Vasodilator substances of the tissues

MONOGRAPHIEN ZUR PHARMAKOLOGIE
UND EXPERIMENTELLEN THERAPIE

HERAUSGEGEBEN VON
PHILIPP ELLINGER

GEFÄSSERWEITERNDE STOFFE DER GEWEBE

VON

J. H. GADDUM

PROF. DER PHARMAKOLOGIE AM UNIVERSITY COLLEGE
LONDON

EINGELEITET VON

H. H. DALE

DIREKTOR DES NATIONAL INSTITUTE
FOR MEDICAL RESEARCH
LONDON

1936

GEORG THIEME / VERLAG / LEIPZIG

Vasodilator substances of the tissues

J. H. GADDUM
Professor of Pharmacology at University College, London

with an introduction by

H. H. DALE
Director of the National Institute for Medical Research, London

and notes for the fiftieth anniversary edition by

F. C. MacIntosh
Department of Physiology, McGill University, Montreal

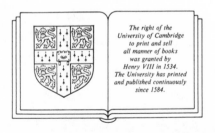

The right of the
University of Cambridge
to print and sell
all manner of books
was granted by
Henry VIII in 1534.
The University has printed
and published continuously
since 1584.

CAMBRIDGE UNIVERSITY PRESS
Cambridge

London   New York   New Rochelle
Melbourne   Sydney

Published by the Press Syndicate of the University of Cambridge
The Pitt Building, Trumpington Street, Cambridge CB2 1RP
32 East 57th Street, New York, NY 10022, USA
10 Stamford Road, Oakleigh, Melbourne 3166, Australia

Originally published in German as *Gefässerweiternde Stoffe der Gewebe* by Georg Thieme/
Verlag, Leipzig, 1936 and © Georg Thieme of Stuttgart (formerly of Leipzig)

First published in English by Cambridge University Press 1986 as
Vasodilator substances of the tissues
English edition © Cambridge University Press 1986

Printed in Great Britain by the University Press, Cambridge

British Library cataloguing in publication data
Gaddum, Sir John
Vasodilator substances of the tissues.
1. Blood-vessels   2. Vasodilators
I. Title   II. Gefässerweiternde Stoffe der Gewebe.   English
612'.18   QP102

Library of Congress cataloguing in publication data
Gaddum, J. H. (John Henry), 1900–1965.
Vasodilator substances of the tissues.
'Originally published in German as Gefässerweiternde
Stoffe der Gewebe by Georg Thieme/Verlag, Leipzig,
1936' – T.p. verso.
Bibliography
Includes index.
1. Vasodilators.   2. Nervous system, Vasomotor.
I. Title. [DNLM: 1. Biological Assay – methods.
2. Neuroregulators – analysis.   3. Vasodilator agents –
physiology. QV 150 G123g]
QP109.G3313   1986   612'.8   86-17534

ISBN 0 521 30860 7

J. H. Gaddum
(from a contemporary photograph)

Contents

Foreword to the 1936 edition

When the German Pharmacological Society met in Wiesbaden in April of 1932, Sir Henry Dale spoke at the session on 'Circulatory actions of body constituents'. His lecture, and the discussion that followed, underlined the importance of a new group of pharmacologically active substances, which have in common that they dilate blood vessels, are formed or released in the tissues, and act close to their sites of origin. I was thus motivated to approach Sir Henry Dale early in 1933 with the request that he should contribute a volume on 'Vasodilator Substances of the Tissues' to the new series of *Monographs on Pharmacology and Experimental Therapeutics*. He accepted, on the understanding that he would work jointly with Professor Gaddum, but the latter's move to Cairo prevented the close collaboration that had been planned. The present book therefore comes primarily from the pen of Professor Gaddum. Sir Henry Dale has limited himself to writing the Introduction and a final section on the anaphylactic symptom complex.

Professor Gaddum's moves, to Cairo and then back to London, have considerably delayed the appearance of the book; but not, I believe, to its disadvantage. For the last two years have brought so much new information to bear on its subject that today's book is quite different from the one that might have been written two years earlier. It is also much more interesting. An especially significant feature of it is that we owe the advances just mentioned in very large degree to its two authors and their intimate colleagues; no one else could have been commissioned to carry out the task.

It will be obvious that the book deals with a number of substances – histamine, acetylcholine, adenosine derivatives, and other materials still unidentified – that constitute a pharmacological group, though chemically they are quite heterogeneous. As we shall see from the story of their discovery and identification, the pharmacological activities that characterize them are remarkably similar, so that a variety of special procedures are required for distinguishing them and separating them from one another. Eventually, more precise information may well stress the differences between them, and permit their assignment to separate pharmacological categories. In the meantime it seems wise not to be committed to giving the group a name of its own.

The present volume has the special virtue of offering the first comprehensive account of the new concept of chemical neurotransmission. It is indeed stimulating to realize how thoroughly the older ideas of 'classical pharmacology', about processes in ganglia and at autonomic and motor nerve endings, have been clarified, and to some extent confirmed and extended; and how, too, the puzzling capability of certain nerves to replace certain others during regeneration, has become almost a corollary of the new and happy recognition that there exist 'cholinergic' and 'adrenergic' nerves.

May the book be a boon to all who are interested in the chemical regulation of the circulation, and in the chemical processes that underlie neurotransmission!

Düsseldorf, September 1935
Philipp Ellinger (translated by F. C. MacIntosh, 1985)

Foreword to the 1986 edition

By 1936, when Gaddum's book came out, the study of the body's short-range chemical mediators had become an important salient of physiological research. It was now evident that there were several, perhaps many, of these agents; that they were highly potent and specific; and that they played key roles in a wide variety of processes, including autonomic neurotransmission.

Like his mentor Dale and his colleague Feldberg, Gaddum had done pioneer work in several sectors of this field. He had read its literature widely and critically, as was his habit, and he was well prepared to offer this first comprehensive account of it. Even more important than his skill as a scientific reporter was his talent for analysis, which enabled him to give a clear outline of the strategies appropriate for further progress. His own research papers illustrated these strategies well. In my judgement, it was Gaddum more than anyone else who showed that pharmacological methods of analysis, properly designed, could reliably detect and measure an active tissue component, even when its concentration was much too low to come within the range of the chemical methods then available. Thus Gaddum and a few others, most of whom were his associates, opened what might be called the 'age of bioassay'. With the invention of new and powerful analytical techniques, bioassay now has fewer practitioners than it once had. But its achievements have been immense, and it still has its triumphs.

Fifty years after Gaddum's book was published, the topics it surveyed have continued their remarkable development, and a minor consequence of that has been a growing interest in their early history. For people with that interest, Gaddum's volume, I believe, is the best single source of information about what happened up to 1935. Having used it in that way myself, I have regretted that it is often unavailable – or even unknown – to many who would have found it helpful. Indeed, I have met scholars who had read Gaddum's research papers of the 1930s but had not heard of his monograph.

Two reasons for that situation are obvious. First, the book, though it was written in English, was published only in German. And second, it went out of print some three years after its publication, in consequence of the outbreak of war. A third possible reason for its neglect may have been its title. As Professor Ellinger explained in the 1936 Foreword, Gaddum did not choose the title, though he accepted it, and over the three years of the book's gestation it became less appropriate than it had been at conception, because of new research findings that could not have been anticipated. The volume as published was still largely concerned with substances that possess vasodilator activity, but by that time vasodilatation *per se* was only one of several major themes.

John Gaddum had the affection as well as the esteem of his scientific colleagues. These feelings extended to his characteristic writing style, which, like him, was modest, terse and sometimes drily humorous. (These qualities survived into the German version by Wilhelm Feldberg.) I enjoyed reading Gaddum myself, and I was pleased when the senior scientists I consulted warmly approved my suggestion that it was time for the monograph to be reprinted, as a historical document in its original language.

One of those colleagues was Sir Arnold Burgen, and he soon told me that he had located Gaddum's typescript in the Library of the Royal Society. Lady Gaddum graciously welcomed the proposal, and so did Lady Todd, Sir Henry Dale's daughter and literary executor. (As noted elsewhere, Sir Henry wrote the book's introductory chapter and a final section on the anaphylactic response. He had planned to be a full co-author with Gaddum but his other duties prevented that.)

The book's copyright had not yet expired: it was held by Georg Thieme of Stuttgart, the successor firm to the original publisher, Georg Thieme of Leipzig. I am grateful to the Stuttgart company for relinquishing the copyright, and to Dr Heinz Kilbinger of the University of Mainz for his part in clarifying the book's status. Finally, and again through the good offices of Sir Arnold Burgen, Cambridge University Press agreed to publish the English version, the Royal Society helped with a grant, and I was invited to supervise the preparation of the typescript for the press.

At that point, as I must now confess, my editorial task became more demanding than I had thought it would be. For I found out what I ought to have ascertained earlier: namely that the German version, as published, was not an exact rendering of the English typescript. It quickly became clear that the author, the translator, and probably others, had invested a great deal of collaborative effort in updating, expanding and otherwise revising the original script, sent to London from Gaddum's department in Cairo, and also that Gaddum had, very sensibly, deferred much of his intended polishing until that penultimate stage of the editorial process. The result of all that was that I found myself with a job for which I was unqualified: I had to try to translate Feldberg's polished German of the printed version back into the sort of polished English Gaddum would have written if he had done all his own editing. This was a daunting responsibility, but fortunately for me Professor Feldberg with typical kindness then came to my rescue, and went through my English text line by line – an operation that took nearly a week, but allowed us both to feel that the final text presented Gaddum's ideas in words that he might have approved, even if they weren't just the ones he would have chosen himself.

Sir Henry Dale's contributions to the present volume needed no such processing. He wrote them in English and they are fine examples of his characteristic literary style. They can now be read for the first time in their original language, and they contain some ideas that their author, so far as I know, did not express elsewhere.

The new Bibliography and Index are anglicized versions of the German ones. In the former, text page references are supplied for each entry. The Table of Contents is a condensed version of the original one, which the publishers and I thought was too detailed to be appreciated by modern readers. I have added some chapter notes, largely on matters that reflect my own interests or prejudices, but partly to explain some of the scientific context of the 1930s that might otherwise seem cryptic. Some of my notes were allowed to grow into short essays. Once or twice I have yielded to the temptation to comment, with the benefit of hindsight, on passages where Gaddum seems to be hospitable to ideas that are no longer popular.

Readers will find that there are remarkably few such passages. More often, the contemporary observations that interested Gaddum were those that we can now recognize as having been seminal. Some of the veteran actors – histamine, acetylcholine, adenosine and its esters – were already on the stage, though not always seen in all their major roles, and we catch sight of some of the coming stars – prostaglandins, neuropeptides, kinins and kininogens, serotonin – most of which have not yet got the names by which they will become famous. (One exception is substance P, which was discovered by Gaddum and von Euler and named by Gaddum and Schild.) Most of these were discovered as pharmacologically active components of tissue extracts, and in some cases it is an interesting exercise for a modern reader to try to identify them from the glimpses that Gaddum provides.

That my share in producing this volume has been a relatively small one should now be obvious. Correspondingly, my debt to the people named above is large. I wish to express my gratitude to them all, and first and foremost to Professor Feldberg. I am particularly pleased that he has consented to add some reminiscences about how the original volume came into being.

My own work on the volume involved secretarial and travel expenses, which arose in part from other studies of a historical kind that I was pursuing at the same time. Most of these were defrayed by a grant from the Hannah Institute for the History of Medicine, and I am grateful to the Institute and its Executive Director, Dr G. R. Paterson, for support and guidance. It is a special pleasure for me to thank the Master and Fellows of Gonville and Caius College, Cambridge, for the privilege of spending the Lent Term of 1985 as a Visiting Fellow there. At Cambridge University Press, Dr Rachel Mellor and Mrs Louise Sanders gave much more than routine help at various stages of the editorial process. And as always, my thanks go to my colleagues in the Physiology Department of McGill University, and to my wife Mary, for their support of my post-retirement activities.

Montreal, April 1986
F. C. MacIntosh

A prefatory note to the 1936 edition

The German edition of this book, which appeared in 1936, was a translation of the English typescript, and I made it while I was working in Dale's laboratory. How did it happen that I was there at the time, and why did I become the translator?

To answer these questions I have to go back to 1925, when I was working in the Physiological Institute of the University of Berlin – and was keen to get married. I must also explain that I come from a closely knit family in which it was the custom to visit relations every weekend. My fiancée thought this a waste of time that could be used for my research. I could not agree more. But how to stop the visits without hurting the feelings of all those nice people?

We found an ingenious way. We would leave Berlin for a long, long honeymoon in England. So after our wedding we went first to Cambridge to work with Langley and Joseph Barcroft, and after that, as I hoped, to London to work in Dale's laboratory. The plan succeeded. When we returned to Berlin two years later the weekly visits had been forgotten.

Langley died about six months after I arrived in Cambridge. He was succeeded by Barcroft and I stayed on. During that time I read Dale's classical paper of 1914 on choline esters, in which he made the fundamental distinction between their muscarinic and their nicotinic actions, as well as his famous 1918 paper on histamine shock. My hope of working with him was strengthened, and I wrote asking permission to come to his lab.; I was happy when he accepted me. That was in 1927, when I spent six months with him. I met him next in April of 1932, in Wiesbaden, at the meetings of the German Congress of Internists and the German Pharmacological Society. Sir Henry – he had been knighted that year – was the invited guest lecturer. He spoke about vaso-active substances of the tissues, but his talk centred on the vasodilator ones. In his Foreword to the 1936 edition of this book, Philipp Ellinger, the editor of the 'Monographs on Pharmacology and Experimental Therapy', tells us that it was that lecture which gave him the idea of approaching Sir Henry and suggesting that he should contribute a monograph to the series, with the title of *Vasodilator Substances of the Tissues*.

During the meeting at Wiesbaden Sir Henry and I had tea together one afternoon in a little forest restaurant, and Sir Henry asked me: 'Feldberg, what do you think of Hitler?' My reply was: 'Sir Henry, you need not worry, he will never win, and if he should, he too will cook with water only' – a German expression which meant that it would all be rather harmless. Sir Henry's only comment was: 'Feldberg, you had better stick to your experiments!'

Not many months later Hitler was Reichskanzler, and within a fortnight the then Director of our Institute in Berlin asked me one morning to see him, and then told me that I must leave the Institute before noon and was not allowed ever to enter it again. My idiotic response was: 'But what about my experiment today?' As a special concession I was allowed to stay until midnight.

In the weeks that followed I felt as if I were paralysed, and I did not know what to do. By chance a colleague told me that a representative of the Rockefeller Foundation was in Berlin; why not try to see him? I succeeded: he was most sympathetic, but said: 'Feldberg, you must understand, so many famous scientists have been dismissed and need our help that it would not be fair to raise any hope of finding a position for a young man like you. But one never knows, let me at least take down your name.' And when I spelt out my name for him, he hesitated and said: 'I must have heard of you'. Turning back the pages of his diary he suddenly said, obviously pleased with himself: 'Here it is; I have a message for you from Sir Henry Dale whom I met in London a fortnight ago. Sir Henry said "Are you going to Berlin? If by chance you should meet Feldberg and he has been dismissed, tell him I want him to come to London at once to work with me." So *you* are all right', the Rockefeller man said warmly, 'there is at least one person I need not worry about any more.'

A few weeks later I was in London. The first question Sir Henry asked me when I entered his office was: 'Feldberg, what do you think of Hitler now?' And I gave him what I think was the only correct answer: 'Can I help it, Sir Henry, if history has made a mistake?'

So that is how I happened to be in London in 1933, at the National Institute for Medical Research. Gaddum was working there, and we soon became friends. One day he brought me the complete typescript of the *Vasodilator Substances of the Tissues* and asked me if I would translate it into German for the series begun by Ellinger. I told him that I did not want to have anything to do with Germany while it was ruled by Hitler, so I was sorry I had to say no. Gaddum understood, and asked a visiting scientist from Germany who had come to work in our lab. to accept the assignment, and he agreed. But a few months later, when the visitor had returned to Germany, he wrote that he was sorry he could not manage it, because he had to spend some months in a Nazi training camp. So Gaddum approached me again and begged me to be his translator. He was afraid that the book might become outdated before it appeared, and he could think of no one else to ask. This time I agreed. Whether I did right or not I don't know, but I think I did, and I am glad now that I was able to help. I did make the condition that my name would not appear on the title page and that there would be no prefatory note by me. Gaddum, however, thanked me in his own Foreword, and that was acceptable.

While I was working on the translation, Gaddum and I discussed the text together chapter by chapter. We made many small alterations, which we did not try to insert into the original English typescript. And of course there were publications to be mentioned that were very recent or had not been available to Gaddum in Cairo. So it was not feasible for the present edition just to reprint the original English typescript. Professor MacIntosh had to compare that version with the published German version of 1936 and then re-translate the latter back into English, a painstaking task which he accomplished beautifully. My own contribution has been to write, fifty years later, this Prefatory Note for the 1936 edition.

National Institute for Medical Research,
Mill Hill, London NW7 1AA,
June 1986
W. S. Feldberg

Preface to the 1936 edition

This book owes much to Sir Henry Dale. I started to write it when I had been working for six years under his inspiration; I wrote most of it on the assumption that it would be published under both our names, and when the work was done he read through my original untidy manuscript and suggested various improvements. For these reasons my opinions, as expressed in this book, are mostly also his opinions. On the other hand the responsibility for any errors, either of omission or of commission, is my own. Some omissions are deliberate. In accordance with the purpose of this series of monographs I have selected for discussion the evidence which seemed to me most interesting and important. A complete 'Handbuch' of this branch of pharmacology would make a much bulkier and more expensive volume.

I am very grateful to my friend and collaborator, Dr Wilhelm Feldberg, who has not only translated the English text into the German tongue, but has also suggested various changes tending to make the book more complete and more accurate. He has himself made important contributions to knowledge in this field, and no-one could be better qualified to undertake this particular piece of translation.

I am also grateful to the editor, Professor Ellinger, who has given more than official help to the work, and to the publishers who have patiently accepted the delays, due to my travels, to which the editor refers in his preface.

London, 26 September 1935
John Gaddum

I

Introduction[1]
(By H. H. Dale)

The origin of this monograph may probably be traced to a discussion at
the Deutscher Kongress für innere Medizin in Wiesbaden in 1932. The title
of that discussion, to which I had the honour of contributing the opening
statement, was *Kreislaufwirkungen körpereigener Stoffe*. Both in that opening
statement and in the ensuing discussion, some reference was made to the
two principles acting on the circulation which can be regarded as true and
typical hormones in the original sense of the word – namely the hormones
of the suprarenal medulla (adrenaline) and of the pituitary posterior lobe
(vasopressin). The discussion was mainly concerned, however, with other
substances than these, not produced in special organs of internal secretion,
but widely distributed in their occurrence throughout the organs and
tissues of the body. Some of these substances rival adrenaline and
vasopressin in the intensity of their action, when they are artificially
injected into the blood stream; on the other hand, there is no reason to
suppose that, under normal physiological conditions, they are distributed
by the blood stream to produce activity at a distance from the site of their
occurrence, according to the original conception of the hormone function.
In the discussion already mentioned, and later in more detail in the Dohme
Lectures delivered in Baltimore in 1933, I drew attention to another broad
difference between the functions of these widely distributed vaso-active
substances, on the one hand, and that of adrenaline and vasopressin on
the other. While, with exceptions in detail, the action of these two
hormones[2] is in the direction of maintaining and intensifying vascular
tone, that of the generally distributed tissue principles is to weaken it; i.e.,
they are essentially and predominantly vasodilator principles. On the same
occasions I have drawn attention to the fact that, while the nervous
control of vascular tone by the central nervous system is predominantly
vasoconstrictor in function, so that provision is made for a general and
sudden increase of that tone, and therewith of the arterial pressure, to
meet physiological needs and emergencies, active vasodilatation by nervous influences appears to be a function localized to particular organs, as

1

in the case of the chorda tympani or the nervi erigentes, or acting through a local, peripheral mechanism, as in the case of the widely distributed axon-reflex dilatation through the terminal branchings of sensory fibres. Active nervous dilatation, therefore, seems to be designed to meet local, metabolic needs, and not for the simultaneous requirements of the whole body. General and simultaneous vasodilatation of nervous origin can only be produced by depression of the vasomotor centres, and we then pass readily from the sphere of physiological adjustment to that of pathological collapse. And I have suggested that there is some analogy between those types of nervous control of vascular tone, and those in which chemical agencies are concerned. Here again the general, simultaneous effects on the whole vascular system, produced by true hormones distributed by the blood, are predominantly concerned with the maintenance or sudden enhancement of vascular tone: while the vasodilator effects of substances widely distributed in the tissues, so far as we have evidence of their physiological function, are localized in the neighbourhood of their release into the tissue fluids. So far as we know, these vasodilator principles subserve no physiological function by distribution in the blood, to cause a general vasodilatation. They can, of course, be made to produce such general effects by artificial injection; but so far as these general vasodilator effects are produced at all under natural conditions, they belong to the domain of pathology, rather than of physiology.

The substances, then, with which this monograph deals are classed together as being natural constituents of the tissues, exhibiting a predominantly vasodilator effect on the vascular system, when they are artificially injected into the blood stream. It was this action which, in each case, first revealed the fact that the substance in question had a potent physiological activity. Of the substances, or groups of substances, which have, hitherto, been chemically identified as responsible for this vasodilator activity in different tissue extracts, none has been new to science. Histamine, choline and its physiologically important acetyl ester, adenosine and its derivatives, were all well known substances, before they were known to have this type of physiological action; and in the cases of histamine and acetylcholine their intense physiological activities had long been recognized and subjected to thorough physiological analysis, long before there was any real evidence for their existence in the body[3] or for their wide distribution in the different organs and tissues. There are yet other physiological actions, and in particular, associations of such, which point to the existence of other powerfully vasodilator substances, which have not yet been chemically identified. Until chemical isolation has been achieved, it is impossible, in any one such case, to be certain that the characteristically associated physiological effects are really due to a single substance. What is quite clear is that regular associations of physiological activities are found in certain tissue extracts, which cannot be explained by any kind of mixture of the chemically known substances; and it seems to me that a good case has been made for the view that there are at least two additional vaso-dilator substances,[4] normally occurring in the tissues, and still await-ing chemical identification.

The question of real interest, with regard to all these substances, is that of the physiological significance of their occurrence in artificial extracts from a wide range of organs and tissues. We have first to consider, in each case, whether the substance which we can isolate from a tissue extract by a chemical procedure, or can even detect and measure in such an extract by means of its characteristic physiological activities, is a natural constit-uent of the living cell protoplasm, or an artificial product of the process of extraction. Strictly speaking we have no evidence, in the case of any one of these substances, that it exists as such in the uninjured, living cells; and it is not very probable that we shall be able to obtain such evidence. In the case of those concerning which our evidence is most complete, such as histamine and acetylcholine, we can say with some assurance that they do not exist, in the tissue from which we can extract them, in a free and diffusible condition. It has been pointed out, for example, that an organ, such as the lungs of most species, may contain sufficient histamine to produce a condition of profound collapse, if it were free to diffuse from the cell protoplasm into the blood. From the same tissue, moreover, an enzyme can be extracted which would destroy the histamine, if the latter were free to diffuse in the protoplasm and to come into contact with the enzyme.

Similar conditions make it even more certain that acetylcholine, which is destroyed with such extraordinary rapidity by the generally distributed cholinesterase, must be held, in the tissue elements from which we can extract it, in some form of combination or physical association which renders it incapable of diffusion, and thereby prevents it from producing its intense physiological effects, and protects it from immediate enzymatic hydrolysis and inactivation.[5] The substances of the adenosine series are in a somewhat different position, in that they are concerned with the cyclical changes of the metabolism of muscle, and probably of other rapidly functioning cells. We must similarly conclude, however, that they are not normally free to diffuse out of the resting protoplasm and to produce, on the neighbouring blood vessels, their vasodilator action, which, though weaker than that of histamine or acetylcholine, is still of a potentially important intensity. In the case of the substance, still of unknown chemical nature, to which the name Kallikrein has been applied, definite evidence has been given of its capacity for the ready formation of an inactive association with a constituent of the blood plasma and the tissue protoplasm, from which it appears to be liberated by treatment with acid.

Occurrence in the living cells in a state of combination or adsorption, which prevents free access to the exterior medium, is not, of course, peculiar to the group of active substances with which we are dealing. It applies, in some sense, not only to true hormones of the type of adrenaline, but even to potassium ions, the mobilization of which from inactivating complexes is coming into increasing prominence, as the probable basis of certain forms of physiological activity.[6] The important questions for our purpose are (1) whether the vasodilator substances which we find in extracts are preformed substances, naturally occurring in the cells, sep-arated from loose, inactivating processes by conservative processes of extraction; (2) whether there is any evidence of their liberation in active, diffusible form as a natural, functional event in the living body; and (3) what can be deduced from the available evidence, concerning the function of each particular substance when liberated.

(1) In favour of preformation, I think it can be said that we have, in the case of most of these substances, if not of all, almost as strong evidence as in the case of any other tissue constituents which biochemical methods can recognize. Histamine or acetylcholine can be extracted from the tissues in which they occur, by the gentlest and most conservative methods. Precautions to prevent changes *post mortem*, or even *intra mortem*, as by rapid freezing of the living tissue, increase rather than diminish the yield. In the case of acetylcholine, in particular, disintegration of the tissue at the ordinary temperature, before extraction, may even cause complete absence of the unchanged substance from the extract. As has been said, there is as good evidence for the existence of these substances in the living cells of many tissues, as for phosphagen[7] or lactic acid in living muscle. And I think it may be expected that, when we know more about them, similar considerations will be found to apply to the cases of the vasodilator substances still unidentified.

Questions (2) and (3) are obviously related, though not identical. If we find evidence of the liberation, in response to stimuli or changes of physiological conditions, of substances having such intense activities as these, it is a natural conclusion that they will produce some physiological effects, and we can proceed to consider what these are, and what is their significance for the normal functional adjustments of the body. Until we can properly assume that they are thus naturally liberated, we cannot begin to form a clear idea of the physiological significance of the activities which they show in artificial extracts. On the other hand, the detection in the living body of actions corresponding to those of one or another of these substances, in the neighbourhood of cells containing them when those cells are subjected to stimulation or injury, may be the main evidence in favour of its natural liberation from such cells. The answer to these two questions must, accordingly, be considered to some extent in parallel; and it is not likely that a definite answer can be given to them, in the case of any of the substances, until evidence from various sources has been accumulated and coordinated over a period of years. It is not surprising, therefore, that satisfactory evidence of the natural liberation of such substances from the cells or tissue elements which yield them to artificial extraction, and of their functional significance when so liberated, is only available, as yet, in the case of these two, of the substances here considered, which have been longest known and most fully studied. By way of illustration of the manner in which the growth of carefully tested knowledge, over a period of many years, leads ultimately to a fairly clear conception of functional significance, it may be useful here to review, very briefly, the very different histories of these two substances, histamine and acetylcholine.

Histamine had earlier been synthesized by Windaus and Vogt, but it was Ackermann's production of it from histidine, by putrefaction, which led to the identification of a substance which Barger and I and Ackermann and Kutscher had independently obtained from ergot extracts. When its activity came to be studied in detail, it became clear that it had an intensely powerful action of the type which Popielski[8] had attributed to his hypothetical 'vasodilatin', producing in different species the central features of the symptom-complex, which, with characteristic variations in the same species, had just acquired a new prominence and interest as the anaphylactic shock. Anomalies in the distribution of the vasodilator component of this complex, and difficulties in reproducing it in artificial perfusion, were ultimately resolved by the discovery that the vasodilator action of histamine was, particularly in the cat, limited to the minutest vessels, including the capillaries. This brought into new prominence the significance of an independent, variable tone of the capillary vessels, which Krogh was simultaneously studying from another point of view. Analysis of the circulatory failure produced by large injections of histamine threw some suggestive light on certain shock-like conditions produced by injury and infection. There was some rather imperfect evidence of the occurrence of histamine itself in certain tissue extracts, but we were still very far from a clear knowledge of histamine as a widely distributed constituent of body cells, or from any evidence that it had a natural, physiological importance. Any views in either direction had still little more than a speculative basis. The whole position was changed, on the one hand by the observations of Lewis and his co-workers, showing that something like histamine was liberated from the cells of the human epidermis when they were lightly irritated and injured, and caused the local changes in the minute blood vessels of the skin; and on the other hand, by the concurrent demonstration in my own laboratory that free histamine can be directly extracted, by the gentlest and most conservative methods, from the fresh tissues of a wide range of organs of the body, and chemically identified. It was possible at once to grasp the meaning of the frequent occurrence of the histamine complex of symptoms, when the tissues were subjected to the general injurious action of a large variety of agents, varying from the ions of heavy metals to the sensitizing antigen in an anaphylactic animal. It was possible, further, to form a clear conception of the normal function of histamine, in the local adjustment of circulation to local injury or irritation. It had required many years to arrive at this point, and there may well be other conditions of the liberation of histamine yet to be revealed; but I think we may feel satisfied that what, for many years, seemed to be scattered items of evidence have come together in the main outlines of a coherent picture.

Acetylcholine, again, had been known as a synthetic curiosity long before Reid Hunt had recognized the intensity of its depressor action. A further long interval elapsed before, having come across it by accident, again as a constituent of an ergot extract, I was led to analyse its activity more fully. I found that its depressor action was largely due to its arteriodilator action; that, in addition, it peripherally reproduced the effects of parasympathetic nerves in a manner resembling the action of muscarine, even matching the facility with which adrenaline had much earlier been shown to reproduce the effects of sympathetic nerves; and that, further, in somewhat larger doses, it stimulated ganglion cells of the autonomic (vegetative) nervous system in a manner comparable to the action of nicotine. Many years later it was shown that it also shared the action of nicotine on voluntary muscles. So far these interesting and striking activities of acetylcholine did not justify any definite ideas as to its possible significance in normal physiology. The idea that adrenaline might somehow intervene in the transmission of the effects of true sympathetic nerves had long ago been tentatively put forward, and it was natural to speculate on the same sort of possibility with regard to acetylcholine and the parasympathetic nerves. There was no evidence, however, of such function in either case, and no evidence that acetylcholine occurred in the animal organism at all. The whole question of the possible function of acetylcholine took an entirely new aspect with the publication, in 1921, of Otto Loewi's now classical demonstration of the humoral transmission of vagus and sympathetic effects to the frog's heart, by the peripheral release of substances having the properties of acetylcholine and of adrenaline respectively. From that point onwards progress has been rapid and continuous. A missing link in the evidence was provided when Dudley and I obtained and chemically identified acetylcholine from an animal organ (spleen of horse and ox). While the chemical nature of the sympathetic transmitter is still unsettled, it is generally accepted that acetylcholine itself is liberated at the peripheral endings of nerve fibres of the parasympathetic system, to transmit their stimuli to the effector cells. And in the last two years a further large extension of the functional significance of acetylcholine has been made possible by investigations in my own laboratory, which have brought into the picture the other, nicotine-like, aspect of its action. The evidence is clear, that acetylcholine is released at the synapse, whenever nervous activity passes from the preganglionic ending to an autonomic ganglion cell. The concept of acetylcholine as transmitter across the synapse of the effect of a single preganglionic impulse, and then immediately disappearing, obviously involves a new set of ideas as to the manner of its action. The same is true, in at least equal degree, of the evidence in favour of its similar action as transmitter of the effect of an impulse reaching the ending of a somatic motor nerve fibre to the end-plate of a striated voluntary muscle fibre. The evidence for this last-mentioned function is not yet as complete as that for transmission at the ganglionic synapse, but it is rapidly becoming so at the time of writing.

This is not the proper place for detailed discussion of the evidence for this wide and still progressive extension of the functional significance of acetylcholine in the body, or of the interesting light which it throws upon long known facts concerning the kinds of nerve fibres, in the whole peripheral system, which can replace one another in artificially-induced regeneration. On these and many other points the later and more detailed sections of this monograph by Professor Gaddum must be consulted; though even there the interval between writing and publication will make it impossible to include all the latest items of the rapidly accumulating evidence. Here I am concerned to lay emphasis on the fact that the two substances, histamine and acetylcholine, which, but a few years ago, could have been regarded as having very similar vasodilator actions in most species, with some differences of incidence and in liability to extinction by atropine, have now been proved to have such fundamentally different functions in normal physiology. Histamine, according to present knowledge, is concerned with strictly localized vascular reactions to irritation or injury, unrelated to nervous control; while acetylcholine now appears to be concerned with the transmission of nervous actions from nerve-ending to effector cell or from one neurone to another, in a large part of the peripheral nervous system. So clear and important is this function that Professor Gaddum, faced with the necessity of providing a special chapter on the chemical transmission of nervous effects, was obliged to complete this section of the picture by dealing with what I have termed 'cholinergic' and 'adrenergic' effects, and accordingly with the true hormone, adrenaline, because of its relation to 'sympathin', although it has no proper place among the *gefässerweiternde Stoffe der Gewebe*.

It may well be that a similarly unforeseen development will yet take place, in our knowledge of the functional significance of each of the other vasodilator principles which have already been detected in tissue extracts. If that should occur, we may be left with no proper reason for considering them together as a natural group of active substances, just as the basis for such an association has already become slender in the cases of histamine and acetylcholine. At present, however, there seems to be still a sufficient ground of convenience, in this attempt to assemble the known facts concerning substances which (with the incidental exception of adrenaline above mentioned) have all been concerned in that 'depressor' action of many tissue extracts, which has so long been recognized, and has only so recently begun to show its physiological importance in widely different directions. Histamine has already had its separate monograph, and acetylcholine will probably have several in the near future. I myself have, on more than one occasion, in special lectures,[9] attempted to summarize knowledge concerning the whole group of substances here concerned, but, of necessity, without any attempt at a full citation of the evidence on which views and suggestions were based. It seemed that a full and detailed review, of the experimental basis of knowledge and theory in this field, would be useful to many. Professor Gaddum and I had originally planned to write it together, but for reasons mentioned by Professor Ellinger this full and intimate cooperation was no longer practicable. He has, accordingly, been responsible for the review of the detailed evidence which forms the main part of this monograph; and I find myself once again, in this introductory chapter, expounding general ideas and principles. In this case, however, Professor Gaddum's work makes the facts and their immediate significance so readily accessible to the reader, that he can check my statements and criticize my views, or formulate his own to replace them.

II

The pharmacological analysis of tissue fluids and extracts: general principles[1]

A. Introduction

The purpose of this book is to examine the nature and the properties of some of the pharmacologically active substances that are widely distributed in the body, and that probably play an important part in the local regulation of tissue activity. The evidence for the physiological significance of these substances is based primarily on the use of pharmacological methods for their detection and assay in tissue extracts, and in other fluids that can be obtained from tissues under various conditions. It may therefore be useful to devote the present chapter to a general account of the methods that have been found useful for this type of pharmacological analysis. It should be remarked at the outset that the methods that have been used by the various investigators are by no means of equal value. In some cases the evidence for an active substance's identity has been vague; in other cases it has been so complete as to leave practically no doubt. A genuinely convincing pharmacological analysis is usually a tedious business, and in the past much confusion has resulted from omitting the precautions that are indispensable for the proper assessment of the outcome of isolated observations. This point is one that becomes even more important when – as is often the case with the theories examined in this book – there is little hope of supporting them with purely chemical evidence.

In special cases it may be feasible to isolate the active substance and thus confirm the pharmacological evidence, but more often this is impossible, owing to the difficulty of obtaining enough material. Pharmacological methods are generally much more sensitive than chemical methods, and for that reason can be used in experiments that would be quite impossible without them. But even when plenty of material is available these pharmacological methods play an important part, because the chemical methods usually involve large losses, and the only way to get a quantitative estimate of the amount of active principle present in the original solution is to use a pharmacological test.

The use of accurate and specific methods for the pharmacological

10

analysis of tissue extracts is therefore of great importance for studying the distribution of active substances, even when their chemical constitution is known. When the chemical nature of an active substance is unknown, the evidence for its existence, and its distribution and its properties, must be based entirely on pharmacological tests.

B. Choice of biological methods

The discovery of a specific test for a given substance depends partly on the discovery of a pharmacological reaction of some animal organ or system specific for that substance, and partly on the choice of the most favourable conditions for carrying out the test. Drugs or special solutions of various kinds may be used to increase the specificity of the test, either by increasing its sensitivity to the substance under study, or by decreasing its sensitivity to other substances. For example various tissues can be made especially sensitive to acetylcholine by means of eserine. Conversely, if an extract contains both histamine and acetylcholine, the blood pressure of a cat can be used as a reagent for the histamine, provided that the cat has received sufficient atropine to render it insensitive to acetylcholine.

Similarly a pharmacological preparation can sometimes be made insensitive to a given substance in the solution to be tested, by administering the substance itself in large quantities.[2] For example if a piece of plain muscle is left in a salt solution that contains a high concentration of histamine, the addition of a small quantity of histamine to the bath has no effect, but the preparation may still respond to certain other active substances whose properties are under study.

C. The quantitative comparison of two solutions

Accurate pharmacological analysis always depends on a quantitative comparison of the action of two solutions, a known and an unknown. Such comparisons are not always carried out as accurately as they should be.

If both solutions owe their activity to the same active principle, the responses they produce should obviously have the same time-course and records of them should be indistinguishable. If this is not so, and the solutions produce records that differ in shape, there may be more than one pharmacologically active substance present, or a single active substance may be present in such different physical association as to retard or accelerate its access to the reactive cells. Under such conditions a strictly accurate quantitative comparison can, of course, not be made. In some cases, it is true, a small amount of impurity may produce minor changes in the shape of the response without invalidating the test; but when the shapes of the responses are recognizably different one must be cautious about interpreting the result.

In making a quantitative comparison the first essential is to be certain that the effects produced are not maximal, and that reasonably small alterations in the dose produce recognizable alterations in the response.

The number of trials required varies according to the test employed and the accuracy needed, but it is seldom possible to have much confidence in the result unless it is based on several trials with each solution. The solutions are usually given alternately, and the doses are adjusted until they produce indistinguishable effects.

In order to be confident about the significance of differences that lie within the range of the uncontrollable variations in the sensitivity of a preparation, it is necessary to carry out a long series of comparisons. In some kinds of test the spontaneous variations are very large, and then a series of comparisons is needed in order to be sure that the result is even approximately correct. The best methods for doing this have been discussed by Burn (1928).[3] Some biological preparations show a tendency to respond to a series of equal doses of a drug with an alternation of larger and smaller responses; and it is therefore sometimes necessary, in order to exclude errors from this source, to alter the order of the injections so that the same dose of one solution is given twice in succession.

The main source of error is the varying sensitivity of the preparation. This can be controlled to some extent by keeping such factors as the depth of anaesthesia, the temperature, the volume of solution injected, and the rate of injection, as constant as possible. It is also usually desirable to follow a rigid time-table since the sensitivity of the reaction is affected by previous injections, and the extent of this effect depends on the time interval.

However, in spite of the utmost care in controlling such factors as these, some preparations may show sudden changes in sensitivity due to unknown causes, and it is never safe to place great confidence in the result unless it is based on a long series of consistent observations.[4]

D. Studies of substances whose chemical constitution is known

Even in the study of known substances, the biological methods used for analysis have often been so unspecific that the observed effects might have been due to any of a variety of substances. In recent years, however, these methods have been greatly improved; and it is now possible in certain cases, in which there is not enough material for chemical isolation, to obtain such convincing evidence for the presence of some active substance that it is almost impossible to doubt its presence in the solution in the concentration indicated by the assay.

As a rule, this evidence depends at least in part on the fate of the active substance when the extract is subjected to various chemical and physical treatments. For example, experiments may be carried out to answer such questions as the following:

(1) Is the substance heat-stable?

(2) Is it stable in the presence of acids, alkalis and other reagents, e.g. oxidizing agents or nitrites?

(3) Does it dialyse through membranes of different porosity?

(4) Is it soluble in the usual solvents (various alcohols, ether, chloroform, acetone, benzene, etc.)?

(5) Is it precipitated or absorbed by various reagents?

(6) Is it rapidly destroyed by autolysis? What enzymes destroy it? (Tests of this kind may be made more specific if purified preparations of enzymes are used, or if the destruction is shown to be inhibited by some substance known to have a specific inhibitory action on a particular type of enzyme.)

The identification of an active principle with a known chemical substance, by any such methods, becomes the more convincing in so far as the comparisons can be made quantitative instead of merely qualitative, for example by showing identical rates of destruction by a particular treatment.

The main source of error in experiments of this type is that the physical and chemical properties of pure substances are often quite unlike their properties in the presence of large amounts of impurities. For this reason it is of the utmost importance to carry out control experiments in which the pure active substance is added to the impure solution and shown to have, under these conditions, the same properties as the active substance originally present. For example, treatment with alkali may remove most of the activity due to histamine in a tissue extract, although histamine in pure solution is quite stable to alkali (Best, Dale, Dudley and Thorpe, 1927).

The evidence for the identification can often be strengthened very considerably by tests of a more strictly pharmacological nature. Either by a suitable choice of the method of preparing the extract, or by applying special procedures of purification, it may be possible to obtain a solution that is approximately pure in the pharmacological sense, though still very impure in the chemical sense. Such an extract, which contains no significant amount of any pharmacologically active substance other than the one under study, can be compared with a pure preparation of the active substance, and a quantitative estimate of the concentration can then be obtained in a number of quite different kinds of pharmacological test. If these quantitative estimates all agree with one another, they provide strong evidence that the active principles are identical. This conclusion is supported by the fact that when an active extract is repeatedly compared with a pure substance other than the one to which it owes its activity, it is generally readily apparent that the results of the different tests do not agree quantitatively with one another. This is true even when exactly the same technique is used in each test. For example, attempts have been made to use potassium or histamine as a standard for assaying the oxytocic activity of extracts of the posterior lobe of the pituitary. These attempts are unsatisfactory because successive tests on the same extract do not give concordant results (Burn and Dale, 1922). Quantitative agreement between the results even of successive tests of the same kind thus provides some evidence for the correct identification of the active principle. When a series of quite different tests is applied, quantitative agreement between their results is of course much more convincing (cf. Chang and Gaddum, 1933).

E. Studies of substances with unknown chemical constitution

Pharmacologically active substances of unknown constitution are of two-fold interest. In the first place it is often important to know something of their properties in order that suitable methods may be devised for removing them from solutions, and thus preventing them from interfering with the results of experiments in which some other substance is of primary interest.

In the second place an unknown substance may have interesting physiological functions of its own. When research has the object of investigating the pharmacological action, and the probable physiological function, of a chemically unfamiliar substance, it is necessary in the first place to have good grounds for believing that the action in question is really due to one substance that is different from any of those already identified.

The number of known constituents of a tissue extract that may produce pharmacological effects is large. Before attributing any effect to some new substance, therefore, one must consider whether the effect might not really be due to any of the following factors acting alone or together:

(1) The reagents used in preparing the extract.

(2) Inorganic salts, such as potassium or calcium. This possibility can be excluded by showing that the effect disappears when the extract is ashed.

(3) The use of a solution whose temperature, tonicity or acidity is either too high or too low.

(4) Histamine, choline esters, adenosine compounds, etc., as well as many other substances that are weakly active, such as bile acids, maleimide, ethylamine and acetic acid (Fleisch, 1933).

(5) Substances that are formed post mortem by autolysis or bacterial action. Thus autolysis liberates large amounts of choline, presumably from lecithin, during the first post-mortem hours; and bacterial action forms histamine and the pressor bases isoamylamine and tyramine by decarboxylation of the corresponding amino acids. In the past these post-mortem changes have caused confusion (Rosenheim, 1909; Barger and Walpole, 1909; Strack, Neubaur and Geissendörfer, 1933). An extreme example of this sort of change is the eventual production of free ammonia.

In considering such possiblities it is particularly important to bear in mind the likelihood that tissue extracts will contain several pharmacologically active substances. For example, if an extract is found to cause a fall of blood pressure in a rabbit after atropine, and if this same extract also stimulates a rabbit's intestine, it is clear that is properties can be attributed neither to acetylcholine alone nor to adenosine alone. For as will be seen from Chapters IV and V, which deal with the pharmacology of these substances, acetylcholine does not lower the blood pressure of the atropinized rabbit, while adenosine, which does have that effect, has an inhibitory rather than a stimulant action on the intestine. The extract might, however, owe its activity to a mixture of the two substances, unless the intestinal preparation has also been treated with atropine.

The possibility must also be kept in mind that two known substances might potentiate each other's action. An effect of that sort was noted by Best, Dale, Dudley and Thorpe (1926) when they studied the effect of a mixture of histamine and acetylcholine on the blood pressure of the cat.

An attempt is sometimes made to exclude the possibility that an observed effect is due to a particular known substance, by showing that the solution still contains activity after it has been subjected to physical or chemical treatment that would be expected to remove or destroy that substance. Such methods must be used with caution. For instance, a method of precipitation that would remove over 99 per cent of a substance when it was present in comparatively high concentration might effect little or no removal when its concentration was too low to be detected by the chemist. Often such concentrations are still high enough to elicit pronounced pharmacological effects. In such experiments it is highly desirable to demonstrate to what extent the substance has been removed, by testing the extract, before and after treatment, with an appropriate pharmacological test that is especially sensitive to the substance in question. Mention has already been made of the converse possibility, namely that activity due to a single substance may be removed from a complex extract by a treatment to which the substance in pure solution is completely stable.[5]

When a solution is found to be inactive in a test that reacts to some known substance, the conclusion is often drawn that the solution does not contain the substance in question. It is desirable that any statement of that kind should be accompanied by an approximate estimate of the least amount of active substance that could have been detected if it had been present. Ideally, one should obtain this estimate by determining the threshold dose of the substance in the same experiment that gave a negative result for the extract. In the absence of such an estimate, the negative result conveys no information, unless the reader himself has some basis for a rough estimate of the sensitivity of the test.

When an extract of a particular organ or tissue has been shown to produce a pharmacological effect that cannot be explained by the presence of any known active substance or substances, a presumption is established that it is due to the presence of a chemically unknown principle. There has been a tendency in such cases to postulate the presence of an active principle that is specific for the organ yielding the extract. Such an assumption is obviously unjustified unless it can be shown that extracts of other organs and tissues do not have the same sort of activity. Neglect of this elementary precaution has had the effect of encumbering the literature with numerous supposed principles named after the organs, often with an unjustified implication of some functional significance attributed to the organs.

When an extract has been definitely freed from known active substances any effect that it still produces must be due to others that are chemically unknown. It is tempting in such cases to attribute the remaining effects to the action of a single unknown substance. Such an assumption, however, can never be strictly justified until the substance has been obtained in a pure state and shown to produce all the actions in question. A degree of support for the assumption may, however, be admitted, if it is shown that the activities of different samples of the extract are present in constant proportion in different tests, in comparison with the corresponding activities of a common standard. Such evidence shows that the different activities are closely and regularly associated, and that various processes of purification do not separate them. It establishes a certain presumption that they are due to one substance; but that presumption disappears as soon as a method of purification is found that effects even a partial separation, or selective destruction, of the different activities.[6] Evidence of this kind has given rise to extensive discussion as to whether the activities of extracts from the pituitary lobe are due to a single principle or to more than one: a line of argument that is equally applicable to many of the cases that will come up for discussion here.

III

Histamine

Histamine or β-iminazolylethylamine is the amine obtained from histidine by decarboxylation. It was first prepared synthetically by Windaus and Vogt in 1907. Its natural occurrence (in ergot) and its pharmacological activity were discovered almost simultaneously by Ackermann and Kutscher (1910) and Barger and Dale (1910). Since then it has been found to be widely distributed in animal tissues.

It will not be possible to refer in this book to all the many publications that deal with this substance. For more complete information the reader should consult the monograph by Feldberg and Schilf (1930), which gives references to nearly all the papers that had appeared up to the time of its publication. A shorter review by Best and McHenry came out a year later (1931).[1]

Synonyms. Histamine was at one time commonly referred to as β-I, an abbreviation of the chemical name β-iminazolylethylamine. Another chemical name is 4-β-aminoethylglyoxaline.[2]

A. Chemical and physical properties

$$\begin{array}{c} CH \\ \diagup \quad \diagdown \\ NH \qquad N \qquad\qquad NH_2 \\ | \qquad\qquad | \qquad\qquad\qquad | \\ C == C —— CH_2 —— CH_2 \\ H \end{array}$$

The constitutional formula of histamine is given above. The molecular weight of the base is 111. Histamine is generally used in the form of the diphosphate (molecular weight 307) or the dihydrochloride (molecular weight 184).

An account of its chemical properties will be found in the book by Guggenheim (1924).

Solubility. Histamine and its salts are very soluble in water, and soluble in alcohol and acetone, but insoluble in ether. The free base is soluble in hot

18

chloroform but the salts are not, so that if an alkaline solution is evaporated to dryness histamine can be extracted with chloroform from the residue; but if the same treatment is applied to an acid solution the histamine is not dissolved by the chloroform.

Stability. Histamine is stable in watery solutions at 100 °C. Solutions that have been sterilized by boiling can be kept at room temperature without loss of activity, but a solution that is not sterile is rapidly inactivated by bacterial growth. In the presence of tissue extracts the stability to heat depends on the pH. If an equal volume of twice-normal acid is added to such an extract the mixture may be boiled for hours without loss of histamine. If histidine is present in the extract prolonged heating may even form some histamine. Ewins and Pyman (1911) obtained a 24 per cent yield of histamine by heating histidine with 20 per cent sulphuric acid to 265–270 °C. On the other hand an extract that has been made alkaline and boiled loses much of its histamine activity. An alkaline solution of pure histamine does not lose activity in this way when it is boiled.

Histamine is inactivated by formaldehyde (Kendall, 1927). According to Best and McHenry (1931) there is an immediate reversible effect involving the loss of about half the activity, followed by a slow irreversible destruction of the histamine. They suggest that the immediate effect is due to the formation of a condensation product.

Like other amines, histamine in weakly acid solutions is destroyed by nitrites (Felix and Putzer-Reybegg, 1932; Gaddum and Schild, 1934). The reaction has an optimal pH between 2 and 6; if the solution is too acid the nitrite is rapidly destroyed, with release of oxides of nitrogen.

Adsorption. Histamine is readily adsorbed on animal charcoal (Saunders, Lackner and Schochet, 1931).

Preparation of histamine. Histamine may be prepared from histidine by decarboxylation (Windaus and Vogt, 1907), by bacterial decomposition (Ackermann, 1910), by acid hydrolysis (Ewins and Pyman, 1911), or by ultra-violet light (Holtz, 1934). It has also been prepared from diamino-acetone (Pyman, 1911; Koessler and Hanke, 1918).

The most convenient way of making histamine is to oxidize histidine with chloramine-T.[3] This forms cyanomethylglyoxaline, which can then be reduced to histamine with sodium and alcohol (Dakin, 1916).

Formation and destruction of histamine by ultra-violet light. It was first shown by F. Ellinger (1928) that solutions of histidine exposed to ultra-violet light acquire pharmacological properties that indicate the presence of histamine. Holtz (1934) confirmed that histamine is actually formed under these conditions when he succeeded in isolating histamine from the irradiated histidine solutions. The identification was established by analysis of the gold and platinum salts and by measurement of the melting points of several salts. The yield is greater if the irradiation is carried out in alkaline solution in the absence of oxygen then in acid solution in the presence of oxygen, but even under the most favourable conditions the yield is less than 1 per cent. The probable reason for this is that the histamine itself is destroyed by ultra-violet irradiation; Ellinger proved this by exposing a pure salt of the base and testing the solution biologically.

There was at one time some difference of opinion about the wave-length of the light that evokes these changes. It is now agreed that pure solutions of histamine or histidine absorb practically no radiation of wave-length greater than 240 mμ. The absorption increases rapidly at shorter wave-lengths, and the absorption curves of the two substances are very similar, except that the sharp rise occurs at a wave-length 2 mμ shorter for histamine than for histidine. As would be expected from these facts it is only light of wave-length less than 265 mμ that either forms or destroys histamine.

When solutions of histidine are irradiated over longer periods a yellow substance is formed, and the solution shows intense absorption at about 313 mμ (Bourdillon, Jenkins and Gaddum, 1930; Ellinger, 1932).

B. Pharmacological properties
1. *Action on the systemic circulation*
(a) *General*

The action of histamine on the peripheral circulation is complex. Under some conditions it appears to act as a vasoconstrictor, under others as a vasodilator.[4] Thus Dale and Laidlaw (1910) could clearly demonstrate that the fall of blood pressure, which is one of the most striking actions of small doses of histamine in the cat, is due mainly to peripheral vasodilatation. This was shown both by plethysmographic experiments and by comparing the latent period of the blood-pressure fall with intravenous as against intra-arterial injection. On the other hand, when the peripheral action of histamine was studied by the older methods of artificial perfusion it was invariably found that histamine produced vasoconstriction. Applied to isolated strips of artery or vein, histamine likewise produces only contraction. By contrast, plethysmographic experiments did not give consistent results: in some experiments histamine caused an increase, in others a decrease, of the leg volume.

The reasons for these variations were studied by Dale and Richards (1918) in plethysmographic experiments on cats. They concluded that the action of histamine on the arteries and veins is always constrictor, but that this action is often insignificant compared with its dilator action on the capillaries. They found that the effects of histamine were closely paralleled by those of small doses of adrenaline, which also dilated the capillaries and constricted the arteries and veins. The dilator action of acetylcholine, on the other hand, appeared to be mainly on the arterioles.

These conclusions received general support from observations on the effects of denervation and ischaemia. When the nerves to the cat's leg were intact, histamine sometimes produced an increase in limb volume and sometimes the opposite effect. These discrepancies were explained on the basis that the capillaries in the first case played a more prominent part in the control of the circulation than in the second. The similarity between the mode of action of histamine and that of small doses of adrenaline was illustrated by the finding that in any one experiment the effect of the two drugs was almost in the same direction.

In some of their experiments Dale and Richards denervated a cat's hind limb completely by section of the sciatic and femoral nerves. This produced dilatation of both arterioles and capillaries. The pads of the foot were then warmer and redder than those of the normal limb. A few minutes later the capillaries regained their normal tone; the pads of the foot became pale, but were still distinctly warmer than normal. At this stage the arterioles were still actively dilated because the vasodilator fibres had been stimulated by the nerve section. Under these conditions the vasodilator effects of histamine and of small doses of adrenaline were particularly well shown in the plethysmographic record, but the action of acetylcholine was diminished. When the denervation was performed 24 hours or more before the experiment, the arterioles were no longer actively dilated and the vasodilator effects of all three drugs were enhanced.

In some of the experiments the circulation to the leg was temporarily interrupted. This dilated both the arterioles and the capillaries, so that no dilator responses could be elicited. The arterioles however appeared to recover their tone sooner than the capillaries, because the dilator effect of acetylcholine reappeared earlier than that of histamine or adrenaline.

In some experiments in which histamine caused constriction in the normal leg it caused dilatation in a leg that had been skinned. It might be inferred from this finding that the vasodilatation was mainly confined to the muscle capillaries; the reddening of the skin of the pads of the feet showed, however, that the skin vessels were also dilated.

In perfusion experiments histamine produced vasoconstriction of the cat's leg unless blood was used as perfusion medium and the capillary tone was maintained with adrenaline. Later Burn and Dale (1926) and Dale and Richards (1927) were able to demonstrate the vasodilator effects of both histamine and adrenaline in the perfused cat limb even when they added no drug to maintain the capillary tone. Since Drinker's artificial oxygenator (Drinker, Drinker and Lund, 1922) was used in those experiments, it is possible that the mechanical action of the apparatus on the blood may have released substances that constricted the capillaries. In experiments of this type the tone of the vessels increased during the first minutes of the perfusion and then stayed high for some time before it finally fell again. While the tone was high histamine caused vasodilatation; later it caused vasoconstriction. If at this stage the vascular tone was restored with adrenaline or pituitary extract, histamine regained its vasodilator action. Barium and ergotoxine,[5] which also caused vasoconstriction, failed to restore the vasodilator effect of histamine, presumably because they raise the tone of the arterioles while adrenaline and pituitary extract raise the tone of the capillaries.

Hemingway and McDowell (1926a, b) and Atzler and Lehmann (1927) have emphasized the relation between the accumulation of acid metabolites and the loss of capillary tone. Hemingway and McDowall perfused the hind limbs of cats with phosphate-buffered salt solution. When the pH of the perfusion medium was 7.6 histamine consistently produced vasodilatation; when the pH was 7.4 the peripheral resistance fell and histamine produced vasoconstriction. The accumulation of acid metabolites is no doubt one of the factors that contribute to the eventual loss of capillary tone in perfused limbs, but it is probable that other factors also play a part.

Experiments in which the vasodilator action of histamine has been directly observed will be described in the next section. The vasodilator effect can also be readily demonstrated either by injecting the histamine intra-arterially and measuring the venous outflow (Feldberg, Flatow and Schilf, 1929) or by employing a method described by Richards and Plant (1915). In this method blood from a cat's carotid artery is pumped into the peripheral end of the femoral artery. The histamine is injected into the blood delivered to the femoral artery and the perfusion pressure is recorded (Gaddum, 1933).

(b) Direct observations of the vasodilator effects of histamine

Macroscopic observations. The indirect observations supporting the theory that the vasodilator effect of histamine is exerted on the capillaries and minute arterioles have been discussed above. This action has often been observed directly. It was first described by Eppinger (1913) as a flushing of the skin in man, and has since been observed by a number of different authors (Sollmann and Pilcher, 1918; Dale and Richards, 1918; Lewis and Marvin, 1927; Feldberg, 1927; and others).

The phenomenon has been described in detail by Lewis and Grant (1924). In their experiments a drop of solution containing 1 mg of histamine diphosphate per cc (a 1:3000 solution of the base) was placed on the skin and the skin was pricked through the drop. After some 20 seconds a triple response appeared, made up of:

(1) *A local redness.* This is most clearly displayed in an arm whose circulation has been arrested. It is then circular and sharply defined, with an ultimate diameter of 3–4 mm.

(2) *A spreading scarlet flush or flare.* This surrounds the local red reaction, has an ill-defined diffuse shape, and is often speckled. It has a diameter of 2–3 cm or more. It is due to the dilatation of arterioles under the influence of an axon reflex. It is absent if the nerves have been paralysed with a local anaesthetic, or cut and allowed to degenerate; but simple section of the nerves without degeneration does not abolish the flare.

(3) *A wheal or local oedema.* This covers exactly the same area as the original local red reaction. The wheal develops slowly, reaches a maximum in 3–5 minutes, and then becomes pale. Later its edges lose their initial crispness and the swelling slowly subsides, to disappear completely in about an hour. The whealing is independent of nerves, since it can be produced when these have degenerated; but the arteriolar dilatation associated with the flare does appear to accelerate the formation of the wheal by increasing the local circulation. Wheal formation is unaffected by raising the venous pressure or applying suction to the skin, but it can be stopped by applying positive pressure (see Lewis, 1927).

These detailed observations form part of an extensive investigation of the reaction of the skin to injury; the subject is discussed further on page 172 *et seq.* The dilator action of histamine on the capillaries can also be observed macroscopically when histamine is applied directly to the peritoneum (Dale and Richards, 1918) or to the conjunctiva (Lewis and Marvin, 1925).

The action is particularly well shown in the rabbit ear after degenerative section of all the vasomotor nerves. If histamine (0.03–0.05 mg) is injected into the vein of the opposite ear one can observe that the central artery and its visible branches are narrowed, while the skin between these branches becomes flushed (Feldberg, 1927; Flatow, 1927). In carrying out this experiment it is important to protect the rabbit from temperature changes due to draughts, etc., so as to avoid changes in vessel calibre during the control period (Grant and Bland, 1933).

Microscopic observations. Hooker (1920, 1921) made observations on the capillaries and venules in the ears of cats under ether. After large doses of histamine both kinds of vessel dilated and became filled with stagnant blood. Rich (1921) observed dilatation of the arterioles, capillaries and smallest veins in the omentum of the cat. Similar observations have been made on human skin (Carrier, 1922), cat mesentery (Florey and Carleton, 1926) and cat skeletal muscle (Eppinger, Laszlo and Schürmeyer, 1928; Hartman, Evans and Walker, 1929). Grant (1930) applied histamine locally to the rabbit ear and observed constriction of small arteries and dilatation of arteriovenous anastomoses.

(c) Action on the systemic circulation in different organs

The vasodilatation due to histamine is not confined to particular tissues but appears to be widespread throughout the systemic circulation. Its actions on the coronary, pulmonary and portal circulations will be considered separately later on.

The flushing of the skin due to capillary dilatation is one of the best known and most striking actions of histamine; however observations by Dale and Richards (1918) suggest that histamine dilates the vessels of skeletal muscle even more strongly. Vasodilatation has also been observed in the conjunctiva, the brain, the intestines, the pancreas (see Feldberg and Schilf, 1930) and in the branches of the hepatic artery (Bauer, Dale, Poulsson and Richards, 1932).

The kidney vessels seem to be exceptional in their response to histamine. Dale and Laidlaw (1910) found that histamine produced a fall of kidney volume in the intact cat, which was so great as to suggest an active constriction of the renal vessels. Other accounts have confirmed this finding. Macgregor and Peat (1933), however, when they perfused a dog's kidney with blood that had been arterialized in the animal's own lungs, found that under these conditions histamine in high concentration produced clear dilatation of the vessels. It seems in any case that the vasodilator action of histamine on the kidney is comparatively weak.

The action of histamine on the cerebral vessels is strongly affected by the anaesthesia. In normal human beings or in cats under amytal, histamine causes dilatation in the brain and raises the pressure of the cerebrospinal fluid, in consequence of its direct action on the vessels within the skull. This action probably accounts for the headache caused by an injection of histamine. In cats or dogs under ether the brain vessels are already dilated, and histamine lowers the cerebrospinal fluid pressure: an effect that appears synchronously with, and is probably due to, the fall of arterial pressure (Lee, 1925; Forbes, Wolff and Cobb, 1929; Pickering, 1933).

(*d*) *Action on the systemic circulation in the organs of different species*
The vasodilator actions of histamine are especially easily seen in the perfused leg of a dog or a monkey. In the cat under ether, histamine also produces an obvious vasodilatation; but with artificial perfusion the vessels it dilates are apt to lose their tone, so that the only change seen is constriction of the larger arteries. Nevertheless, as has already been mentioned, the vasodilator action (and also that of adrenaline) can be demonstrated in this situation also, if precautions are taken to maintain the tone of the capillaries.

In the rabbit under ether small doses of histamine are ineffective, and larger doses (0.05–1 mg) produce a pure rise of arterial pressure. In the earlier stages of anaesthesia large doses produce a fall of blood pressure due to pulmonary vasoconstriction (Dale and Laidlaw, 1910). Perfusion experiments in rabbits have always revealed only a vasoconstrictor action of histamine; but as already noted, the vasodilator effect of histamine on the capillaries can be seen in the unanaesthetized animal by direct observation of the conjunctival or ear vessels (Lewis and Marvin, 1927). If the animal is anaesthetized with chloralose instead of ether the vaso-dilator action can be recorded as a fall of blood pressure (Feldberg, 1927). In the guinea-pig the vasoconstrictor action of histamine predominates, and in general no sign of vasodilatation is observed.

In rats under ether large doses of histamine produce a fall of blood pressure, presumably due to peripheral vasodilatation, as in the cat or dog (Voegtlin and Dyer, 1924).

These differences in the vascular responses of different species probably depend on whether the dilator or the constrictor effect occupies the larger part of the vascular tree. In rabbits the capillaries are dilated by histamine while the small arterioles are constricted. In dogs both the small arterioles and the capillaries are dilated and only the large arteries are constricted. In the other animals mentioned, the position in the vascular tree where the constrictor action changes to dilator lies between these two extremes. The correctness of this explanation is evidenced by the observation that histamine in the dog still causes vasodilatation in experiments in which the perfused area is limited to the mesenteric artery and its branches, whereas in the cat with the same procedure the result is vasoconstriction (Burn and Dale, 1926).

Histamine has little or no action on the vessels of the frog (Clark, 1924; Grant and Jones, 1929; Feldberg and Schilf, 1930). In the tortoise it produces a fall of blood pressure, attributed to vasodilatation (Sumbal, 1924).

2. *Action on the lymph*

Dale and Laidlaw (1911) found that 2 mg of histamine increased the lymph flow in the thoracic duct of a dog from about 0.3 cc to 1 cc per minute.

Since the lymph in the thoracic duct comes mainly from the liver, constriction of the hepatic outflow veins is probably partly responsible for this phenomenon in the dog. But the general capillary-dilator effect of histamine must also play a part, as shown by the occurrence of a similar increase in lymph flow from the dog's foot in response to histamine (Haynes, 1932).

3. *Action on the heart, coronary vessels, pulmonary circulation and portal circulation*

(a) *Heart*

Dale and Laidlaw (1910) found that in cats and rabbits small doses of histamine increased both the force and the frequency of the beat (positive inotropic and chronotropic action), whether the heart was observed *in situ* or perfused. Larger doses weakened the heart. On the dog or cat heart–lung preparation, Fühner and Starling (1913) and Rühl (1929) also found that large doses (0.5–2 mg) weaken the heart. The left side of the heart is affected more than the right, since there is a larger rise of pressure in the left auricle than in the right. Smaller doses, according to Feldberg and Schilf (1930), increase the force and frequency of the beat in a heart–lung preparation.

Many other investigators have found that histamine increases the force and frequency of the beat of the isolated heart. Such observations were made on the hearts of cats, dogs, rabbits and guinea pigs (Einis, 1913; Rothlin, 1920b; Abe, 1920; Gunn, 1926; Oppenheimer, 1929). The beat of both auricle and ventricle is increased.

Reports on the action of histamine on the frog heart are contradictory. Stimulation certainly occurs under some conditions, which are not well-defined; higher concentrations (1:100) are inhibitory (see Feldberg and Schilf, 1930).

Table 1. *Action of histamine on the coronary vessels*

Animal	Action	Author
Cat	Vasodilatation	Gunn (1926)
Dog	Vasodilatation	Rühl (1929)
Tortoise	Vasodilatation	Sumbal (1924)
Rabbit	Vasoconstriction	Dale and Laidlaw (1910)
Guinea pig	Vasoconstriction	Viotti (1924)

The effect of histamine on the heart *in situ* has recently been discussed by Dixon and Hoyle (1930). They showed that the alterations of auricular pressure seen after histamine result from its action on other organs rather than from its direct effect on the heart itself.

(b) Coronary vessels

As Table 1 illustrates, the coronary vessels of different species respond differently to histamine. The differences correspond to those observed in the peripheral vessels of the greater circulation.

The contrast in the responses of the cat and rabbit has been shown with especial clarity by Gunn (1926). He perfused the hearts with Ringer's solution, and found that histamine caused vasoconstriction in the rabbit and vasodilatation in the cat. Histamine also caused coronary dilatation in the dog heart–lung preparation (Rühl, 1929). According to Feldberg and Schilf (1930) this coronary dilatation can be evoked by doses of histamine that are too small to affect the beat, and they conclude that it must represent a direct action on the vessels.

Isolated strips of coronary artery from man, ox, horse, pig or dog are contracted by histamine (Barbour, 1913; Rothlin, 1920a; Anrep, 1926; Cruickshank and Rau, 1927). Such observations probably have little or no relation to the regulation of coronary flow *in vivo*.

(c) Pulmonary circulation

Histamine constricts the pulmonary vessels of cats, dogs and rabbits. Dale and Laidlaw (1910) who first observed this effect in the cat supposed that it represented constriction of the arterioles. Inchley (1923), however, devised a method of perfusing the arteries and veins separately with Ringer's solution, and showed that both were constricted by histamine. Later (Inchley, 1926), he found that isolated veins in general, and pulmonary veins in particular, are more sensitive to histamine than are isolated arteries; he concluded that the action of histamine is principally on the veins. Mautner (1923) held the same view, and supported it with experiments in which the lung volume of dogs with intact circulation was recorded simultaneously with pressures in the pulmonary artery and the left auricle.

Gaddum and Holtz (1933) perfused the lungs of dogs and cats with blood, and found that the vasoconstrictor action of histamine was accompanied by small changes in lung volume; these changes depended on whether the perfusion was at constant flow or at constant pressure. They came to the conclusion that histamine under these conditions constricted both arteries and veins, so that the lung volume remained almost constant, though not perfectly so.

Isolated rings of pulmonary vessels contract in response to histamine (see Feldberg and Schilf, 1930).

(d) Portal circulation

The effect of histamine on pressure and flow in the portal vein has been studied frequently. In such experiments these variables may be affected by changes in the general arterial pressure or in either splanchnic or hepatic vascular resistance. The effect of drugs on the hepatic vascular resistance varies with the animal species; so it is not surprising that contradictory results have been obtained. (See pages 25–6 for the action of histamine on the mesenteric vessels.)

When vascular tone is high, histamine dilates the hepatic artery branches in the dog, cat or goat. This presumably corresponds to its effect on capillaries in other tissues. It causes some constriction of the branches into which the portal vein divides on entering the liver (Bauer, Dale, Poulsson and Richards, 1932).

In addition, histamine has in some animal species a powerful and significant action on the outflow of blood from the liver (*Lebersperre*). This appears to reflect its action on a special muscular mechanism that controls the opening of the hepatic veins into the vena cava. The mechanism was first brought to notice by the Vienna school (Mautner and Pick, 1915, 1922, 1929; Lampe and Mehes, 1926; Baer and Roessler, 1926). These authors showed that histamine produces a great swelling of the liver in the dog, attributable to an action on the hepatic veins. This conclusion has been confirmed by Bauer, Dale, Poulsson and Richards (1932), who used an improved experimental procedure in which both the hepatic artery and the portal vein of a dog were perfused simultaneously with blood. They showed that the site of the altered resistance was near the openings of the hepatic veins into the vena cava. In contrast to the situation in the dog, they found no evidence at all for such a mechanism in the cat or the goat. The rise in portal pressure, which Feldberg (1928) observed in the cat when he injected histamine directly into that vein, was probably due to a direct action on the veins into which the portal divides on entering the liver.

A possible significance of this action of histamine on the veins of the liver is that small quantities of histamine may be absorbed from the intestine during digestion and carried to the liver, where by restricting venous outflow they may increase the amount of blood in the liver itself and in all the viscera drained by the portal vein. Within limits, such congestion might be of advantage during digestion (cf. Dale, 1929).

4. *Action on smooth muscle*
(a) *General*

Histamine produces a contraction of nearly all smooth muscle. The uterus and intestine of some species are particularly sensitive and contract in response to minute concentrations of histamine; but both these organs have been found to relax if they remain in contact with histamine in high concentration for a sufficient time, the ultimate effect being inhibition of the muscle's spontaneous activity. With the rat uterus and the frog intestine this inhibitory effect of large doses is commonly the only one seen. Histamine also strongly contracts the bronchi and the retractor penis, with a weaker action on the spleen (Dale and Laidlaw, 1910). It is also a weak stimulant of the gall bladder (Brugsch and Horsters, 1926), the ureter (Rothmann, 1927; Gruber, 1928) and the urinary bladder (Adler, 1918; Macht, 1926).

The response of the pupil to histamine must be interpreted in the light of the drug's ability to discharge adrenaline from the suprarenals. Its direct action is to dilate the pupil in man, dog, cat and frog, and to constrict it in the rabbit (Matsuda, 1929; Hadjimichalis, 1931).

The action of histamine on arterial and venous smooth muscle has been discussed under the heading of the circulation (page 20 *et seq.*). Anaesthetics weaken that action (Katz, 1929).

None of these actions is specifically blocked by atropine in dosage high enough to abolish the effect of acetylcholine, but in the case of some tissues, e.g. intestine (Feldberg, 1931) or bronchi (see Baehr and Pick, 1913), histamine has little or no effect if the tissue is already depressed by a high concentration of atropine. Whether this failure of the heavily atropinized tissue to respond to histamine represents a genuine antagonism must be regarded as doubtful.

(b) Uterus

The action on the uterus was the first pharmacological action of histamine to be observed. Dale and Laidlaw (1910) obtained a definite contraction of the isolated horn of the virgin guinea pig uterus when they added histamine in a concentration of 1:250000000. A tenfold higher concentration always produced a strong response. The uterus becomes less sensitive in pregnancy. The sensitivity is increased by a sudden fall, and decreased by a sudden rise, in the tonicity of the solution in which the uterus is bathed (Dale, 1913b).

The uteri of cat, dog and rabbit also contract when treated with histamine, but are less sensitive than that of the guinea pig. The isolated uterus of the rat is rhythmically active when it is isolated, and this rhythm is usually suppressed by histamine (Guggenheim, 1913). The uteri of some exceptional rats respond to histamine by contracting, but inhibition is much more commonly observed. This test is sometimes used to confirm the pharmacological identification of a substance when other tests have suggested that it is histamine. With the mouse uterus, different workers have obtained apparently conflicting results, but their observations were reconciled by Abel and Macht (1919), who found that small doses cause contraction and large doses inhibition, while intermediate doses had no effect at all.

Histamine stimulates the human uterus. Jäger (1913a, b) came to the conclusion that histamine injected subcutaneously or intramuscularly during labour increased the pains and hastened delivery. Guggenheim (1914) and Quagliarello (1914) showed that histamine will contract an isolated strip of human uterus. The most interesting demonstration of the effect of histamine on the human uterus, however, was that of Bourne and Burn (1927), who recorded intrauterine pressure during labour. They found that the subcutaneous injection of 2 mg of histamine base caused a series of violent contractions of the uterus, which went on for up to 45 minutes. Dilatation of the cervix did not occur during this time, and the histamine did not appear to accelerate labour.

Moir (1932) found that intramuscular histamine evoked a slight contraction of the human puerperal uterus; histamine by mouth had no effect.

(c) *Intestine*

Histamine contracts the isolated intestine of the cat and usually increases its spontaneous activity (Dale and Laidlaw, 1910). The longitudinal muscle of the distal end of the guinea pig ileum is especially sensitive to histamine (Guggenheim and Löffler, 1916a; Kendall and Shumate, 1930), and will sometimes respond to a concentration of 1:1 000 000 000. It can be used as the basis of a test for histamine which is at least as sensitive as, and distinctly more rapid than, the test on the isolated uterus of the guinea pig. Histamine also contracts isolated intestinal segments from the cat, dog, rabbit, rat or hen, but not from the frog. In high concentration (1:20000) it inhibits the frog intestine (see Feldberg and Schilf, 1930). The intestine of the cat is more sensitive than that of the rabbit or rat but less sensitive than that of the guinea pig.

The rectal caecum of the hen is particularly sensitive to histamine (Barsoum and Gaddum, 1935a).

When histamine is applied in high concentration (1:75000 to 1:500) to the cat or the rabbit intestine, the muscle usually gives an immediate strong contraction, which however does not last long and is followed by loss of tone and absence of rhythmic movements. When the concentration is in the range 1:750000 to 1:75000 there are several periods of contraction and relaxation at intervals of 1–4 minutes (Olivecrona, 1921).

When the ganglion cells are removed from a piece of cat intestine, the muscle still responds to histamine by contracting, but it is much less sensitive than before (Gasser, 1926; Esveld, 1928). Histamine reduces the threshold of the guinea pig intestine for electrical stimulation (Bishop and Kendall, 1928).

In the intact animal, intravenous histamine increases the movements of the entire intestinal tract. This response has been studied in man with X-rays and in animals after the abdomen has been opened under anaesthesia. It is, however, best observed in an unanaesthetized rabbit which has been provided with a celluloid window in its abdominal wall. An injection of 50 μg of histamine then causes a strong increase of peristaltic activity, especially in the small intestine, with visible rapid propulsion of the intestinal contents. The colon becomes active later, and eventually, after several injections, faeces appear at the anus (Dale and Feldberg: see Feldberg and Schilf, 1930).

The motor response of different parts of the intestine has also been recorded in the intact animal by means of the 'filling method' described by Babkin (MacKay, 1930; Tidmarsh, 1932).

(d) Bronchial musculature

Guinea pigs are in many ways specially sensitive to histamine, and this is particularly true of their bronchial muscles. The most striking effects of an intravenous injection of histamine, which are described under the general effects of large doses (page 36 *et seq.*), are all due to bronchospasm. Histamine also constricts the bronchi in rabbits, dogs, cats and rats, but in these animals its effects on the circulation are relatively more prominent (see Feldberg and Schilf, 1930).

The bronchoconstrictor action of histamine is also seen in spinal animals (Dale and Laidlaw, 1910; Koessler and Lewis, 1927) and in perfused lungs (Baehr and Pick, 1913; Hanzlik and Karsner, 1920; Bartosch, Feldberg and Nagel, 1932a, b; Daly and Schild, 1935, etc.) Also responding with contraction are the isolated bronchial strips from the dog, cow, pig or guinea pig, but only when the concentration is rather high (Titone, 1914; Quagliarello, 1914; Macht and Ting, 1921; Epstein, 1932). These isolated strips, which necessarily come from large bronchi, are apparently rather insensitive to drugs.

In the dog denervated bronchi may be less sensitive than normal bronchi, but this does not prove that the bronchoconstriction is mainly of central origin (Houssay and Cruciani, 1929); its main cause is undoubtedly the direct action of histamine on the bronchial smooth muscle.

(e) Gall bladder

Histamine produces contraction of the isolated gall bladder of the dog or the guinea pig (Brugsch and Horsters, 1926; Erbsen and Damm, 1927; Halpert and Lewis, 1930). It also produces contraction of the gall bladder in the intact animal, but the effect is inconstant and may be secondary to the fall of blood pressure or to the passage of gastric content into the intestine (Ivy and Oldberg, 1928).

5. Action on glandular secretion

(a) General

The injection of histamine causes the secretion of saliva, pancreatic juice, bronchial mucus and tears (Dale and Laidlaw, 1910). It stimulates gastric secretion (Popielski, 1920a).[6] It appears to have no effect on the secretion of milk or sweat. It has little or no action on the secretion of mucus by intestinal glands (Florey, 1930).

Its actions on the salivary glands and the pancreas are antagonized by atropine (see Feldberg and Schilf, 1930).

(b) Salivary secretion

In some types of experiment the salivary secretion evoked by histamine may depend partly on impulses carried by the chorda tympani (Popielski, 1920) and partly on adrenaline discharged from the suprarenals, but it is not wholly accounted for by these factors. Even after atropine and ergotoxine a small secretory effect of histamine can be detected (MacKay, 1929). The effect is partly a true secretion and partly the expression of saliva from the ducts by contraction of their smooth muscle (Stavraky, 1931).

When repeated doses of histamine are given the gland becomes fatigued and is then insensitive to both histamine and adrenaline (Fröhlich and Pick, 1912; MacKay, 1927a). Adrenaline may have the effect of diminishing the secretory action of histamine (MacKay, 1929).

The secretory action of histamine is small unless the chorda has been stimulated. After stimulation of the chorda its action is increased, but the increase does not occur if the animal has received atropine (MacKay, 1927a). Since atropine abolishes the secretory but not the vasodilator actions of the chorda, it is clear that the increase in the response to histamine is not due merely to the local vasodilatation resulting from chorda stimulation. The effect probably has a complex explanation (see Babkin and McLarren, 1927). After the injection of pilocarpine, on the other hand, the action of histamine is reversed and it becomes inhibitory. This is probably a secondary effect, related to the action of these drugs on blood flow through the gland (MacKay, 1927b).

Saliva secreted under the influence of histamine has a higher content of organic matter and a lower content of ash than saliva produced by stimulation of the chorda (Stavraky, 1931).

(c) Pancreatic secretion

Lim and Schlapp (1923) found that histamine produced no secretion of pancreatic juice after the pylorus had been tied off, and concluded that its action on pancreatic secretion was secondary to the entrance of gastric juice into the duodenum. Molinari-Tosatti (1928) found in contrast that histamine still evoked secretion after both the pylorus and the bile duct had been tied off. The mechanism of the histamine-induced secretion is thus still unclear.

(d) Gastric secretion

The gastric secretagogue action of histamine has been demonstrated in the dog, cat, rabbit, guinea pig, duck and frog (Keeton, Koch and Luckhardt, 1920) as well as in man (Carnot, Koskowski and Libert, 1922). It can be seen in a gastric pouch whose nerve supply has been completely removed and allowed to degenerate (Ivy and Javois, 1924). It involves an increased secretion of water and acid, but not of pepsin (Vineberg and Babkin, 1931). There can be no doubt that it is due to a direct action of histamine on the gland cells.

This action appears to be remarkably resistant to fatigue (Lim, 1924; Lim and Liu, 1926). It only appears when a sufficient concentration of histamine is maintained in the blood over several minutes. This is usually achieved by injecting the histamine subcutaneously, but the result is the same if it is given by slow intravenous infusion. It is ineffective if it is given in a single large intravenous dose, presumably because it disappears too rapidly from the blood (Gutowski, 1924).

Histamine is almost certainly the active constituent in extracts of gastric mucosa that produce the gastric secretagogue that Edkins attributed to 'gastrin' (Sacks, Ivy, Burgess and Vandolah, 1932; Gavin, McHenry and Wilson, 1933).[7]

The gastric secretory response to histamine has been widely used clinically as a test of the secretory capacity of the gastric mucosa. It has been found that histamine is often more effective than a test meal and can evoke a secretion when the latter is ineffective. Some investigators have therefore given histamine instead of a test meal, while others have used both and have distinguished a special type of case in which there is a response to histamine but not to a test meal. The histamine test now has a large literature, which will not be discussed here; for a fuller account see Feldberg and Schilf (1930) and Best and McHenry (1931).

Addendum: development of gastric ulcers after histamine

The hypersecretion of acid gastric juice induced by histamine can lead to the development of gastric ulcers. This action was described for cats by McIlroy (1928) and for rats by Büchner, Siebert and Molloy (1929); Harde (1932) observed a similar effect in mice and guinea pigs. The ulcers appeared only if the animals were allowed to fast; apparently histamine is not otherwise ulcerigenic (Bücher, Siebert and Molloy, 1929; Strömbeck, 1932; Brummelkamp, 1934). According to Brummelkamp ulcers can be produced in rats without histamine treatment if they are alternately fed for one day and fasted for two days, water being given throughout. The presence of free acid is critical for ulcer formation.

McIlroy did experiments that point in the same direction. He produced surgical defects of the pyloric mucosa in cats, and gave them subcutaneous histamine every second day. The defects did not heal in these animals as they did in his control animals; instead they became larger, and one of the animals suffered a perforation. O'Shaughnessy (1931) observed ulcers in cats not only after injecting histamine subcutaneously, but also after injecting it directly into the gastric mucosa. Ulcers so produced might resemble chronic ulcers seen in man.

6. *The relation of histamine to the suprarenals*

Dale and Richards (1918) drew attention to the antagonism between the actions of histamine and adrenaline on the capillaries. When these vessels are dilated by histamine their tone may be restored by adrenaline. They supposed that one of the functions of the suprarenals might be to maintain the tone of the capillaries. Later Dale (1920a) showed that if a cat's pupil was sensitized to adrenaline by denervation, the injection of histamine caused it to dilate. He attributed this finding to secretion of adrenaline induced by the action of histamine on the suprarenals. Extirpation of the suprarenals in the cat produced effects that had some features in common with histamine shock, and adrenalectomized animals were especially sensitive to histamine. This effect, however, is presumably due to the loss of the suprarenal cortex rather than the medulla. In confirmation of this view it has been found that suprarenalectomy lowers the resistance of rats to histamine and that their resistance can be restored to normal by injection of bovine cortical hormone (Marmorston-Gottesmann and Perla, 1931; Wyman, 1928).

Dale's observations on the pupil were confirmed and extended by Kellaway and Cowell (1923). They showed that the effect disappeared when the suprarenals were extirpated. The mechanism may be partly reflex, since a fall of blood pressure may evoke a reflex secretion of adrenaline (Anrep and de Burgh Daly, 1924); but the direct action of histamine on the suprarenals certainly plays an important part in the reaction, since Kellaway and Cowell still obtained dilatation of the pupils in response to histamine after they denervated the suprarenals.

Burn and Dale (1926) obtained further evidence for the same mechanism by studying the effect of histamine on the blood pressure. They had noticed that when the blood pressure of a cat was low the primary depressor effect of a histamine injection was complicated by a secondary rise of pressure accompanied by acceleration of the heart. This rebound of the blood pressure was more striking if the cat's circulation was restricted by evisceration. It was abolished by removing the suprarenals, and reversed by injecting enough ergotamine to reverse the action of adrenaline. Thus there could be no doubt that the pressor effect of histamine was secondary to the liberation of adrenaline.

Szczygielski (1932) demonstrated this effect of histamine on the suprarenals by injecting small doses (0.05–0.3 µg) into the central stump of the coeliac artery in an eviscerated cat. This injection caused a rise of blood pressure in cats but not in rabbits. The action was not abolished by atropine.

This output of adrenaline must be a complication in many kinds of experiment. MacKay (1929), for example, encountered it in her study of the action of histamine on the salivary glands. She found that adrenaline weakened the secretory action of histamine, and by removing the suprarenals she regularly succeeded in obtaining a clear secretory response from cats whose glands had previously seemed refractory to histamine. It is interesting in this context that McCarrison (1924) found that long-term administration of histamine to young rats increased the size of the suprarenals.

7. General effects of large doses; effect on blood composition

The effects of large doses of histamine can in general be interpreted as an intensification and prolongation of the various effects seen in more transient form with small doses. The condition produced by these large doses bears many points of resemblance to traumatic shock and other pathological states. It will be convenient to devote a special section to the discussion of histamine shock.

When a large enough dose of histamine is injected, it causes death either from its action on smooth muscle (rabbit, guinea pig) or from its action on the capillaries (cat). In the former case the action is weakened by anaesthesia; in the latter case it is often increased.

Guinea pigs. The intravenous injection of 50 µg of histamine into an unanaesthetized guinea pig of 200–250 grams is followed by a typical series of symptoms which end with death through bronchospasm within 3–4 minutes. Usually the animal begins to struggle and gasp before the injection is completed. It makes chewing movements and tries to breathe but cannot. The gasps become rarer, and the cyanosis more intense. Urine is passed, and peristalsis in seen through the abdominal wall. The nose continues to make respiratory movements for a short time after respiration has otherwise ceased. The heart continues to beat after even that movement has ceased. When the chest is opened after death, it is found that the lungs are fixed in distension and that the right heart is filled with blood.

Rabbits. The lethal dose with intravenous injection is 0.3–0.6 mg per kg (Oehme, 1913; Leschke, 1913). The symptoms are like those seen in the guinea pig, but the heart stops before the respiration, and death is due to obstruction of the pulmonary circulation, leading to acute distension of the right ventricle (Dale and Laidlaw, 1910).

Rats and mice. They are very insensitive to histamine. Large doses lead to death apparently as a result of fall in blood pressure. The lethal dose is about 250 mg per kg (see Feldberg and Schilf, 1930).

Cats. The action of large doses of histamine was studied in detail by Dale and Laidlaw (1919).

If a dose of 1 mg per kg or more is injected into an anaesthetized cat, there is first a rapid immediate fall of blood pressure due to pulmonary constriction. This fall is succeeded by a temporary rise due to a general increase of arterial tone. The blood pressure then falls again slowly to a low level, from which it may never recover. During this last period the condition of the animal resembles traumatic shock. The central feature of this histamine shock is a great diminution of the circulating blood volume. This is due partly to the accumulation of blood in dilated capillaries and partly to escape of plasma through the capillary walls. The importance of the latter factor is shown by the fact that the concentration of cells in the circulating blood rises by 30–50 per cent.

In the unanaesthetized cat the injection of a large dose of histamine leads to vomiting, defaecation, salivation, laboured breathing and partial narcosis. The pupils are constricted, but dilate if the animal is disturbed.

Dogs. The general picture of intoxication in the dog differs from that in the cat chiefly on account of the great increase in hepatic venous resistance, which leads to an accumulation of blood in the portal circulation and thus intensifies the fall of systemic arterial pressure. The lethal dose is about 3 mg per kg.

Man. The subcutaneous injection of 2–8 mg produces definite effects (Kehrer, 1912; Eppinger, 1913; Jäger, 1913a, b; Schenk, 1921; Fenyes, 1930, and others). The intravenous injection of 0.001 to 0.2 mg has similar consequences (Harmer and Harris, 1926; Best and McHenry, 1931). The effects include flushing, palpitations of the heart, vomiting, asthma and a profound fall of blood pressure, as well as a great increase in the secretion of gastric juice (see page 33).

An intravenous injection is followed by a characteristic series of sensations, the intensity of which is much diminished when the injection is given slowly. 'A rapid injection of 0.015 mg of the base may produce a very definite metallic taste, a feeling of increased temperature, particularly in the face and neck, a feeling of tightness in the head, and a headache which may last for several hours. With larger doses (0.15–0.2 mg injected within 10 seconds) a very distinct metallic taste was the first symptom noted…This persisted for only a few minutes. The flush was felt

Table 2. *Alterations in blood composition after histamine*

Erythrocytes	+ Cat	Dale and Laidlaw (1919)
	+ Dog	
	0 Rabbit	Underhill and Roth (1922)
Leucocytes	− Cat	Dale and Laidlaw (1919)
Viscosity	− Dog	Waud (1928)
Surface tension	− Guinea pig	La Barre (1926a, b)
Alkali reserve	− Dog	Underhill and Ringer (1921)
	− [a]	Hiller (1926a, b)
	− [a]	La Barre (1926a, b)
	− [a]	Boyd, Tweedy and Austin (1928)
	+ Rabbit	Katzenelbogen (1929)
	+ Man	Martin and Morgenstern (1932)
Clotting time	−0 Dog	Dale and Laidlaw (1911)
	0 Rabbit	Zunz and La Barre (1926)
	+0 Rabbit	Smith (1920)
Chloride	− Dog	Drake and Tisdall (1925)
	− Man	Saradjichvili and Rafflin (1925)
Non-protein nitrogen	+ Dog	Hashimoto (1925)
	+ [a]	Drake and Tisdall (1926)
Serum protein	− Cat	Derrer and Steffanutti (1930)
Cholesterol	− Dog	Cornell (1928)
	+ Dog	Tangl and Recht (1928)
Sugar	+ Dog	Chambers and Thompson (1925)
	+ Guinea pig	La Barre (1926c, d, e)
	+ [a]	Katzenelbogen and Abramson (1927)
	+ Rabbit	Menten and Krugh (1928)
Lactic acid	+ Dog	Chambers and Thompson (1925)
Calcium	0 Guinea pig	La Barre (1926c, d, e)
	+ Dog	Kuschinsky (1929)
Inorganic phosphate	+ Dog	Chambers and Thompson (1925)
	− Guinea pig	La Barre (1926c, d, e)
Lipase	− [a]	Steffanutti (1930)
Oxygen	− Cat	Rühl (1930)
	− Man	Tang (1932)

+ = rise of concentration; − = fall of concentration; 0 = no change.
[a] The failure to state the species is a shortcoming of the original.

5 to 10 seconds later and was noted by observers a few seconds after that. The sensation of tightness in the head was very uncomfortable and was accompanied by ocular disturbances' (Best and McHenry, 1931). In many people asthma is a prominent feature of the response.

Various suggestions have been made as to the significance of the changes in blood composition shown in Table 2. The rise in the erythrocyte count is presumably a direct consequence of the escape of plasma from the circulation. The changes in alkali reserve and chloride concentration are probably due to the secretion of acid (or alkaline) liquids by the stomach (or pancreas). The rise in blood sugar and lactic acid may be secondary to the release of adrenaline evoked by histamine. The fall in oxygen is probably due to vasomotor and other changes in the lungs, which interfere with the passage of gases through the alveolar wall.

C. The natural occurrence of histamine: its extraction, estimation, isolation and distribution

1. Introduction

The first indication of the natural occurrence of histamine came in 1910, when Ackermann and Kutscher found that it was produced by the bacterial decomposition of histidine. In the same year it was found in extracts of ergot. Such extracts strongly contracted the isolated uterus, and one of the methods for estimating their activity was based on that action (Kehrer, 1907). This effect could not be accounted for by the active principles that were known at that time. Histamine was isolated from such extracts simultaneously by Barger and Dale (1910) and by Kutscher (1910). Barger and Dale identified it correctly, and they confirmed the identification by comparing it with a sample of the histamine that Ackermann had obtained from the putrefaction of histidine. In his original publication Kutscher considered that the base he had isolated from ergot was closely related to histamine but not identical with it, because of an apparent difference in the pharmacological actions of the two preparations.

Since that time histamine has been isolated from many different sources. The most convincing evidence for its natural occurrence in fresh extracts of animal tissues was obtained by Best, Dale, Dudley and Thorpe (1927). They showed that it was probably responsible for many of the unknown effects of tissue extracts; in particular, it was probably the main active constituent of Popielski's 'vasodilatin'.[8]

2. The preparation of extracts

Tissue histamine appears to be very stable; the minced tissue can be allowed to stand for hours without detectable loss or formation of histamine. Autolysis eventually leads to the disappearance of the base, but the process is too slow to have much effect in less than 24 hours at room temperature. On the other hand histamine may be either destroyed or formed in minced tissues or extracts if time is allowed for bacterial growth to occur.

Histamine in the tissues is only very loosely combined with other materials. It can be extracted by a number of simple procedures:

(a) The minced tissue is placed in 95 per cent alcohol (2–4 cc per gram of tissue). After standing for 24 hours the extract is filtered, and the tissue is again extracted with 60 per cent alcohol for a further 24 hours. The combined filtrates are concentrated *in vacuo* until the alcohol is removed. The extract is defatted by shaking with ether, and when the ether has been removed *in vacuo* the extract can be used in biological tests. This is the method employed by Best, Dale, Dudley and Thorpe (1927); similar methods have been used by other authors. Best and McHenry (1931) found that about 60 per cent of the total histamine was recovered in the first alcoholic extract and 30 per cent in the second; a third extraction recovered a further 5 per cent.

(b) A more drastic procedure was employed by Hanke and Koessler (1920c), with hydrochloric acid as extractive. As modified by Best and McHenry the procedure is as follows:

Add 150 cc of 10 per cent HCl to 20 g of tissue. Heat at 95 °C for an hour or less; this treatment disintegrates the tissue completely, but is insufficient to form histamine from histidine (Best and McHenry, 1931). Evaporate the mixture to dryness *in vacuo*. Add successively two portions each of 50 cc of 95 per cent alcohol and remove this along with some of the acid by distillation. Suspend the residue in water and neutralize the remaining trace of acid with 20 per cent NaOH. Filter through paper and wash paper and flask repeatedly with distilled water. The filtrate is used for biological testing.

(c) Histamine can be removed quantitatively from minced tissues by electro-dialysis (MacGregor and Thorpe, 1933a, b). A cell is constructed having three water-filled compartments separated by parchment paper or cellophane. The minced tissue is placed in the central compartment; the other compartments contain the anode (carbon) and the cathode (nickel). A current of 0.2–0.5 ampere is applied, which carries the histamine to the cathode. Samples are withdrawn at 15-min intervals from the cathodal chamber and tested with the Pauly reaction; when two successive samples give the same colour the extraction is taken to be complete. The fluid from the cathodal chamber is neutralized and used for the biological test. Special arrangements are required for keeping the solutions cool. As the whole procedure takes less than 2 hours, it is faster than the methods described above. The results agree with those obtained by method (a) and are about 30 per cent below those obtained by method (b).

(d) Tissue histamine can also be extracted by trichloroacetic acid (Gaddum and Schild, 1934; Barsoum and Gaddum, 1935a), and no doubt by a variety of other procedures.

3. The detection and estimation of histamine

The isolation of histamine from tissue extracts inevitably involves large losses and can only succeed when large amounts of starting material are available. For this reason a number of simpler and more sensitive methods of estimating histamine have been described and developed into quantitative procedures.

(a) Chemical methods

Methods based on the Pauly reaction. These depend on the formation of coloured compounds resulting from the linkage of the imidazol ring of histamine with diazo compounds. For the test as described by Pauly (1904) diazobenzenesulphonic acid (*p*-phenyldiazosulphonate) is prepared by adding potassium nitrite to a solution of sulphanilic acid in hydrochloric acid. The solution to be tested is made strongly alkaline with sodium carbonate, and the diazobenzenesulphonic acid is added. A positive reaction shows as a dark cherry-red colour, which retains its tint on dilution and changes to orange when the solution is made acid. The reaction is not specific for histamine; histidine gives the same result. Tyrosine also gives a red colour but this becomes yellow on dilution and bronze-yellow on acidification.

Koessler and Hanke (1919a; see also Hanke and Koessler, 1920, 1924c) developed a quantitative method for determining histamine in fluids that contain large amounts of impurities. Their method is based on the Pauly reaction and involves complicated procedures for removing impurities that interfere with the reaction. An alcoholic extract is taken to dryness and hydrolysed with HCl; HCl and ammonia are removed by distillation; histamine is then precipitated as the phosphotungstate. After decomposition of the phosphotungstate the histamine is extracted from an alkaline solution with amyl alcohol, the histamine remaining in the aqueous layer. The histamine is then extracted from the amyl alcohol with sulphuric acid and precipitated with silver in the presence of baryta. The silver salt is decomposed with HCl, and the preparation is dried and extracted with methyl alcohol or hot chloroform. The Pauly reaction with diazobenzenesulphonic acid is now applied to the purified preparation and the colour is compared with that of standard mixtures of methyl orange and congo red. Tables are given for estimating quantities of histamine ranging between 1 and 50 µg.

One might expect that such a procedure would involve large losses, but Koessler and Hanke found that they could with their method recover quantitatively 10 mg of histamine that had been added to 40 g of casein. According to other authors these methods do involve losses when they are applied to organic material. Gerard (1922) was unable to obtain quantitative recovery of added histamine. Best, Dale, Dudley and Thorpe (1927) found that histamine was not extracted by amyl alcohol from purified extracts of liver even when it was undoubtedly present. They also found that extracts often failed to give the Pauly reaction even after histamine had been added to them. Hanke and Koessler had recognized this possibility, but considered that their procedure removed the materials that interfere with the reaction. Thorpe (1928) was unable to recover more than 93 per cent of the histamine from the phosphotungstate precipitate.

There is little evidence as to whether the results obtained with Koessler and Hanke's procedure agree with those obtained by biological methods. A certain lack of faith in the procedure has arisen, not from any proof that it is unreliable as carried out by Koessler and Hanke, but mainly from observations incidental to other investigations. If the biological tests had not been simpler and more sensitive, this chemical test would, no doubt, have been much more widely used than it has been.

Gebauer-Fuellnegg (1930) modified the Pauly reaction by using *p*-nitro-aniline instead of diazobenzenesulphonic acid. The coloured substance produced with this reagent by histamine or histidine can be extracted by shaking the watery solution with butyl alcohol, and the red colour persists in the butyl alcohol. It the test is applied to tyrosine the butyl alcohol extract is coloured brown.

A rapid assay based on this reaction has been used by MacGregor and Thorpe (1933b); they matched the colour in the colorimeter of Rosenheim and Schuster (1927), which employs tinted Lovibond glasses. They did not attempt to remove impurities, and they emphasize that the test used in this manner is not specific, and gives a much more intense colour than would be given by the histamine present in the solution.

Methods based on other colour reactions. Histamine and histidine give a reddish colour with bromine (Knoop, 1908). In the modification of this test proposed by Hunter (1922a, b), excess bromine and pigments were removed by shaking the solution with chloroform. Knoop's reaction has not been adapted for quantitative estimation. It is less sensitive than the Pauly reaction.

Zimmermann (1929) has described a test that is said to be specific for histamine, but is not sensitive enough to detect quantities less than 0.5 mg. The extract is mixed with cobalt nitrate and alkali under anaerobic conditions in a Thunberg tube. If histamine is present a violet colour develops.

(b) Biological methods

None of the biological tests that have been used to detect histamine in tissue extracts is as specific as some of the tests used for acetylcholine. Commonly, other substances present in the extracts have actions so like those of histamine that special precautions are necessary to exclude their contribution to the response on which the determination is based.

Choline and acetylcholine may be present, but their effects can generally be excluded by using atropine. Pharmacological tests for histamine are however likely to give results that are less sharp when atropine is present, so this drug is used only when it is necessary.

Adenosine derivatives and the still unidentified depressor substances kallikrein and substance P have some actions that resemble those of histamine, but they are destroyed when the solution is boiled with acid. For this reason methods of extraction that involve boiling with acid, like those of Koessler and Hanke (described above), are likely to be especially specific; but it has to be remembered that prolonged and intense heating with acid can convert histidine into histamine (Ewins and Pyman, 1911; see page 19).

The presence or absence of histamine in posterior pituitary extracts can be demonstrated by adding an equal volume of 2N NaOH, letting the mixture stand at room temperature for two hours, and then testing it on the guinea pig uterus. Such alkali treatment destroys the oxytocic principle but not histamine (Dudley, 1920).

As further support for the pharmacological identification one can apply tests that depend on the chemical and physical properties of histamine. Though histamine is stable at high temperatures in most acids, it is unstable in nitric acid, and it is of course destroyed by ashing. It is dialysable; it is soluble in alcohol, but insoluble in ether. Tests based on these properties must be used cautiously because the properties may be profoundly modified by the presence of impurities. One should always do control experiments in which histamine is added to the extract.

The most convincing evidence for the pharmacological identification of histamine is obtained when an extract can be compared quantitatively with pure histamine by a number of quite different methods and gives estimates that agree quantitatively (Bartosch, Feldberg and Nagel, 1932a, b, 1933; see also Gaddum and Schild, 1934). To achieve this, it may be necessary to apply treatments to the extract that will remove impurities. Some *in vitro* preparations can be made insensitive to histamine by adding large amounts of histamine to the bath, without loss of their sensitivity to other drugs. This phenomenon can help to confirm the identification of an unknown active material with histamine (Barsoum and Gaddum, 1935a).

Blood pressure. The most generally useful test for the determination of histamine in extracts depends on the fall of blood pressure that it produces in a cat when given by vein. The cat is usually anaesthetized with ether.

Injections are made at regular intervals of 3 min, and the drum is stopped between injections to facilitate comparison. Histamine in doses of 0.3–1 μg and extract are given alternately, and the dose of the extract is adjusted until its effect matches that of the histamine. With appropriate modification the method can be used with other anaesthetics, e.g. chloralose or urethane.

Histamine also lowers the blood pressure of the dog, and raises that of the rabbit under ether anaesthesia. These animals are less sensitive than cats, but may be used in confirmatory tests.

Uterus. Dale and Laidlaw (1912) described a satisfactory method for using the guinea pig uterus for quantitative estimations. Burn and Dale (1922) have modified the method. Histamine must be used as standard. The test is not specific, since pituitary extracts, choline esters, adenosine derivatives and a number of other substances can also cause contraction. The stimulant action of choline esters can be abolished by atropine.

The uteri of other animals are generally much less sensitive. The rat or mouse uterus is generally inhibited by histamine and may offer a useful confirmatory test. Adenosine derivatives also inhibit the rat uterus, but acetylcholine stimulates it.

Intestine. Guggenheim and Löffler (1916a) advocated the use of the Magnus preparation of the guinea pig intestine for the estimation of histamine. Its reaction is faster than that of the uterus, and it is more sensitive. Adenosine derivatives inhibit the intestine. The stimulant action of choline esters can be abolished by atropine (cf. Forst and Weese, 1926). The rectal caecum of the hen is especially sensitive to histamine (Barsoum and Gaddum, 1935a). The cat intestine, and also the rabbit intestine, can be used in confirmatory tests, but both are much less sensitive.

Bronchi. If 50 μg of histamine is injected intravenously under local anaesthesia into a guinea pig weighing 200–250 g, it produces the typical signs of bronchial spasm ending in death (see page 36). The reaction has been used as a confirmatory test for histamine. A method by which the effect of 5 μg of histamine on the bronchi can be recorded has been described by Koessler, Lewis and Walker (1927). In a second method, described by McDowall and Thornton (1930) and used by Thornton (1931), the bronchi are perfused from the trachea, the lungs being scarified to allow escape of fluid. In a third simple method, employed by Bartosch, Feldberg and Nagel (1932a) and by Daly and Schild (1935), the pulmonary vessels are perfused, the lungs are subjected to rhythmic inflation, and broncho-constriction is recorded with the aid of a lever attached by a thread to the edge of the lung.

Gastric juice. According to Popielski (1920a) 3.3 µg of histamine per kg evokes a secretion of gastric juice in the dog.

Human skin. If a drop of histamine solution (1:1000 to 1:100000) is placed on the skin and the skin is pricked through the drop, the typical response described on page 23 is produced. The test may be made more sensitive if a small accurately measured volume of the solution is injected intradermally by means of a micrometer syringe of the type described by Trevan (1925).

Suprarenals. The adrenaline-releasing action of histamine may be used for estimating its concentration (Szczygielski, 1932; Bartosch, Feldberg and Nagel, 1932b).

4. *Methods for isolating histamine*

The method used by Barger and Dale (1910, 1911) for isolating histamine from ergot and from the mucosa of the small intestine followed the procedures in general use for isolating bases. Impurities were removed from the watery extract with tannic acid; the bases were precipitated as phosphotungstates, and then fractioned by the silver nitrate method of Kossel and Kutscher. The histamine was contained partly in the histidine fraction, which precipitates from neutral solution, and partly in the arginine fraction, which precipitates in the presence of baryta. The bases were extracted from these precipitates with hot absolute alcohol, and histamine picrate was obtained on the addition of picric acid to a watery solution.

This method was modified in various ways by Best, Dale, Dudley and Thorpe (1927). The tissues (lung and liver) were first extracted with acid alcohol. After the alcohol was removed by evaporation *in vacuo* and fat was removed with ether, lead acetate was used to precipitate impurities. Lead was removed from the filtrate with H_2S, and the bases were precipitated as phosphotungstates and fractionated with silver. After decomposition of the silver salts with H_2S a large excess of solid barium hydroxide was added and the histamine was extracted in a Soxhlet apparatus from the paste thus formed. After removal of the alcohol and barium, picric acid was added, and histamine picrate crystallized out.

In the course of this work each of the successive fractions was tested physiologically for histamine on the cat's blood pressure, and it was found that there was no important loss at any stage. The investigators were thus able to show that all, or practically all, the histamine-like activity of the original extract was in fact due to histamine. (This did not, of course, exclude the possibility that other vasodilator

Table 3. *Publications in which the isolation of histamine is described*

Histamine exposed to putrefaction	Ackermann (1910)
Ergot extract	Barger and Dale (1910); Kutscher (1910)
Soya beans	Yoshimura (1910)
Intestinal mucosa (ox)	Barger and Dale (1911)
Faeces (man)	Eppinger and Guttmann (1913)
Urine (parathyroidectomized dog)	Koch (1913)
Gastrointestinal mucosa (dog) ⎫ Pituitary (commercial preparation) ⎬	Abel and Kubota (1919)
Lung ⎫ (ox) Liver ⎬	Best, Dale, Dudley & Thorpe (1927)
Skeletal muscle (horse)	Thorpe (1928)
Spleen (ox, horse)	Dale and Dudley (1929)
Heart (ox)	Thorpe (1930)

substances, such as adenosine compounds and substance P, were present. They could more readily have been detected by using the rabbit's blood pressure for testing.)

Thorpe (1928) used the same method to isolate histamine from voluntary muscle, except that he extracted the phosphotungstates with acetone and discarded the insoluble fraction. He found that this removed a great deal of potassium and increased the specificity of the silver precipitation, so that nearly all the histamine came down in the histidine fraction.

Adenosine triphosphate is known to be present in voluntary muscle in such high concentration that it must account for most of the vasodilator activity of muscle extracts prepared with trichloroacetic acid. In Thorpe's work this compound was presumably excluded at the outset owing to its insolubility in alcohol.

5. *Distribution of histamine*

A list of papers dealing with the tissue distribution of histamine has been published by Best and McHenry (1931). In some cases the evidence depended on chemical isolation, in others on biological tests which were often carried out in a way that made them quite unspecific. A list of the papers that deal with the isolation of histamine is given in Table 3.

Table 4 gives an idea of how histamine is distributed in the tissues of a number of species.[9] It has been compiled from various sources, but mainly from the paper by Thorpe (1928), in which alcoholic extracts of

Table 4. *Concentration of histamine in different tissues (mg per kg)*

Organ	Species					
	Man	Horse	Ox	Dog	Cat	Rabbit
Blood[10]	0.04	—	—	0.05	0.01	10
Voluntary muscle	—	1.4; 1.1	4	7	—	—
Intestinal smooth muscle	—	7.5	—	35	—	—
Bladder	—	7.8	—	—	—	—
Heart	—	—	{ 9.4; 18; 9.6; 8.1	4	—	—
Lung	—	35	44; 75	16.5	—	—
Liver	—	2.5; 6.6	5.4; 5	33	—	—
Pancreas	—	1.6; 3	—	—	—	—
Spleen	—	7.5	—	—	—	—
Testis	—	—	1.8	—	—	—
Ovary	—	—	9	—	—	—
Parotid gland	—	5.3; 6.7	—	—	—	—
Submaxillary gland	—	0.6; 0.5	—	—	—	—
Thyroid	—	0.5	—	—	—	—
Skin { Epidermis	24	—	—	—	—	—
Skin { Cutis	4	—	—	—	—	—

various tissues were compared with histamine on the cat's blood pressure. Atropine was not used in those tests, and it is possible that part of the effect was due in some cases to choline or other vasodilator substances. Further estimates may be obtained from the papers by Harris (1927), Thorpe (1928), Best and McHenry (1930) and Barsoum and Gaddum (1935a).

D. The absorption, excretion and metabolism of histamine

1. Absorption

Histamine injected subcutaneously or intraperitoneally is slowly absorbed. When it is applied to the peritoneum it can be seen to produce vascular changes. When it is applied to the surface of the skin it is normally not absorbed unless the skin is pricked; it can, however, be made to pass through the skin by electrophoresis (Ebbecke, 1922; Hosoye, 1931, and others). Koessler and Hanke (1924) found that histamine put into a guinea pig's mouth was absorbed through the oral mucosa. Absorption from the cerebral ventricles is very slight (Heller and Kusonoki, 1933).

(a) Absorption from the intestinal tract.

The question of how far histamine is absorbed from the intestine has been studied in detail, because of its probable physiological importance.[11] Histamine has been found to be normally present in the intestinal contents; presumably it is continually being absorbed under physiological conditions, although this has not yet been proven by direct experiment. The problem has been approached from two directions.

(1) A large amount of histamine has been put into the intestinal lumen, and its fate has been studied either by measuring the rate at which it disappears, or by observing its effect in the animal, or by looking for it in the animal's organs.

(2) Attempts have been made to demonstrate an excessive absorption of histamine in intestinal obstruction. That this may happen is theoretically quite possible, but it has not been proven. The issue will not be discussed further here; for a summary of the literature consult the review by Best and McHenry (1931).

E. Mellanby (1915) studied the fate of histamine in the cat's intestine by introducing it into an intestinal loop some 25 cm long and then withdrawing the content at intervals to assay it for histamine. Since the intestine contains bacteria that can form histamine, and also, probably, bacteria that can destroy it, Mellanby thought it important to know whether these factors could affect his findings, so he did control experiments in which the circulation to the loop was shut off. In the control situation some of the histamine did disappear from the large intestine, but under his normal experimental conditions he found that all the changes in the histamine content within the small intestine could be accounted for by absorption alone.

Koessler and Hanke (1924) used a simpler method of carrying out the same control. They removed some of the intestinal content and determined its histamine level before and after incubation at body temperature *in vitro*.

Mellanby found that the absorption was fastest at the lower end of the ileum. He could detect no absorption of histamine from the large intestine. Later investigations have confirmed the rapid absorption from the lower ileum, but have shown that there is also some absorption from the stomach, the duodenum and the colon (Lim, Ivy and McCarthy, 1925; Meakins and Harington, 1923; Koskowski and Kubikowski, 1929; Florey, 1930).

Mellanby found, further, that the rate of absorption increased when the intestinal content was made alkaline, and decreased when it was made acid. It was fastest when the intestine was empty, and was slowed by the presence of foodstuffs, Ringer's solution, bile or bile salts, concentrated $MgSO_4$ solution or morphine, It was diminished by interfering with the circulation. Intravenous injection of glucose appeared to accelerate the absorption.

MacKay (1930) and MacKay and Baxter (1931) obtained results that disagree on some points with those of Mellanby. They introduced large amounts of histamine into the dog or cat small intestine and found effects on the blood pressure only when the medium was acid. According to Mellanby's findings the presence of acid should inhibit absorption.

Kendall and Varney (1927) used the contractile response of an intestinal loop as an index of the histamine absorbed from its lumen. They applied histamine to the mucosal side of an isolated segment of guinea pig intestine. A slow contraction occurred after an interval of 30 to 60 seconds; this delay was absent when the histamine was applied to the serous side. Weakly acid histamine solutions applied to the mucosa elicited no response.

The reaction of the whole animal can also be used as an index of the absorption of histamine from the bowel, but it has been found that very large amounts of histamine may disappear from the intestine without producing much effect on the animal. Koessler and Hanke (1924), for example, introduced 500 mg of histamine dihydrochloride into the stomach of a dog and found that it disappeared at the rate of about 2.2 mg per minute. In another experiment 100 mg of the same salt was introduced into the stomach of a guinea pig, and its rate of disappearance was 0.5 mg per minute. If the histamine had been injected intravenously at these rates the dog would certainly have been very severely intoxicated and the guinea pig would have been killed. In fact, the dog seemed to be quite unaffected and the guinea pig showed only slight discomfort. Thus it appears that histamine is detoxicated when it is absorbed from the stomach or intestine, but the site of the detoxication is not known with certainty. Some experiments of Meakins and Harington (1922) on cats equipped with Eck fistulas suggest that the liver may act as a trap for histamine. Ivy and Javois (1924), on the other hand, could find no difference in tolerance between normal and Eck-fistula dogs; and later evidence has made it unlikely that the liver plays a major role in the detoxication of histamine. It contains less histaminase than many other tissues, and perfused livers inactivate histamine much more slowly than do perfused kidneys (Best and McHenry, 1930).

Thus the fate of the histamine absorbed from the intestine is still unknown. Only a small part of it can be recovered from either the liver or the intestinal wall (Koessler and Hanke, 1924). It has been suggested (Popielski, 1920a) that it is destroyed by the intestinal mucosa itself, a suggestion that is supported by the finding that this tissue is especially rich in histaminase.

Histamine does, however, produce significant effects on the animal if enough of it is introduced into the intestine. In experiments on cats Meakins and Harington (1922) observed that blood pressure fell and breathing became irregular. If the intestinal circulation was interrupted for 5–15 minutes the mucosa was irreversibly damaged and no absorption occurred. Florey (1930) observed a continuous fall of blood pressure when he introduced histamine solutions into the rectum and then tied off the anus. Ivy and McIlvain (1923) and Ivy, McIlvain and Javois (1923) found that histamine (1 mg per cc) placed in the small intestine of a dog produced an increased secretion of gastric juice. The secretion of gastric juice is a particularly good indicator in such experiments because it can respond to a low level of histamine maintained for a long time. Fenyes (1930) gave patients 2–12 mg by mouth and observed increased gastric secretion and slight haemoconcentration. MacKay and Baxter (1931) saw a fall of blood pressure after introducing histamine (10 mg) in acid solution into the cat duodenum.

2. Excretion

Pharmacological tests have failed to detect histamine in the urine of animals into which it has been injected (Dale and Laidlaw, 1910; Oehme, 1913; Guggenheim and Löffler, 1916a). MacGregor and Peat (1933), however, were able to detect it in the urine when a large amount of it had been added to the blood perfusing an isolated kidney. The levels of histamine in the blood must have been higher in these experiments than ever occurs in a normal animal. Apparently, therefore, histamine is not excreted as such by the kidney. It should be noted that Koch (1913) reported the isolation of histamine from the urine of a parathyroidectomized dog.

Yoshida (1931) supposes that when histamine is injected subcutaneously, it is secreted into the gastric juice formed in response to the injection. He found a substance in gastric juice that resembled histamine in its pharmacological action. His findings require confirmation.

The possibility that histamine is excreted into the intestine has not been investigated.

3. Metabolism
(a) Mammalian tissues

The only metabolic change that histamine is known to undergo in mammalian tissues is destruction by an enzyme, histaminase[12] (Best, 1929; Best and McHenry, 1930).

Histaminase was found in relatively large amounts in the kidney and in the walls of the small and large intestine; in smaller amounts in blood, muscles, spleen, lungs, suprarenals and bladder. The amount in the liver was too small to be measured reliably, and the enzyme was apparently not present in the heart, the skin or the wall of the stomach. It was not excreted in the urine. Beef kidney is an excellent source of histaminase. Best and McHenry described the preparation of a stable dry powder from kidney, which can be used as a source of active enzyme.

Histaminase has an optimum pH between 6.8 and 8.0 and an optimum temperature of about 38 °C. It is inactivated at 60 °C. Sodium and potassium ions have no effect on its activity; ammonia and sulphate are slightly inhibitory, while calcium and citrate ions inhibit it strongly. Potassium cyanide is inhibitory at concentrations as low as 0.0005 M. This observation, and the fact that oxygen accelerates the reaction, suggest that histamine is oxidized by histaminase, but the reaction products are not known.

McHenry and Gavin (1932) found that histaminase at pH 7.2 releases one molecule of ammonia for each molecule of histamine destroyed, which indicates that deamination is involved. Tyramine is unaffected by histaminase.

Injected histamine disappears very rapidly from the blood. This is already apparent from the fact that a cat under ether can receive a series of small doses at 3-minute intervals for a long time with no lasting fall of blood pressure or reduction of sensitivity. A slow infusion of histamine is similarly well tolerated, and it is possible in this way to give at least 20 times the dose that would have caused death if given in one injection (Dale and Laidlaw, 1910; Oehme, 1913; Guggenheim and Löffler, 1916a).

The fate of the histamine that disappears from the blood has been studied by adding histamine to the blood used to perfuse different organs of the dog (Best and McHenry, 1930; MacGregor and Peat, 1933). It was found that large amounts of histamine were removed from the blood by the kidneys, the lungs, the liver and the extremities, but not by the heart. Over a perfusion period of $\frac{1}{2}$ to $5\frac{1}{2}$ hours a large proportion of the added histamine could not be recovered from the tissue, and must therefore have been destroyed. Presumably histaminase was responsible for the destruction, for the kidney, which is especially rich in this enzyme, was also the tissue which destroyed histamine most rapidly, while the heart was inactive in both respects.

MacGregor and Peat found in control experiments that their method of perfusion produced no change in the natural histamine content of the tissues, unless they added histamine to the perfusion, in which case the tissue's histamine content usually rose slightly. MacGregor and Peat think that the natural histamine store of the tissues is in some way inaccessible to histaminase. Whether or not this is so, it is evident that histaminase is a very effective agent for keeping the histamine level of the tissues constant.

The origin of this natural histamine is unknown. Lewis (1927) has pointed out that the phenomena of reactive hyperaemia suggest that some histamine-like material is continually being produced by the tissues. This may well be so, but at present we have no direct evidence of any process by which histamine is newly formed in the tissues.

(b) Bacteria

Formation of histamine. In Ackermann's original work (1910) on the bacterial formation of histamine by decarboxylation of histidine, the organism responsible for the reaction was not isolated. Mellanby and Twort (1912) isolated a small gram-negative bacillus that acted on histidine to form a substance that they identified as histamine by pharmacological tests. In the same year Berthelot and Bertrand (1912a, b) isolated a different bacillus, which resembled the _B. pneumoniae_ of Friedländer, from human faeces. This bacillus decarboxylated histidine, tyrosine and tryptophan.

Koessler and Hanke (1919–1924) published a series of papers on the production of toxic amines by bacteria. These papers have been summarized in a further report by Koessler, Hanke and Sheppard (1928). They qualified the finding of Mellanby and Twort (1912) in that they found that decarboxylation occurred only in an acid medium, and they proposed that amine formation is a defence reaction that keeps the medium from becoming too acid. In all they examined 223 organisms. Nine of these produced histamine, and all nine belonged to the coli-typhus group. Five produced tyramine, and none of the five belonged to that group. Out of 94 organisms from that group, 67 formed unknown substances that could cause contraction of bronchi and isolated arteries.

Roske (1928) found that _B. coli_ produced amines from peptone only in an acid medium. If buffer was added to prevent the medium from becoming too acid no amines were formed: a confirmation of the finding by Koessler and Hanke.

Destruction of histamine. It has often been observed that non-sterile solutions of histamine lose their activity sooner than sterile solutions. The process appears not to have been studied in detail, but it seems clear that histamine is readily attacked and destroyed by a variety of common bacteria and moulds.

E. The physiological significance of histamine

The liberation of histamine has been thought to play a part in a number of different processes that occur in the body. It is probable that in most of these cases histamine is not the only substance liberated. We shall discuss these phenomena in separate sections, under the following headings:

(1) The normal regulation of blood flow in resting and active tissues (pages 168 *et seq.*)

(2) The response to mild local injury, and the general shock that results from more extensive injuries and burns (pages 172 *et seq.*)

(3) The anaphylactic symptom complex (pages 179 *et seq.*).

IV

Acetylcholine

The acetyl ester of choline was synthesized by von Baeyer (1867), but the first hint of its physiological importance was the discovery by Hunt and Taveau (1906) that it has great pharmacological activity. In more recent years it has been found to be present in many animal tissues, and to play an important part in the activity of many nerves. Our knowledge of its functions has been growing rapidly while this book has been in preparation.

References to studies on different aspects of acetylcholine function will be found in the publications of Riesser (1931), Hirschberg (1931), Kroetz (1931), Alles (1934) and Villaret, Justin-Besançon and Cachera (1934),[1]

A. Chemical and physical properties

$$\overset{\displaystyle OH}{\underset{\displaystyle |}{(CH_3)_3 \equiv N}} - CH_2 - CH_2 - O(CH_3CO)$$

The formula of acetylcholine is given above. The molecular weight of its chloride is 181.5.

A summary of its chemical properties is given by Guggenheim (1924). Acetylcholine may be prepared by heating choline with acetyl chloride or acetic anhydride to 100 °C for several hours in a sealed tube in the absence of water. Under these conditions the choline is quantitatively acetylated, and the acetyl chloride or acetic anhydride can be removed by vacuum distillation. The chloroaurate of acetylcholine can easily be obtained in pure crystalline form from the acetylated choline. A neutral solution of acetylcholine chloride can be obtained from the chloroaurate by shaking an aqueous solution with finely divided metallic silver until the solution is decolorized. The gold and excess chloride, as metallic gold and silver chloride, can be removed by filtration (Dudley, 1929).

Acetylcholine and choline form a double chloroplatinate that is less soluble than either choline chloroplatinate or acetylcholine chloroplatinate. The double salt crystallizes from a solution of the mixed chloroplatinates, and its preparation offers a convenient method of separating acetylcholine from a large excess of choline. The chloroplatinates can also be decomposed with metallic silver (Dudley, 1929, 1931).

Solubility. Acetylcholine is readily soluble in water and alcohol but is insoluble in ether.

Stability. In the presence of alkali acetylcholine is rapidly hydrolysed to choline and acetate. It is also destroyed by strong mineral acids. The pH of maximum stability is 3.9. Hoffmann (1930) boiled a solution at that pH for 12 hours and found by titration that only 3.9 per cent of the original acetylcholine was hydrolysed. At pH 4.0 the rate of hydrolysis was doubled. Solutions that have been made neutral to congo red and sterilized are stable for many days in the cold; however when acetylcholine is dissolved in a slightly alkaline bicarbonate solution like Locke's solution it loses much of its activity in the course of a day. If an equal volume of N sodium hydroxide is added to a tissue extract that contains acetyl-choline, most of the acetylcholine will be destroyed within 10 minutes at room temperature (Dale and Dudley, 1929; Velhagen, 1931a).

Adsorption. Acetylcholine is adsorbed by animal charcoal, but not so completely as histamine (Galehr and Plattner, 1928; Engelhart and Loewi, 1930; Saunders, Lackner and Schochet, 1931).

B. Pharmacological properties

1. *General*

The actions of acetylcholine are qualitatively similar to those of choline. The presence of the acetyl group may increase the activity a thousand-fold or more and may slightly alter the response, but it does not change its general nature. The actions of choline are similar to those of other quaternary ammonium bases and have been comprehensively described in the Heffter *Handbuch* (1923). The summary given below is limited to a discussion of experiments on acetylcholine itself; the actions of other bases are mentioned only incidentally.

The actions of acetylcholine, like those of other quaternary ammonium bases, can be conveniently divided into two classes (Dale, 1914a):

(1) '*Muscarine actions*'. These are like those of true muscarine from *Amanita muscaria* (Kögl, Duisberg and Erxleben, 1931) and may with some justice be summarized as reproducing those of the cranial and sacral autonomic nerves.[2] They are all abolished by small doses of atropine but are unaffected by nicotine. Drugs that have similar or identical muscarine actions include pilocarpine and arecoline.

(2) '*Nicotine actions*'. These may be summarized as stimulation, followed

by paralysis, of sympathetic ganglia[3] throughout the body. They are abolished by large doses of nicotine but are unaffected by small doses of atropine. The release of adrenaline and the contracture of voluntary muscle, actions which acetylcholine and related substances display under certain conditions, must also be classified as nicotine actions.

The action of eserine, which prevents the hydrolysis of acetylcholine in the body, will be discussed later (see page 83 *et seq.*). Here it is sufficient to note that an injection of eserine greatly increases an animal's sensitivity to acetylcholine. This sensitization presumably affects every action of acetylcholine, so it is unnecessary to mention all the different types of experiment in which it has been observed.

2. Action on the systemic circulation

Acetylcholine has a variety of strong vasomotor actions. Depending on the circumstances it can cause either vasoconstriction or vasodilatation.

(a) Mammalian vessels

Muscarine action (*vasodilatation*). A clear fall in blood pressure can sometimes be elicited by a dose of less than 0.01 µg (Hunt and Taveau, 1906).[4] This effect of very small doses is due mainly to a direct vasodilator action, and like that of muscarine is abolished by atropine (Dale, 1914a). The vasodilator action can easily be demonstrated on the perfused limb preparation; it differs from histamine vasodilatation in that a different segment of the vascular tree is involved. Acetylcholine dilates relatively large arterioles, while the dilator action of histamine is confined to the smallest arterioles and the capillaries (Dale and Richards, 1918; cf. pages 20 *et seq.*). These observations have been confirmed by direct observations on human skin vessels (Carrier, 1922) and on the retina (Villaret, Schiff-Wertheimer and Justin-Besançon (1928). Carrier did observe some capillary dilatation but this was probably a passive consequence of the widening of the arterioles. In the rabbit ear acetylcholine dilates not only the small arteries, but also the arteriovenous anastomoses that are present in that tissue (Grant, 1930).

The distribution of the vessels that respond in this way to acetylcholine was examined by Hunt (1918) in plethysmographic experiments. He concluded that the principal effect was on the blood vessels of the skin, but well-marked dilatation was also apparent in vessels of the spleen, the penis and the submaxillary gland. There was comparatively little vasodilatation in voluntary muscle and in kidney; there is, however, no doubt that the vessels of these latter tissues are also dilated (for muscle, see

Hartman, Evans and Walker, 1929; for kidney, see Hamet, 1926, and MacGregor and Peat, 1933).

The vasodilator effect of acetylcholine on perfused mesenteric arteries (Dale and Richards, 1918) is replaced, in the intact animal, by vasoconstriction, which is due to an action on the mesenteric ganglia (Feldberg and Minz, 1932). Owing to this action and to the marked vasodilatation in the skin and other vascular beds, the effect of an intravenous injection of acetylcholine in the whole animal is to diminish blood flow through the splanchnic bed (McMichael, 1933).

The action of acetylcholine on the coronary, pulmonary and liver vessels will be considered below (see pages 61 *et seq.*).

Nicotine action (*vasoconstriction*). Larger doses of acetylcholine (for example 0.1 mg) produce vasoconstriction in the cat. This action, however, is only apparent under special conditions, or when atropine has been given previously to abolish the intense vasodilator effect of acetylcholine (Dale, 1914a). This reversal of the acetylcholine effect by atropine can be observed also in the absence of an anaesthetic (Gruber, 1929).

In a cat under chloralose the injection of 1 mg of acetylcholine produces a steep fall of blood pressure, but after atropine this effect is almost completely lost and is replaced by a pressor action. The pressor action is particularly well displayed if the experiment is done on a pithed cat under atropine (Feldberg and Minz, 1931). A similar vasoconstriction is seen after nicotine; and correspondingly, the pressor action of acetylcholine is abolished by large doses of nicotine, and can therefore be classified as a nicotine effect. This pressor action is presumably due to stimulation of sympathetic ganglia and of the suprarenal medulla. The effect is brought about in several ways:

(1) Acetylcholine causes a discharge of adrenaline from the suprarenals. Thus Dale (1914a) found that removal of the suprarenals diminished the pressor effect in the cat. Feldberg and Minz (1931) confirmed this observation and studied the matter further. They showed that the action of acetylcholine was exerted directly on the suprarenals, and was not secondary to an action on the central nervous system or on ganglion cells: in support of that conclusion, the pressor effect was larger and occurred more rapidly if the acetylcholine were injected directly into the aorta through the central stump of the coeliac artery, rather than intravenously. The released adrenaline produced a definite dilatation of the pupil: and repeated injections reduced the amount of adrenaline that could subsequently be extracted from the gland.

The action of acetylcholine on the suprarenals is greatly weakened by nicotine. Atropine has a double action on the suprarenals. If it is given together with acetylcholine, the latter does not produce its usual pressor effect. This action of atropine is transient; a few minutes later acetylcholine acts in the usual way. On the other hand acetylcholine still has a small residual pressor action after nicotine, and this is permanently abolished by an injection of atropine (Feldberg, Minz and Tzudzimura, 1934).

Siehe (1934) found that the action of acetylcholine on the suprarenals could still be obtained three weeks after section of the splanchnic nerves and removal of the abdominal sympathetic chain.

(2) After removal of the suprarenals the injection of a large dose of acetylcholine, 5–10 mg, will still raise the blood pressure of the atropinized cat. This phenomenon is discussed, with full references to earlier work, in a second paper by Feldberg and Minz (1932). It is largely due to splanchnic vasoconstriction and is greatly diminished or abolished by evisceration.

If acetylcholine (2 mg) is injected through a cannula in the abdominal aorta, so that it reaches the splanchnic circulation immediately, it produces a large rise of blood pressure even in the atropinized cat deprived of its suprarenals. This action is abolished by large doses of nicotine, which implies that it is due to an action of acetylcholine on the sympathetic ganglia, and Feldberg and Minz accept that interpretation. As further evidence for it they showed that the effect was greatly reduced by removal of the solar plexus and the inferior mesenteric ganglion. They could also produce a rise of blood pressure by painting a 5 per cent solution of acetylcholine on these ganglia.

(3) There is some reason to believe that large doses of acetylcholine may also produce vasoconstriction by a peripheral action, seen on tissues that contain no ganglion cells. In the perfused ear of the rabbit, though small doses cause vasodilatation, large doses regularly cause vasoconstriction (Hunt, 1918; Bartosch and Nagel, 1932). According to Bartosch and Nagel this vasoconstrictor action is abolished by nicotine. Hirose (1932) injected acetylcholine directly into the femoral arteries of cats and dogs and measured femoral venous outflow. The injection caused a brief increase followed by a prolonged diminution of the flow. When the injection was made into the renal, mesenteric or pulmonary artery, it caused pure vasoconstriction. In perfused lungs obtained from rabbits (von Euler, 1932) or cats or dogs (Gaddum and Holtz, 1933), large doses of acetylcholine are vasoconstrictor, though smaller doses can be vasodilator.

It seems likely in all these cases that acetylcholine must have been acting on tissues that contain no ganglion cells; the point, however, is one that has not been studied in much detail.

(b) Frog vessels

According to Doi (1920) acetylcholine lowers the frog's blood pressure. The effect must have involved peripheral vasodilatation because it was accompanied by an increase in limb volume.

Amsler and Pick (1920) perfused the limbs and splanchnic vessels in the frog with salt solutions. The addition of acetylcholine to the perfusion fluid caused vasoconstriction; however, this was replaced by vasodilatation when the perfusion fluid was made slightly alkaline (Teschendorf, 1921) or when the ratio of potassium to calcium was elevated (Voss, 1926).

3. Action on the lymph

Acetylcholine applied locally has been found to increase lymph flow from the dog's paw, probably in consequence of the increased blood flow. The action was much less striking than that of histamine (Haynes, 1932).

4. Action on the heart and other parts of the circulation
(a) Heart

Muscarine actions. Acetylcholine has several actions on the heart, which are the same as those of vagal stimulation, and which are abolished by atropine. Thus it slows the rate and diminishes the force of the beat (negative chronotropic and negative inotropic action), as found by Dale (1914a) on the cat; it slows conduction from auricle to ventricle (negative dromotropic action), as found by Clark (1926a) on the frog; and it also shortens the refractory period, as found by Wedd and Fenn (1933) on the turtle's auricle. These different effects are not always equally well shown, even in preparations from the same species. In one preparation the slowing may be especially prominent, while in another the weakening of the heart may be the only obvious effect (Dale, 1914a).

Isolated frog hearts have been widely used as test-objects for acetylcholine. The usual response is a simple diminution of the force of the beat. The action has been made the basis of a quantitative study of the relation between dose and effect (Clark 1926a, 1932). Strips of ventricle were suspended in Ringer's solution and stimulated with single electric shocks at a constant frequency. When acetylcholine was added to the bath it diminished the amplitude of the beat, and this diminution, expressed as a percentage, was taken as a measure of the effect. The curve relating the effect so measured with the concentration of acetylcholine resembled the dissociation curve of oxyhaemoglobin. The form of the curve is compatible with the notion that there is a reversible monomolecular reaction between acetylcholine and its receptive substance in the tissue.

The action of a combination of acetylcholine and atropine in experiments of this kind can be expressed by the formula

$$K \times \frac{concentration\ of\ acetylcholine}{concentration\ of\ atropine} = \frac{y}{100-y}$$

where y represents the effect and K is constant. The size of the effect is thus determined by the ratio of the concentration of the two drugs, and the effect of a raised concentration of atropine can be overcome by using a still higher concentration of atropine. The formula is found to describe the facts accurately enough over the middle range of doses. It becomes inaccurate when the concentration of atropine is low, since it suggests that when no atropine is present any concentration of acetylcholine whatever would give a maximal effect. More recent but less complete studies of this question have been published by Kahlson (1932) and by von Beznák (1934).

In mammals, stimulation of the vagus or administration of vagomimetic drugs can affect the ventricle only as a secondary consequence of an action on the auricle. Thus Cullis and Tribe (1913) found that muscarine and pilocarpine had no action on the ventricle of perfused hearts of cats and rabbits after the A–V bundle had been cut; and Leetham (1913) found that these drugs had no action on isolated strips of ventricle. These observations were confirmed by A. S. Dale (1930), using arecoline and acetylcholine. Contrariwise, the frog's ventricle is quite sensitive to acetylcholine, as shown by Clark's results, discussed above. It is interesting in this connection that extracts of auricle and ventricle of the frog heart show equivalent concentrations of acetylcholine, while in the rabbit the ventricle contains very little acetylcholine compared to the auricle (Engelhart, 1930; Vartiainen, 1934).

The strength of the acetylcholine effect on the heart is much influenced by changes in the ionic composition of the medium. If this fluid is slightly alkaline the action of acetylcholine on the isolated rabbit auricle is sharply diminished (Andrus, 1924). This cannot be entirely due to the instability of acetylcholine in alkaline solution, for alkalinity has the same effect when choline is used instead of acetylcholine. (Choline is stable in alkaline solution.) Clark (1927) and Davis (1931) have reported similar results with perfused hearts from frogs and rabbits. In view of the evidence, to be discussed later, that the vagus acts on the heart by liberating acetyl-choline, it is significant that the action of the vagus on the heart is also diminished if the medium in contact with the heart is made alkaline (Andrus, 1924; Clark, 1927).

Clark (1927) found that the effect of acetylcholine on the frog heart was weakened by raising either the potassium or the calcium concentration of the medium and/strengthened by lowering the potassium concentration. Davis (1931) could not confirm the latter observation; his results were in the opposite direction. Clark's observations are rather surprising, because a high enough concentration of potassium has an effect on the heart that superficially resembles that of acetylcholine, though it is not influenced by atropine. Howell (1906) found that vagal stimulation became more effective if the medium in contact with the heart of a frog or terrapin contained potassium in excess, and that this effect was seen even when the concentration of potassium was insufficient to produce inhibition by itself. In the absence of potassium the vagus ceased to act. References to other work on this subject will be found in the paper by Frank, Nothmann and Guttmann (1923).

The effect of acetylcholine on the frog heart is increased if chloroform, ether or chloral hydrate is present in the medium in low concentration; but higher concentrations of these anaesthetics abolish the effect al-together (Rydin, 1924).

Acetylcholine inhibits the snail heart, and it appears that the effect is not antagonized by atropine (Jullien and Morin, 1931).

Acetylcholine inhibits the embryonic chicken heart after only 70–90 hours of incubation, and the action is abolished by atropine. The cardiac nerves do not develop for another day or two, so they are not required for the response (Plattner and Hou, 1931). Armstrong (1935) has criticized this study, arguing that the acetylcholine concentrations used were high. He found that acetylcholine sensitivity first appeared in the heart of *Fundulus* when the nerve cells developed.

Nicotine actions. In the atropinized cat acetylcholine increases the frequency of the beat. The effect is probably mainly due to a nicotine-like action on the stellate ganglion. It is abolished by nicotine and diminished by extirpation of the ganglion (Feldberg and Minz, 1932).

(b) *Coronary circulation*

Information about the action of acetylcholine on the coronary vessels is scanty and conflicting. In the dog and the cat the vagus nerve contains vasoconstrictor fibres for these vessels. Their effect is abolished by atropine, but larger doses are required than for blocking the action of the vagus on the heart rate (Anrep, 1926).

There is some evidence that acetylcholine also can cause vasoconstriction. It has been shown to contract isolated strips of coronary artery (Eppinger and Hess, 1909; Lange, 1932). Wedd and Renn (1933) obtained coronary vasoconstriction in the perfused rabbit heart with high concentrations of acetylcholine (1:50000). On the other hand Smith, Miller and Graber (1936), who used a similar technique, observed coronary dilatation, which was abolished by atropine. Narayana (1933) worked with the dog heart–lung preparation and found that small doses of acetylcholine added to the blood in the venous reservoir had no effect on coronary flow. This is hardly surprising because the drug was probably hydrolysed before it reached the heart.[5] On the other hand 6 mg of acetylcholine produced coronary dilatation but failed to slow the heart.

In the tortoise both acetylcholine and vagal stimulation produce coronary dilatation, which atropine abolishes. These effects could be directly observed in an intact heart, and were also recorded from a perfused heart (Sumbal, 1924).

(c) Pulmonary circulation

Acetylcholine has been found to constrict the pulmonary vessels of rabbits (Antionazzi, 1931; von Euler, 1932) and of cats (Hirose, 1932). Franklin (1932) found that acetylcholine caused relaxation of pulmonary arterial strips from dogs and contraction of pulmonary venous strips. Gaddum and Holtz (1933) perfused dog and cat lungs with blood and observed vasodilatation after small (3–10 µg) doses and constriction after larger (100 µg) doses. Both effects were abolished by atropine. From their observations on lung volume they concluded that in cats large doses constrict both the arteries and the veins; in dogs the constriction involves mainly the arteries. They were unable to determine the site of the vasodilatation.

(d) Portal circulation

Fleisch (1931) perfused the mesenteric veins and the colic vein of cats *in situ* and measured the perfusion rate. He found dilatation with low concentrations of acetylcholine and constriction with higher concentrations. Franklin (1926) could detect no action on the mesenteric veins of sheep.

The action of acetylcholine on the liver vessels of the dog is weak. There is some evidence that it may dilate the branches of the hepatic artery (Bauer, Dale, Poulsson and Richards, 1932). Large doses constrict the hepatic branches of the portal vein in the cat by a direct action, and smaller doses have a similar effect by stimulating sympathetic ganglia (McMichael, 1933).

5. *Action on the eye*
(a) *Iris*

Acetylcholine can act directly on the iris to constrict the pupil; this action is abolished by atropine. It may also cause dilatation, which may be secondary to the release of adrenaline; but when the effect results from the injection of a small dose into the carotid artery it is due mainly to an action on the superior cervical ganglion and is abolished by extirpating that structure (Lipschütz and Schilf, 1931).

Acetylcholine in very high concentration can constrict the pupil of the frog's isolated eye (Hadjimichalis, 1931). The constriction may perhaps result, like that produced by related drugs, from direct stimulation of the sphincter iridis (Poos, 1927).

In fishes the nervous control of the pupil is different from that in mammals. In *Scyllium, Mustelus, Trygon* and *Uranoscopus* the pupil dilates on stimulation of the third cranial nerve. In *Uranoscopus* it has also been shown to be constricted by stimulation of the sympathetic nerve. Adrenaline, acetylcholine, pilocarpine and eserine all dilate the pupil in some species and constrict it in others, but there is no obvious relation to the two kinds of nerve supply. Nevertheless the actions of adrenaline are always antagonized by ergotoxine, and those of acetylcholine, pilocarpine and atropine are antagonized by atropine. In *Uranoscopus* acetylcholine and sympathetic stimulation both constrict the pupil, and the effects of both are prevented by atropine (Young, 1930, 1932).

(b) *Intra-ocular pressure*

If the head of a dog is perfused with blood and acetylcholine (10 μg) is injected, the intra-ocular pressure rises. The effect is partly secondary to vasodilatation and partly due to contraction of the extrinsic muscles of the eye (Colle, Duke-Elder and Duke-Elder, 1931).

6. *Action on smooth muscle*

The actions on the smooth muscle of the blood vessels and the iris have already been described (see pages 56 *et seq.* and 63).

(a) *Alimentary canal*

Acetylcholine in very low concentration increases the tone of isolated strips of longitudinal or circular smooth muscle from any part of the alimentary tract, but it has comparatively little action when injected intravenously (Dale, 1914a). Preparations of the small intestine of the rabbit have been favoured in testing for acetylcholine, but the intestines of other animals are also extremely sensitive to it. The small intestine of the guinea pig, for example, responds to concentrations (1:100000000) similar to those used in work on the rabbit intestine (Guggenheim and Löffler, 1916b; Fühner, 1916). The small intestine of the mouse is apparently still more sensitive and has the advantage as a test-object that it is insensitive to histamine. The chicken intestine is also fairly sensitive to acetylcholine (Wedum and Gebauer-Fuelnegg, 1932; Kahlson, 1934; Barsoum and Gaddum, 1935a).

Kuroda (1923) has studied the action of acetylcholine and other drugs on strips of smooth muscle taken from various levels of the alimentary canal of the dog. All these muscles were relatively insensitive to acetylcholine when compared with the corresponding preparations from other animals. Strips of longitudinal muscle from the stomach were the most sensitive. In each case the response was abolished by atropine but not by nicotine or curare.

Acetylcholine has been used, apparently with success, to treat paralytic ileus in human patients (Abel, 1933). It produces a weak contraction of the frog stomach (Fühner, 1918a).

The intestine of the chicken embryo acquires acetylcholine sensitivity between the 7th and the 10th day of incubation (Suma, 1931).

The intestine of the tench (*Tinea vulgaris*) has two muscular coats, an outer one composed of striated muscle and an inner one composed of smooth muscle. With electrical stimulation there is a quick twitch of the striated muscle followed by a slow contracture of the smooth muscle. This double response can be evoked by acetylcholine. Curare blocks the response due to the striated muscle and atropine blocks the part due to the smooth muscle. If the muscle is soaked for some hours in a 1:10000 acetylcholine solution, electrical stimulation produces only the slow contracture, the quick twitch being absent (Frey, 1928; Méhes and Wolsky, 1932).

(b) Uterus

The uterus is also stimulated by acetylcholine but it is less sensitive than the intestine. In this respect acetylcholine contrasts with histamine, which acts more strongly on the uterus than on the intestine.

The rat uterus, which is relaxed by histamine, is excited by acetylcholine (Dale, 1914a). Feldberg (1931) found that acetylcholine in high concentration can relax the isolated uterus of the cat.

The action of acetylcholine on the uterus is blocked by atropine (Führner, 1916) or by large doses of nicotine (Rydin and Backman, 1925). The sensitivity of the muscle is raised by the addition of serum albumin to the bath (Fröhlich and Paschkis, 1926). Acetylcholine stimulates the hen's oviduct (McKenney, Essex and Mann, 1932).

Like other drugs, acetylcholine acts more strongly on the pregnant than on the non-pregnant human uterus (Murakami, 1931).

Kitakoji (1931) has found that acetylcholine also causes contraction of the gallbladder and the sphincter of Oddi.

(c) *Other organs containing smooth muscle*

The urinary bladder and the retractor penis of the dog are stimulated by acetylcholine. The bronchioles are constricted (Dale, 1914a), and this action is antagonized by atropine (Dirner, 1929; von Euler, 1932). The splenic capsules of the dog and the rabbit are stimulated by acetylcholine, and this action is intensified by eserine and abolished by atropine; the human spleen is inhibited (Vairel, 1933; Saad, 1934).

(d) *The effect of nerves on the responses of smooth muscle to acetylcholine*

If a strip of intestinal muscle from a rabbit is freed from all the cells of Auerbach's and Meissner's plexuses, it still exhibits rhythmic contractions and responds to acetylcholine and other drugs (Gasser, 1926). Van Esveld (1928) thinks that the peripheral nervous syncytium, which survives the removal of the plexuses, may be responsible for the responses. It should be noted, however, that the hen's amnion has no nerves at all, but is readily stimulated by acetylcholine (1 : 7 000 000) and by eserine, and that these responses are blocked by atropine (Baur, 1928).

The effect of degenerative nerve section has been examined on segments of the dog's stomach or small intestine. In a preliminary aseptic operation a 'nerve-free' gastric pouch of the Bickel type was prepared, or alternatively the nerve supply of an isolated loop of jejunum was severed. Later these tissues were excised and used for the preparation of isolated strips which were tested with acetylcholine. It was found that the muscles still retained their sensitivity after several weeks of deprivation of their nerve supply (Misuda, 1923; Kuroda, 1923). On the other hand the mucous glands in these denervated segments lost their ability to respond to acetylcholine within a few weeks. They were still functionally active however, for they could be stimulated with histamine (Mitsuda, 1923; Suda, 1924). It is thus clear that the nerve supply of a smooth muscle is not necessary for the muscle's response to acetylcholine.

7. *Action on striated muscle*

It is typical of nicotine and the drugs allied to it that they produce a slow contracture of certain striated muscles. The literature on this subject has been summarized by Gasser (1930). The following investigations have been carried out with acetylcholine.[6]

That choline can exert this peculiar action on frog muscles was first established by Boehm in 1908. Later Riesser (1921) showed that acetylcholine acts in the same way. In the case of frog's muscle the reaction is little if at all affected by degenerative section of the motor nerves (Simonson, 1922; Rehsteiner, 1927).

Mammalian striated muscle on the other hand does not normally respond in this way, except when very large doses of acetylcholine are applied (Wachholder and von Ledebur, 1932). However, if the motor nerves to such muscles are divided the muscles acquire, in about a week, an extraordinary sensitivity to acetylcholine, which can now bring about a very powerful slow contracture (Frank, Nothmann and Hirsch-Kauffmann, 1922, 1923; Feldberg, 1932).

The behaviour of avian muscles is intermediate between that of frog muscles and that of mammalian muscles. Usually they are fairly sensitive to the drugs of this group, but their sensitivity is not greatly increased by denervation (Langley, 1905). The muscles of reptiles, in many cases, are normally sensitive to acetylcholine; those of fishes, in contrast, are insensitive (Rückert, 1930).

The muscles of various invertebrates also react to acetylcholine (see Riesser, 1927, 1933). For the striated dorsal muscle of the leech, see page 72.

The muscles of a mammalian foetus normally respond to acetylcholine in much the same way as the frog gastrocnemius; this property is not lost until a few days after birth (Rückert, 1930a).

Different muscles of the same species vary widely in their response to acetylcholine. Thus the rectus abdominis of the frog is particularly sensitive and has frequently been used as a test-object for acetylcholine (see page 74). Its response is a slow contracture lasting many minutes. The sartorius, on the other hand, responds to high concentrations of acetylcholine with a relatively brisk contraction that lasts less than half a minute. Other muscles, the gastrocnemius for example, contain fibres giving either type of response. In this case the fibres that give the tonic reaction are often grouped together in a bundle that lies close to the nerve's point of entry and can be dissected free from the rest of the muscle (Sommerkamp, 1927).

It has long been known that the injection of nicotine causes the frog to adopt a clasping attitude (*Umklammerungshaltung*). Wachholder and Ledebur (1930) have shown that acetylcholine induces the same type of response, and have suggested that the muscles used in this attitude have special properties because they are used to maintain a tonic contracture over a long period in the mating season. Similarly the muscles that a tortoise uses to withdraw its head and limbs within the shell are extremely sensitive to acetylcholine.

Certain mammalian muscles also normally respond with a contracture when acetylcholine is applied to them. Rückert (1930a) found that acetylcholine evokes a contracture of the diaphragm of the adult mouse, and Duke-Elder and Duke-Elder (1930) observed a similar reaction in the extrinsic ocular muscles of the dog. According to Wachholder and von Ledebur (1932) most mammalian muscles will contract if sufficiently high concentrations of acetylcholine are used. The most sensitive muscles are those that have to maintain tone over long periods. There is no strict correlation between a muscle's acetylcholine response and its colour, but 'red' muscles tend to give the contracture more readily than 'white' muscles. Both in frogs and in mammals that hibernate, the muscles that do not react to acetylcholine in the summer tend to acquire this property in the winter.

Rückert's (1930b) analysis of why some muscles react differently from others led him to conclude that it is the phylogenetically young muscles that are sensitive to acetylcholine.

The action of acetylcholine on voluntary muscle clearly has to be classed as a nicotine action. Drugs that produce this effect are those that produce the same general effects as nicotine: cytisine, nitrosocholine, hordenine methyl iodide, tetramethylammonium and trimethylsulphonium salts. Many drugs whose actions mimic those of parasympathetic nerve stimulation – for example pilocarpine, muscarine, arecoline and tetramethylammonium salts – do not have this effect (Dale and Gasser, 1926).

The action of acetylcholine on frog muscle is prevented by procaine, curare, atropine and calcium, but not by adrenaline (Riesser, 1922; Simonson, 1922; Hess, 1923). The antagonism between acetylcholine and atropine on the frog rectus abdominis has been the basis of a precise quantitative study by Clark (1926b).

Wachholder and Matthias (1933) have investigated the action of salt solutions of different composition on the acetylcholine sensitivity of frog muscles. The sensitivity rose if potassium were raised somewhat, or if calcium were reduced; but if potassium were raised to the point where the muscle became unresponsive to electrical stimulation it also no longer responded to acetylcholine. Phosphate increased sensitivity. Small amounts of alkali made the sartorius more, and the rectus abdominis less, sensitive. Acid had the opposite effect.

The influence of atropine and of adrenaline on the acetylcholine contracture depends on the experimental conditions. The denervated gastrocnemius of the cat, with its circulation intact, will respond with a contracture to acetylcholine even after large doses of atropine. But if a piece of cat diaphragm, denervated 10 days previously, is excised and suspended in salt solution, it can be shown that the acetylcholine contracture is prevented by atropine (Dale and Gaddum, 1930).

Adrenaline has a double effect in mammalian experiments. If acetylcholine is injected at a time when adrenaline is still circulating, it fails to evoke a contracture. This has been attributed to an effect of adrenaline on the capillaries, making them impermeable to acetylcholine. But the direct effect of adrenaline on the muscle, as seen for example in experiments on the isolated diaphragm preparation, is to intensify and prolong the action of acetylcholine (Dale and Gaddum, 1930).

The contractures elicited by the stimulation of vasodilator nerves will be discussed in a later section devoted to this topic (see page 139 *et seq.*).

8. Action on glandular secretion

Acetylcholine causes secretion of saliva, tears, pancreatic juice and sweat (Dale, 1914a). Atropine abolishes these effects. Acetylcholine also increases bronchial secretion (Schilf, 1932). The action of acetylcholine on the salivary glands has been studied by a number of investigators during the last few years in connection with the function of cholinergic nerve fibres in the chorda (see pages 133 *et seq.*; cf. also Anochin and Anochina-Ivanova, 1929).

Acetylcholine increases the secretion of water and acid in the human stomach, but its action in that respect is feeble compared with that of histamine, and is sometimes completely lacking. (For references, see Gebhart and Klein, 1933.)

Degenerative section of all the nerves supplying the stomach (Kuroda, 1923) or the small intestine in the dog (Mitsuda, 1923) causes the glands to become insensitive to acetylcholine, but not to histamine.

9. Action on the respiratory centre

Hunt and Taveau (1906) produced complete apnoea by injecting 4 μg of acetylcholine into an anaesthetized rabbit. Fühner (1916) however saw no effect on respiration in the rabbit unless he used doses large enough to cause a profound fall of blood pressure. In experiments on unanaesthetized dogs, Gruber (1929) noticed a slight increase in respiratory depth and frequency. Such effects might be secondary to the effects of acetylcholine on the blood pressure or the bronchi. Feldberg and Minz (1932) observed transient deepening and acceleration of the breathing when they injected 0.1 mg of acetylcholine into an atropinized dog. Later experiments by Bouckaert and Heymans (1935) have shown that this results from an action on the carotid sinus. Dikshit (1933b) found that 0.05 to 1 µg of acetylcholine injected into the third ventricle of a cat produced depression or even complete stoppage of respiration lasting several seconds. In a few exceptional cases breathing was accelerated by acetylcholine. These effects must have been due to a direct action on the brain.[7]

10. Metabolic actions of acetylcholine

Depending on the circumstances, acetylcholine can either raise or lower the blood sugar of a rabbit. After destruction of the nervous system it causes glycogen to accumulate in the muscles of a cat after these have been fatigued; evisceration does not prevent the effect. There is no evidence that acetylcholine affects the secretion of insulin by the pancreas (Ernould, 1931).

According to Tutkewitsch (1929) acetylcholine has a slight and probably unspecific action on the blood lipids.

Acetylcholine given by mouth resembles choline in that it causes an increase in the amount of fat in the liver (Best and Huntsman, 1932), but its effect is no greater than that of the choline it contains.

11. Action on ciliated epithelium

According to Plattner and Hou (1931), acetylcholine, choline, pilocarpine, arecoline and adrenaline all increase ciliary movement in pieces of epithelium excised from a frog's pharynx.[8] Atropine abolishes the effects of all these substances.

12. General effects of acetylcholine; lethal doses

Some data on the lethal dose of acetylcholine are given in Table 5.

Man appears to be very insensitive to acetylcholine, presumably because of its rapid destruction in human blood (see page 82). Ellis and Weiss (1932) found that the rate of acetylcholine infusion had to exceed 90–140 mg per minute before toxic effects ensued. Govaerts and van Dooren (1931) injected as much as 500 mg intravenously without observing any effects. Carmichael and Fraser (1933), however, found that an intravenous injection of 10–30 mg caused a transient slowing of the heart and a fall of blood pressure. The effect was enormously increased by a

Table 5. *Lethal doses of acetylcholine in different species*

Species	Route	Dose (mg per kg)	Author(s)
Mouse	Intravenous	30	White and Stedman (1931)
Mouse	Subcutaneous	310	Hunt and Taveau (1910)
Mouse	Subcutaneous	100	Fühner (1932)
Rabbit	Intravenous	0.2–0.3	Justin-Besançon (1929)
Frog	Subcutaneous	200	Fühner (1932)

preceding subcutaneous injection of eserine. A strong local vasodilation was seen if the acetylcholine was injected into the femoral artery. A subcutaneous or intramuscular injection of 50 mg had no appreciable effect, and no action resulted from an oral dose of 2 g (Justin-Besançon, 1929).

C. The natural occurrence of acetylcholine

1. *General*

The natural occurrence of choline esters was first suspected by Reid Hunt (1899, 1901). He found that the vasodilator action of suprarenal extracts from which he had removed the adrenaline could not all be attributed to choline since the extracts were active after atropine. In the light of later evidence it seems likely that these atropine-resistant effects were due to histamine or adenosine compounds. Hunt succeeded in obtaining residues that were 50 to 100 times as active as choline in lowering the blood pressure, but he was unable to isolate any choline ester (Hunt and Renshaw, 1934).

The first proof for the natural occurrence of acetylcholine, like that for histamine, came from studying the properties of ergot extracts. Some of these extracts proved to have effects that could not be ascribed to known active principles: they lowered the blood pressure, inhibited the heart and stimulated the isolated rabbit intestine, and all these effects were antagonized by atropine (Dale, 1914b). The finding that the active material was unstable in alkali suggested that it might be an active choline ester of the sort studied by Hunt and Taveau (1911); and Ewins (1914) was successful in isolating acetylcholine from the extracts.

The presence of acetylcholine in mammalian tissues has been established by Dale and Dudley (1929), working with extracts of horse and ox spleen. In this case also the first hint of the presence of acetylcholine was that the extracts had an almost incredibly powerful choline-like activity which they lost on being made alkaline. Later work has made it probable that most mammalian tissues contain some acetylcholine, though in many cases the amount present is very small.

2. Preparation of extracts for identification of acetylcholine

Dale and Dudley (1929) have discussed the conditions necessary for obtaining acetylcholine from extracts of horse spleen. The spleen may be removed from the animal and kept intact for several hours without change in its content of acetylcholine, but as soon as it is cut up or minced the acetylcholine quickly disappears. For maximum yield, therefore, it is important to mince the tissue in an extraction medium that will immediately arrest the process that destroys acetylcholine.

The medium most generally used for making such extracts is ethyl alcohol. Some hydrochloric or sulphuric acid is usually added, since acetylcholine is known to have maximum stability in weak acid. The method worked out in Loewi's laboratory (Engelhart, 1930) is as follows: the tissue is extracted with 0.01 N HCl in alcohol; the extract is filtered and the filtrate is dried on a water bath at 90 °C; after the residue has been washed with ether, the purification with acid alcohol is repeated, and the purified material is taken up in Ringer's solution. This treatment is said to destroy *Acceleransstoff* (Engelhart, 1931).

Chang and Gaddum (1933) made a comparison of several methods and concluded that trichloroacetic acid, which had already been used by Galehr and Plattner (1928), was a better extractive than alcohol. In the procedure they recommend, the tissue is disintegrated while it is immersed in 10 per cent trichloroacetic acid (2 cc per g of tissue). After standing for an hour the extract is filtered and the acid is removed with ether. The extract is then concentrated, neutralized and tested on the frog's rectus abdominis muscle or on the leech.

When the acetylcholine concentration is low, the extract may contain enough potassium to invalidate the test. The use of alcohol as extractive excludes some of the potassium. It is always possible to test whether potassium is present in interfering amounts: one ashes the extract and sees if adding a solution of the ash to the appropriate amount of acetylcholine alters the effect of the latter.

When Chang and Gaddum applied Engelhart's extraction method to the horse spleen they found that the final extract contained only about 3 per cent of the activity present originally, as estimated in an extract made with trichloroacetic acid. The alcoholic extraction if done carefully does extract all, or nearly all, the acetylcholine; but when such an extract is taken to dryness the pH must be carefully adjusted to about 3.9. Engelhart's method cannot be applied without modification to all tissues. Plattner

(1934) reports, however, that the method gives satisfactory results in his hands.

Kahlson (1934) has used metaphosphoric acid to prepare extracts for acetylcholine estimation. This procedure has the advantage that the pH of the filtrate is approximately that at which acetylcholine is most stable.

3. Detection and estimation of acetylcholine

The following table summarizes a number of tests that have been used.

Table 6. *Tests for detection and estimation of acetylcholine*

Test-object	Approximate acetylcholine concentration in µg per l	Remarks
Leech (in eserine)	2	Specific but slow
Isolated rabbit auricle (in eserine)	4	Non-specific
Frog heart, Straub preparation	10	Non-specific
Mouse intestine	10	
Rabbit intestine	20	Non-specific
Frog rectus abdominis (in eserine)	20	Specific and quantitative
Cat blood pressure	20	Non-specific
Denervated gastrocnemius, cat	100	Specific

The figures given for the concentration of acetylcholine represent the approximate concentration needed for a good response in an average test; there is naturally a good deal of variation from test to test. For isolated organ tests, the concentrations stated are those actually present in the bath; usually a small volume of a more concentrated solution is added. For injections, the concentration is stated on the assumption that the injected volume will be about 0.5 cc.

Presented below is a more detailed account of the special features of some of these tests, and of the methods that may be used to confirm the pharmacological identification of acetylcholine.[9]

Leech (Hirudo medicinalis). Except in high concentration acetylcholine has no action on the normal leech muscle, but when the muscle has been soaked for 15 minutes or longer in a weak solution of eserine it becomes very sensitive to acetylcholine (Fühner, 1918b, c). The use of this preparation for the detection of acetylcholine was recommended by Minz (1932); Chang and Gaddum (1933) and Feldberg and Krayer (1933) provide further information about it. It provides a very sensitive and specific test which has been employed in many interesting observations.[10]

A longitudinal strip of the body wall is taken from the dorsal side of the anterior end of the leech and its inner surface is cleaned by careful dissection. The strip is suspended in Locke's solution diluted 1:1.4, which contains eserine in a concentration of about 5 mg per litre. If this concentration produces a strong contraction of the leech, a lower concentration must be used. The strip is attached to a lever which writes on a smoked drum. The tension depends on the size of the muscle strip, but is usually 2–3 g; the magnification is about five-fold; when the strips are prepared by splitting the dorsal muscle lengthwise a tension of about 1.2 g is appropriate. The volume of the bath generally used is 20 cc, but a 2 cc bath works equally well. It is kept at room temperature and stirred by a slow stream of air. When acetylcholine is added to the bath the contracture starts slowly and may take many minutes to attain a maximum. When the fluid in the bath is replaced with fresh Ringer's solution the muscle relaxes slowly; the process can be accelerated to some extent by stretching the muscle mechanically. If room temperature is too low, under 18 °C, the bath must be warmed or the muscle will not relax properly. The muscle is ready for a second application of acetylcholine when it is fully relaxed and the writing-point has traced a horizontal line for several minutes. The procedure is tedious; but when it has been found that the solution under test produces responses that have about the same shape as the responses to acetylcholine, the determination can be accelerated, and an exact result achieved, by working to a strict time-table and allowing the active solution to be in contact with the muscle for a pre-determined time of 2 or 4 minutes (see Chang and Gaddum, 1933; Vartiainen, 1934).

The leech also contracts in response to choline, but only when high concentrations are used. The sensitivity of the leech to other unstable choline esters is increased by eserine in the same way as is its sensitivity to acetylcholine. Butyrylcholine, for example, produces a contracture after eserine that roughly matches the one produced by an equimolar concentration of acetylcholine. The leech preparation is unaffected by large amounts of dog's blood, or by quite high concentrations of adrenaline, histamine, adenosine or most other known tissue components. Only potassium is an exception. If KCl is added to the bath at a concentration of about 0.5 mg per cc, it produces relaxation. Such an amount of potassium in a fluid under test for acetylcholine, therefore, will partly antagonize the effect of the latter. With higher concentrations of potassium the relaxation is followed by a contracture. This action of potassium does not significantly affect the result, except when extracts are made of tissues containing less than about 0.2 µg per g of acetylcholine. Tissues often contain large amounts of intracellular potassium, and when the cells are broken up in the preparation of the extract, enough potassium can be released to interfere with the result if the extract is not applied in considerable dilution. In such cases most of the potassium can often be precipitated with alcohol, but the precipitation is never complete. A small increase in the potassium content of the physiological salt solution in which the leech muscle is suspended, for example by replacing the frog Ringer's solution with diluted Locke's solution, will make the muscle more sensitive to acetylcholine, perhaps by improving the resting relaxation.

The response of the leech muscle to acetylcholine is unaffected by atropine, but it is abolished by prolonged contact with nicotine or curare. The muscle is easily damaged by alcohol or trichloroacetic acid, substances that are often used in preparing the extract, so one must be sure that they are completely removed before the test is made.

Frog rectus abdominis muscle. Chang and Gaddum (1933) have used this muscle, in eserine solution, for estimating acetylcholine in tissue extracts. The rectus test is similar in many ways to the leech test, and most of what has been said about the latter applies here too. A somewhat higher concentration of eserine, 10 mg per l, is needed than for the leech, but even so it is less sensitive to acetylcholine. On the other hand it is just as insensitive to other materials in the extract that might affect the result. Potassium can affect the result of either test, but while it weakens the effect of acetylcholine on the leech it produces a contraction of the rectus and so can strengthen the response to acetylcholine. The rectus can be used as the basis of a rough quantitative test for potassium if a solution of ashed extract is applied to it. Chang and Gaddum preferred the rectus to the leech because they found that the test was quicker and gave a more precise quantitative estimate of the acetylcholine in a tissue. Kahlson (1934) also used the rectus test, but found that the result could be made inaccurate by sensitizing substances in the tissue extracts.[11]

Frog heart (Straub method). This is a sensitive preparation for the estimation of acetylcholine. Witanowski (1925), working in Loewi's laboratory, found that simple alcoholic extracts of various tissues contained a substance that acted like *Vagusstoff* on the frog heart and might be identical with acetylcholine. In later studies from the same laboratory the extracts were prepared by means of the more elaborate method described above and were tested on the frog heart in comparison with acetylcholine (Engelhart, 1930). A similar procedure was used by Plattner (1926), by Plattner and Krannich (1932), and by others. It has been criticized on the ground of low specificity by Chang and Gaddum (1933), von Beznák (1934) and Vartiainen (1934). Tissue extracts contain many substances that can affect the frog heart. Extraction with absolute alcohol according to the procedure described above would probably remove most of the potassium, but it must be supposed that such extracts contain other pharmacologically active materials, some of which will stimulate the heart while others, adenylic acid for example, will inhibit it. In either case, the presence of such substances in significant quantity will destroy the accuracy of the test.

Intestine. Acetylcholine causes a contraction of the longitudinal muscle of the rabbit duodenum immersed in Tyrode's solution. This is a very quick and accurate test for acetylcholine (Dale, 1914a), but unfortunately it is too non-specific to be valuable unless acetylcholine is present in high concentration compared with other active substances. The intestine is inhibited by adenosine compounds, and it is stimulated by a so far unidentified substance present in extracts of intestine, brain and prostate. This latter material, called 'substance P', will be discussed on page 108.

The isolated guinea pig intestine has also been used for the estimation of acetylcholine, but it is likewise unspecific, being stimulated by histamine in low concentration and inhibited by adenosine. On the other hand the isolated intestine of the mouse is a very suitable reagent for acetylcholine, since it is very insensitive to histamine and very sensitive to acetylcholine (Wedum and Gebauer-Fuelnegg, 1932; Kahlson, 1934).

Solutions that contain eserine cannot be tested on the intestine, because eserine causes contraction.

Other tests. The blood pressure of the rabbit under ether can be used to assay acetylcholine in the presence of histamine; but both adenosine and substance P produce a blood-pressure fall in the rabbit that is almost indistinguishable from that produced by acetylcholine (Chang and Gaddum, 1933).

The blood pressure of the cat is lowered by very small doses of acetylcholine, Either ether or chloralose may be used as anaesthetic, and the sensitivity can be increased by evisceration and by injection of eserine (0.1–0.15 mg per kg); however, the test is not specific. Brown and Feldberg (1935b) found that cats that had been kept for several hours under chloralose were particularly sensitive.[12]

The rabbit auricle preparation, isolated by the method of Clark (1931) as modified by Trevan (Trevan, Boock, Burn and Gaddum, 1928) is very sensitive to acetylcholine when treated with a sufficient quantity of eserine. This test too is unspecific.

The cat gastrocnemius, sensitized by previous denervation and by eserine, is a specific but not very sensitive reagent for acetylcholine (Chang and Gaddum, 1933).

4. *Pharmacological differentiation from other substances*

The following special tests have been used to distinguish between effects due to acetylcholine and those due to other substances.

(1) If the activity of a solution is much increased by eserine, this fact by itself justifies the conclusion that the effect is not due to choline, or to any substance other than acetylcholine that is known to occur in tissues.

(2) The activity should disappear rapidly when the solution is mixed with blood, but if eserine has been added to the blood this destruction should be greatly slowed or prevented.

(3) The active material should be unstable in alkaline solution. If a portion of the solution is mixed with 2N NaOH, and left for 10 minutes or more and then neutralized, any acetylcholine present is destroyed, while choline is unaffected.

(4) The choline content of the extract can be determined by acetylating it and then comparing it pharmacologically with acetylcholine. From the result of this test it can be calculated whether the amount of choline originally present was sufficient to account for the activity originally observed.

(5) The active material should withstand boiling for several minutes in a solution made weakly acid to congo red.

(6) Many of the effects of acetylcholine are antagonized by atropine, others by nicotine. The material under test should react to these drugs in the same way as acetylcholine itself.

(7) When the activity of the extract is compared quantitatively with that of acetylcholine by means of different pharmacological tests, the same answer should be obtained in every test.

The last of these tests provides stronger evidence for the pharmacological identification of acetylcholine than do the other tests. It differentiates acetylcholine from most other choline esters, while the other tests do not (Chang and Gaddum, 1933).[13]

5. *Methods for isolating acetylcholine*

In the procedure used by Ewins (1914) for the isolation of acetylcholine from ergot, an extract made with dilute alcohol was concentrated, and after removal of the alcohol the watery solution was treated with mercuric chloride, which precipitated only some of the impurities. After removal of the mercury the filtrate was concentrated and extracted with absolute alcohol. Acetylcholine and choline could then be precipitated with mercuric chloride from the alcoholic solution. Choline was crystallized as choline acid tartrate from alcoholic solution; when the filtrate was treated with platinic chloride, an insoluble chloroplatinate crystallized out and was identified by Ewins as the chloroplatinate of acetylcholine. Later work by Dudley (1929) makes it probable that it was really the double chloroplatinate of choline and acetylcholine.

Boruttau and Cappenberg (1921) used a similar method for extracting acetylcholine from shepherd's purse (*Capsella bursae pastoris*).

Several modifications of the method were introduced by Dale and Dudley (1929). The original alcoholic extract was concentrated to dryness and extracted with absolute alcohol, then filtered. The alcohol was removed from the filtrate, the residue was taken up in water, and the aqueous solution was treated with mercuric chloride, which precipitated impurities. Acetylcholine was then precipitated by adding an excess of sodium acetate; when the precipitate was shaken with cold water, the acetylcholine remained in the slightly soluble fraction. This fraction was decomposed with H_2S, concentrated to dryness and extracted with alcohol, and the alcoholic solution was then treated directly with alcoholic platinic chloride.

The precipitate obtained in this way consisted mainly of the chloroplatinate of choline and acetylcholine. The double chloroplatinate crystallizes out of a watery solution of the mixed chloroplatinates (Dudley, 1929, 1931) as it is less soluble than the chloroplatinate of either base and is particularly insoluble when choline is present in excess. It was identified by its melting point, by elementary analysis (C, H and Pt), by its physiological activity, and by the fact that its activity was exactly doubled by complete acetylation.

Kapfhammer and Bischoff (1930) introduced an improved method for isolating acetylcholine. Its basis is that acetylcholine forms an insoluble reineckate in the presence of reinecke acid or a reinecke salt. (Reinecke acid is $[(NH_3)_2Cr(CNS)_4]H$.)

In this procedure alcohol is first removed from an alcoholic extract, and the extract is then treated with tricholoroacetic acid and filtered. The acid is removed with ether and the aqueous solution is treated with reinecke salt. The precipitate is washed with ice water, alcohol and ether and dried in the desiccator. It is then extracted with acetone, and the acetone-soluble fraction, which consists in part of the reineckates of choline and acetylcholine, is treated with silver, which precipitates silver reineckate and liberates the bases. With this method Kapfhammer and Bischoff obtained astonishingly large amounts of acetylcholine from ox blood, and later from other organs. Several other investigators have failed to confirm their result with blood, as discussed in the following section. Nevertheless there is no doubt that the use of reineckate represents a great advance in methods for the isolation of acetylcholine.

6. *The distribution of acetylcholine*

A quantitative measure of the concentration of acetylcholine in a tissue can only be obtained by a biological test. The results thus obtained for the presence of acetylcholine in a variety of tissue extracts are summarized in Tables 7 and 8.

As already mentioned, the situation has been complicated by the announcement (Kapfhammer and Bischoff, 1930; Bischoff, Grab and Kapfhammer, 1931a, b) that unexpectedly large quantities of acetylcholine can be chemically isolated from blood. Later a similar claim was made for other tissues (Bischoff, Grab and Kapfhammer, 1932). Their findings, however, are not generally accepted. It has been pointed out in this connection that the intravenous injection of acetylcholine, in the quantity these authors claim to have obtained from 1 cc of blood or 1 g of muscle, would produce a severe physiological response; and also that acetylcholine is rapidly destroyed when it is added to blood. Such facts do not in themselves render the conclusions of Bischoff, Grab and Kapfhammer incredible, because it is possible that acetylcholine is combined with other chemical groups, or is present as a precursor, in such a way that it is neither physiologically active nor liable to destruction. Indeed it is generally agreed that extracts of certain tissues can contain enough acetylcholine to have a pronounced effect if they are applied to another piece of that tissue.

The real reason why the results of Bischoff, Grab and Kapfhammer were so surprising was their assertion that various tissues contained up to 100 times as much acetylcholine as had hitherto been supposed. Their first report was on ox blood, and owing to the importance of their conclusions a number of attempts have been made to confirm them. Vogelfänger (1933) did get confirmatory results, but all the other findings have been contradictory (Wrede and Keil, 1931; Dale and Dudley, 1931; Dudley, 1933; Chang and Gaddum, 1933; Gollwitzer-Meier, 1934; Kahlson and Romer, 1934; Ettinger and Hall, 1934; Loach, 1934).

These studies were undertaken with the special purpose of confirming the conclusions of Bischoff, Grab and Kapfhammer with respect to blood, and in several of the studies the technique described by them was followed in every detail. Because of the preponderance of contradictory findings, those of Bischoff, Grab and Kapfhammer will not be tabulated here.[14]

Table 7. *Reports on the presence of choline esters in tissues*

Hunt (1899, 1901)	Suprarenals
Witanowski (1925)	Heart, nerve, striated muscle
Plattner (1926)	Heart
Dale and Dudley (1929)	Horse and ox spleen
Engelhart (1930, 1931)	Heart, eye
Plattner and Krannich (1932)	Striated muscle
Plattner (1932)	Striated muscle
Henderson and Roepke (1932)	Salivary glands
von Beznák (1932, 1934)	Salivary glands
Minz (1932)	Suprarenals
Velhagen (1932)	Eye
Chang and Gaddum (1933)	Numerous tissues
Chang and Wong (1933)	Human placenta
Gollwitzer-Meier and Bürgel (1933)	Skin
Dikshit (1933a)	Brain
Kahlson (1934)	Spleen, striated muscle, bowel
Plattner (1934)	Various tissues
Feldberg and Schild (1934)	Suprarenal cortex and medulla
Barsoum (1935)	Nervous tissue
Brown and Feldberg (1935b)	Sympathetic ganglion

Table 7 lists the major studies that have dealt with the question of the presence of choline esters in animal tissues other than blood.

The values in Table 8 are estimates of tissue acetylcholine content, obtained by different methods. They reveal no consistent discrepancies, but it is of course not possible to state with confidence than in every case the active material was acetylcholine itself.[15]

Although acetylcholine is not present in normal plasma, it may appear when eserine has been given, in the blood of the submaxillary vein (Babkin, Stavraky and Alley, 1932) or in the portal blood coming from the stomach or small intestine (Feldberg and Rosenfeld, 1933; Dale and Feldberg, 1934a). These findings are discussed more fully later (see pages 117–20). Acetylcholine was detected in concentrations of 0.005–0.5 µg per cc; under these conditions the acetylcholine content of the plasma was roughly equal to that of the corpuscles.

The presence of large amounts of acetylcholine in the spleens of oxen and horses is surprising. The spleens of most laboratory animals contain practically none. The only other tissue in which similarly large quantities of acetylcholine have been found is the human placenta. Chang and Wong (1933) have suggested that this acetylcholine may act on the uterus in normal labour. Apart from these two exceptional cases, the highest

Table 8. *Estimates of the acetylcholine content of different tissues, taken from the papers selected in Table 7*

	mg per kg		mg per kg
Blood corpuscles	0.01–0.09	Sympathetic ganglion	10–20
Blood plasma	0	Suprarenal cortex	0.1
Striated muscle	0.07	Suprarenal medulla	0.45
Smooth muscle, stomach	0.7	Lung	0.13
Intestine	2.4	Liver	0.11
Urinary bladder	1.2	Pancreas	0.1–0.2
Uterus	0.4	Spleen (horse and ox)	4–30
Heart, auricle	1.4	Spleen (dog)	0
Heart, ventricle	0.1–0.3	Kidney	0
Brain	0.4	Testes	0
Somatic nerve	1.7–2.5	Skin	0.1
Vagus nerve	5–10	Human placenta	15–133
Sympathetic nerve	1–4		

concentrations of acetylcholine are in the trunks of cholinergic nerves and in the tissues supplied by such nerves. Thus the distribution of tissue acetylcholine conforms roughly to expectation; but there are some minor anomalies, and the correlation between acetylcholine content and cholinergic nerves does not seem to be a very close one.

Engelhart (1930) found, as he had anticipated, that in mammalian hearts the auricle had a much higher acetylcholine concentration than the ventricle. In the suprarenals, the high concentration in the medulla as compared with the cortex can be related to the cholinergic innervation by the splanchnic nerve. The choline concentration is higher in the cortex than in the medulla (Feldberg and Schild, 1934).

D. Acetylcholine metabolism

1. *General*

Little is known with certainty about the sources of the body's acetylcholine. Some findings by Le Heux (1919) suggest that the intestine can esterify choline directly in the presence of acetate or other fatty-acid salts to form the corresponding choline ester.

Abderhalden and his colleagues (1925) and Ammon and Kwiatkowski (1934) have reported experiments that indicate that cholinesterase can synthesize acetylcholine if choline and acetic acid are supplied in high concentration. From recent observations by von Beznák (1934) it appears that minced heart tissue from the frog can manufacture acetylcholine if enough eserine is present to hinder its breakdown. Chang (1934) obtained similar results with placental extracts. Ammon and Kwiatkowski (1934) on the other hand, state that eserine inhibits the synthesis of acetyl-choline.[16]

Our information about acetylcholine breakdown is more complete. Acetylcholine is destroyed more rapidly in solutions to which minced tissue has been added than it is in pure solutions. As early as 1914 Dale suggested that the evanescence of its effect when injected into the circulation of a living animal might well be attributed to destruction by an esterase. The correctness of this view was shown by the experiments of Loewi and Navratil (1926a), when they confirmed the existence of such an esterase by showing that watery extracts of the frog heart have the power of destroying the physiological activity of acetylcholine, and showed that the extract was inactivated by heating to 56 °C and by ultraviolet irradiation. They showed, further, that the original activity could be fully regenerated by acetylation, from which they concluded that the chemical change produced by the enzyme is hydrolysis of acetylcholine into choline and acetic acid. Later work has confirmed this conclusion.

The cleavage of acetylcholine by blood has been studied rather more extensively by Galehr, Plattner and their co-workers (1927, 1928). The reaction is so fast that in human blood at 40 °C the destruction is almost complete in 15 seconds. The activity was mainly due to the corpuscles; serum was much less active. The time-course was exponential, with the velocity of the reaction increasing by a factor of 1.3 for each 10 degree rise of temperature over the range 11 to 40 °C. They arrived at the conclusion that the destruction was due to adsorption of acetylcholine, and that the reason its rate was affected by various factors was secondary to surface alterations; thus the reaction was slower in weakly acid media, and that was attributed to a diminution of the active surface. They found that acetylcholine was also destroyed by adsorption on charcoal, and concluded that its destruction by blood was due to a similar unspecific adsorption and not to the action of an enzyme. Exposure to heat and ultra-violet irradiation, which destroyed the heart esterase of Loewi and Navratil, did not affect the action of blood on acetylcholine.

The whole problem was thoroughly reinvestigated by Engelhart and Loewi (1930), who showed that all the evidence is compatible with destruction by an esterase. The resistance of the blood enzyme to heat and irradiation is due to the presence of protective substances in human blood; the esterase of the frog heart is similarly protected if it is added to mammalian serum. Matthes (1930) also showed that the cleavage of acetylcholine was entirely explicable on the basis of an esterase.

Purified preparations of cholinesterase have been obtained from horse serum by Stedman, Stedman and Easson (1932). They measured the activity of their preparation by adding a large excess of substrate and adding alkali continuously so as to keep the pH constant. Under these conditions the amount of substrate split was directly proportional to time. Although these purified preparations acted powerfully on acetylcholine and butyrylcholine, their action on methyl butyrate was very weak compared with that of the original serum. Thus their findings greatly strengthened the view that the destruction of acetylcholine is due to a specific cholinesterase.[17] The action of horse serum on methyl butyrate must be mainly due to a different enzyme: pure cholinesterase, they suggested, would have no effect on that ester.

Ammon (1933) has described an especially convenient method for the quantitative estimation of cholinesterase. The acid liberated by the hydrolysis of acetylcholine acts on bicarbonate, and the CO_2 evolved is measured in the Barcroft–Warburg apparatus (see also Stedman and Stedman, 1933).

2. The distribution of cholinesterase

The presence of cholinesterase was demonstrated by Loewi and Navratil (1926) and by Clark (1927) in the frog heart, by Abderhalden and his colleagues (1925) in the intestine of the horse and pig, and by Engelhart and Loewi (1930) and Matthes (1930) in blood.

Galehr and Plattner (1928) and Plattner and Bauer (1928) compared the hydrolytic activity of the blood of different species. These were in order: man (highest activity), pig, ox, dog, horse, rabbit, frog, cat. According to Feldberg and Rempel (1933) the activity of guinea pig blood is even lower than that of cat blood. Galehr and Plattner express the opinion that the reason why some species are particularly resistant to the pharmacological effects of acetylcholine is that their blood is especially rich in cholinesterase. Stedman, Stedman and White (1933) have measured serum cholinesterase concentration in a series of animals, and found that they could only harmonize their results with those obtained earlier for blood by supposing that the serum and corpuscular levels are independent of each other.

The order in the list above probably gives a general idea of the relation between the different species, but it should not be supposed that the cholinesterase content in the blood of a given species is a constant quantity. There is much more reason to believe that different animals of the same species, or for that matter the same individual under different conditions, can differ widely in this respect (see Govaerts, Cambier and Van Dooren, 1931). The possible influence of external conditions, hormones, etc. on tissue cholinesterase has not yet been investigated.

Plattner and Hintner (1930a) have studied the distribution of cholinesterase in the different tissues. Their data show some degree of correlation with those obtained later by Chang and Gaddum (1933) for the distribution of acetylcholine. Thus the acetylcholine equivalent (apparent acetylcholine) of intestine, salivary glands, suprarenals, brain and uterus was more than 0.3 μg per g, and these tissues contained enough cholinesterase to halve the acetylcholine concentration in less than 10 minutes under the chosen experimental conditions. Lung, ventricle, skeletal muscle and kidney all gave lower values for both acetylcholine equivalent and cholinesterase content. On the other hand blood, pancreas, liver and spleen failed to exhibit this correlation, all being rich in esterase and poor in acetylcholine equivalent. The reverse situation – high acetylcholine equivalent with low cholinesterase – was not encountered.

It would be much easier to discuss the cholinesterase results if there were some generally accepted method of expressing the concentration of the enzyme. One difficulty that arises here is that some investigators use a large excess of enzyme (pharmacological tests) while others use a large excess of acetylcholine (chemical tests). In the first case the reaction is approximately exponential, and a cholinesterase unit would naturally be defined as the amount of enzyme necessary to halve the acetylcholine concentration in a given time under standard conditions; whereas in the second case the unit would be defined as the amount required to destroy a given quantity of acetylcholine in a set time under standard conditions. The relationship between the results obtained by the two sorts of method is still unknown, and cannot be calculated without further data.

3. The action of eserine and other drugs on cholinesterase[18]
Eserine is a specific inhibitor of cholinesterase, which explains why it increases the effects of active choline esters in every case in which it has been tried. It has only a slight and unspecific potentiating action on the effects of more stable drugs such as choline, pilocarpine, muscarine and tetraethylammonium salts (Loewi, 1912; Fühner, 1918b, c; Dale and Gaddum, 1930; Feldberg and Vartiainen, 1934). Its influence on the responses to choline esters were first observed by Hunt (1915a), who also found (cf. Hunt, 1934) that it strengthens the actions of acetyl- and

methylcholine on the heart and the pupil. Hunt showed later (1918) that eserine strengthens all the circulatory actions of acetylcholine in the same way. 'It strengthens the depressor action and the bradycardia before atropine and the pressor action after atropine.'

Fühner (1918) studied the effect of various drugs on the dorsal muscle of the leech. Equal quantities of choline and acetylcholine produced roughly equal contractions of the muscle. But after the muscle had been soaked for 20 minutes in a 1:100000 solution of eserine the action of acetylcholine was increased a million-fold; in other words the same result could be produced by a millionth of the original concentration. The actions of choline and pilocarpine were unaffected by the treatment. Fühner thought that these findings could be explained by supposing that acetylcholine was normally hydrolysed by the tissue so rapidly that it could not exert its full effect, and that this hydrolysis was inhibited by eserine.

Loewi and Navratil (1926a) showed that the effects on the frog heart of *Vagusstoff* and acetylcholine were greatly prolonged by eserine, while the effects of choline and muscarine were not. They demonstrated by *in vitro* experiments that the explanation was inhibition of the esterase. Similar results were obtained by Engelhart and Loewi (1930) and by Matthes (1930) when they added eserine to blood, and by Plattner and Hintner (1930a) when they added it to extracts of various organs.

The inactivation of cholinesterase by eserine is a slow process. It takes more than a quarter of an hour to become complete when the eserine concentration is low, but can apparently proceed more rapidly at higher concentrations. The quantitative relationship between eserine concentration and degree of inhibition indicates that the process gradually advances until a point of equilibrium is reached. Matthes (1930) has shown that the reaction is indeed reversible, since the enzyme is re-activated when eserine is removed by dialysis.

Stedman and Stedman (1931) have put forward an attractive theory of the mechanism of the inhibition. They suggest that eserine, which is itself a stable ester, forms a compound with the enzyme in the same way as a choline ester, and thus blocks its action. The theory is supported by the finding that eserine inhibits at least one esterase besides cholinesterase, and that esters that are chemically related to eserine also have this action.

Cholinesterase is also inhibited by miotine and other urethanes that are chemically related to eserine and have similar pharmacological properties (Matthes, 1930; White and Stedman, 1931; Aeschlimann and Reinert, 1931). It is also inhibited by high concentrations of ergotamine and ergotoxine, by various narcotics, by fluorides and quinine, but not by atoxyl. Magnesium, calcium and barium are weakly inhibitory (Loewi and Navratil, 1926; Matthes, 1930; Plattner and Galehr, 1928; Plattner and Hou, 1930). Ammon (1933), who confirmed many of these findings, showed further that muscarine also is an inhibitor.

E. The physiological significance of acetylcholine

The physiological significance of acetylcholine will be discussed in Chapter VII, which deals with the liberation of active substances by nerves (see page 111 *et seq.*).

V

Adenosine compounds

A. Chemical and physical properties

The chemical structure of muscle adenylic acid is:

Adenine   +   d-Ribose   +   Phosphoric acid

Muscle adenylic acid

The purine base adenine linked with the pentose d-ribose forms the nucleoside adenosine. Adenosine linked with phosphoric acid forms the nucleotide adenylic acid. Adenylic acid was first recognized as a product of the partial hydrolysis of nucleic acids, but only a small part of the adenylic acid of the tissues is found in that combination; nearly all of it is normally linked to pyrophosphoric acid to form adenosine triphosphate or adenyl pyrophosphate (Fiske and Subbarow, 1929; Lohmann, 1929).

The adenylic acid obtained by hydrolysis of yeast nucleic acid differs from muscle adenylic acid, but probably only in the attachment of the phosphate group to the ribose. Levene (1931) gives a comprehensive account of the chemical properties of these compounds, so it will suffice here to give a brief summary of the properties that are of special interest to pharmacologists.

Stability. The action of heat on aqueous solutions of adenosine compounds depends on pH. In alkaline solution adenosine is very stable: a solution of it can be boiled for 30 minutes in N NaOH without apparent loss of physiological activity. On the other hand it is unstable in acids, which hydrolyse it to adenine and pentose; if it is heated to 100 °C for 5 minutes

Table 9. *Molecular weights of adenosine compounds*

Adenine	135
Adenosine	267
Adenylic acid (as monohydrate)	365
Adenosine triphosphate (adenyl pyrophosphate)	507
Barium salt of adenosine triphosphate	885

in strongly acid solution it is almost completely inactivated. This behaviour in the presence of alkalis and acids can be used to test whether the physiological effects of extracts are due to adenosine compounds or to other substances. Histamine, acetylcholine and other vasodilator substances are partially or completely destroyed in a tissue extract that has been made strongly alkaline and then boiled, whereas histamine is very stable in acid solution.

If adenyl pyrophosphate is boiled in neutral solution, the pyrophosphate is split off and adenylic acid is liberated. If it is boiled in the presence of N HCl, the pyrophosphate is split off more rapidly, and the action is practically complete in 7 minutes; the adenyl pyrophosphate is converted into adenine, pentose and phosphoric acid (Lohmann, 1931a, b), and its physiological activity is lost. If adenylic acid is heated to 145 °C for 2 hours in strong ammonia, it is hydrolysed into adenosine and phosphoric acid.

Neutral solutions of adenosine compounds are stable at 0 °C for several days.

Solubility. Adenosine compounds are practically insoluble in ether. Adenosine itself is soluble in physiological saline solution in a concentration of about 4 mg per cc.

Chemical precipitants. Like other purine bases, adenosine compounds are precipitated by silver from acid solutions. Adenyl pyrophosphate can be separated from adenylic acid on the basis of the much lower solubility of its barium and calcium salts. Adenylic acid is precipitated by uranyl acetate, whereas adenosine is not. On the basis of these facts Ostern and Parnas (1932) have proposed a test for free adenosine. The tissue extract is treated with uranyl acetate and the filtrate is assayed on the frog heart.

B. Pharmacological properties of adenosine compounds

1. Action on the heart

The pharmacological activity of adenosine compounds was discovered and studied in some detail by Drury and Szent-Györgyi (1929).[1] Except where the contrary is stated, the information presented here has been taken either from their original paper or from the later one by Bennet and Drury (1931).

The effects of the adenosine compounds on the heart have certain resemblances to the effects of vagal stimulation. They slow the heart, and they weaken the auricular beat while having no direct action on the mammalian ventricle. On the other hand their action is quite unaffected by atropine, and they dilate the coronary vessels.

Frequency. Adenosine and adenylic acid slow the heart (negative chronotropic action). This can be demonstrated in the dog by injecting 20 mg intravenously. The cat, rabbit and guinea pig hearts are less sensitive. The slowing, as mentioned above, is due to a direct action on the heart and is unaffected by atropine. In the cat there is also a secondary slowing; it reaches its maximum some 80 seconds after the injection, and is of vagal origin, being abolished by atropine (Drury and Szent-Györgyi, 1929). Adenylic acid also slows the heart in man (Honey, Ritchie and Thomson, 1930; Rothman, 1930).

In the isolated frog heart (Straub preparation) the bradycardia is sometimes the only visible effect of adenosine (Ostern and Parnas, 1932). This response, which can be elicited with doses as low as 2 µg, has been made the basis of a sensitive and apparently specific test for adenosine compounds. The slowing is also well shown by isolated heart muscle from turtle, guinea pig, rabbit, dog, and rat (Wedd and Fenn, 1933).

Force of the beat. In the auricle of the normal dog with intact circulation the beat is weakened by adenosine and adenylic acid (negative inotropic action). This effect is due in part to the slowing of the heart, but not entirely, since the compounds have the same action when the heart's rate is kept constant by electrical stimulation. They do not have such an action on electrically stimulated strips of ventricle.

The negative inotropic effect is well shown by the isolated hearts of a number of animals (Wedd and Fenn, 1933). The beat of the Straub isolated frog heart preparation is often weakened by adenosine, but in some hearts the negative chronotropic action is more prominent. Lindner and Rigler (1930) found that the calcium salt of adenosine triphosphate differed from adenosine and adenylic acid, in that it strengthened, even in very low concentration, the beat of the hypodynamic frog heart.

Conduction. In the guinea pig adenosine and adenylic acid produce complete heart block (negative dromotropic action). In the dog and cat the heart block is usually incomplete, because the beat is slowed so much that the conducting tissues have more time to recover; but if the heart is driven electrically at constant frequency complete block may occur. In any case there is slowing of conduction from auricle to ventricle. Adenosine also causes heart block in man (Honey, Ritchie and Thompson, 1930).

Excitability. The threshold for electrical stimulation is unaltered (absence of bathmotropic action).

Coronary vessels. Adenosine compounds are powerful coronary dilators.[2] In the isolated rabbit heart adenosine is about 20 times as active as sodium nitrite in increasing coronary flow. The same dilator action is seen on the isolated hearts of the guinea pig, the cat and the dog (Wedd, 1931; Rigler and Schaumann, 1930). This coronary dilator action may be responsible for the increase in force and regularity of the beat that has sometimes been observed to follow the injection of adenosine compounds.

Coronary vasodilatation also occurs in birds and reptiles (Rigler, 1932).

2. Action on the blood vessels

Adenosine compounds dilate the vessels of the systemic circulation, presumably in consequence of a direct action on the muscular coat of the smaller vessels. This may be the main cause of the fall of blood pressure seen in the intact animal after an intravenous injection. It has been demonstrated in experiments in which a dog's leg was perfused with blood. It can also be shown as an increase of flow when a rabbit's ear is perfused with Locke's solution. No vasodilatation is seen in the isolated perfused kidney; rather, there is vasoconstriction and reduced urine flow after large doses (100 mg).

Adenosine compounds constrict the perfused pulmonary vessels of the rabbit (Bennet and Drury, 1931). The perfused lung of the cat responds with vasodilatation to small doses (5 µg); with larger doses (0.25 mg) the initial vasodilatation is followed by vasoconstriction. The pulmonary vessels of the dog react in the same way but are less sensitive (Gaddum and Holtz, 1933).

Adenosine compounds dilate the perfused vessels of the frog (Rigler, 1932). The coronary dilator action of adenosine compounds has been described in the section on the action on the heart.

3. Action on smooth muscle

The adenosine compounds have been found to inhibit every kind of smooth muscle on which they have been tested, with the exception of the guinea pig uterus. Thus adenosine and adenylic acid inhibit the tone and contractions of the cat and rabbit small intestine, relax the gall bladder, and dilate the bronchioles. They produce contraction of the guinea pig uterus (Bennet and Drury, 1931; Deuticke, 1932a); however the uteri of the cat, the dog and the rat are inhibited, like other kinds of smooth muscle (Barsoum and Gaddum, 1935a).

4. *Leucocytosis*

Sodium nucleinate, adenylic acid, adenosine and adenine all increase the number of neutrophil leucocytes in the blood. The effect seems to be due to stimulation of the bone marrow. The leucocytosis is sometimes preceded by a leucopenia. A local accumulation of neutrophil leucocytes also appears in the conjunctiva if adenosine or adenylic acid is instilled into the rabbit eye for a long time (Fox and Lynch, 1917; Neymann, 1917; Doan, Zerfas, Warren and Ames, 1928; Reznikoff, 1929; Bennet and Drury, 1931). Adenine and guanine are used clinically for the treatment of agranulocytosis (Reznikoff, 1930).

C. The natural occurrence of adenosine compounds

1. *General*

Adenosine compounds are rapidly destroyed during tissue autolysis; for studies of their distribution it is important to inhibit autolysis as quickly as possible after death. The steps taken by different authors to achieve this have varied widely.

It is probable that the autolytic changes proceed relatively slowly before the tissue is minced. A sufficiently accurate estimate of the adenosine compounds in small samples of tissue can thus be obtained by putting the tissue in 10 per cent trichloroacetic acid and mincing it below the surface of the liquid.

2. *Estimation*

Quantitative chemical methods. Buell and Perkins (1928) have described a nephelometric method for the determination of adenylic acid. Lohmann (1928a, 1928b, 1931c) has developed a method for the quantitative estimation of adenosine triphosphate, based on measuring the phosphoric acid set free by hydrolysis in N HCl within 7 minutes at 100 °C. Schmidt (1928) describes an alternative method, and Ferdmann (1933) one of even greater specificity.

Pharmacological methods. Some of the pharmacological methods for detecting and estimating adenosine compounds are much more sensitive than the chemical methods, and are therefore suitable for use when only small quantities of material are available.

Bennet and Drury (1931) have used the duration of the heart block in the intact atropine-treated guinea pig as a measure of the adenylic acid

Table 10. *Summary of reports on the isolation of adenosine compounds*

Substance	Source	Authors
Adenine	Pancreas	Kossel (1885)
Adenosine	{ Yeast hydrolysate	Levene and Jacobs (1909)
	{ Urine	Calvery (1930)
	{ Yeast hydrolysate	{ Levene (1918)
		{ Jones and Kennedy (1919)
		{ Thannhauser (1919)
	Blood	{ Jackson (1923, 1924)
		{ Hoffman (1925)
Adenylic acid		{ Zipf (1931b)
	Skeletal muscle	Embden and Zimmermann (1931)
	Heart muscle	{ Pohle (1929a)
		{ Drury and Szent-Györgyi (1929)
	Brain	Pohle (1929b)
	Kidney	Embden and Deuticke (1930)
	'Lacarnol'	For literature see Zipf (1931)
Adenosine triphosphate	Skeletal muscle	{ Lohmann (1929)
		{ Fiske and Subbarow (1929)
		{ Barrenscheen and Filz (1932)
	Erythrocytes	Meyerhof and Lohmann (1932)
Other polyphosphates	Heart muscle	{ Lohmann (1931a, b, c)
		{ Embden: see Deuticke (1932b)

in their tissue extracts. About 0.1 mg of adenosine will elicit this response, and the amount can be reduced to 0.02 mg by cooling the animal. A simpler method, and one which requires even smaller tissue samples, has been described by Ostern and Parnas (1932): it uses the isolated frog heart.

The depressor action of the adenosine compounds has also been used as the basis of a method for their assay. Here it is particularly important to exclude the possibility that other materials contribute to the effect. If atropinized rabbits under ether are employed, histamine and choline esters need not be considered. Kallikrein and substance P can be removed by boiling the extract with alkali. Adenosine compounds are the only known constituents of extracts that lower the blood pressure of the atropinized rabbit under ether and that can withstand boiling for 10 minutes in the presence of N NaOH.

Recent studies by Barsoum and Gaddum (1935a) have shown the rectal caecum of the fowl to be a sensitive test-object for adenosine.

3. Isolation and distribution

Table 10 gives a brief survey of the reports by authors who have isolated adenosine compounds from organic material. The methods employed have

Table 11. *Estimates of the concentration of adenosine compounds in tissue.* Results calculated as mg of adenyl acid per kg.

Organ		Biological tests		Chemical tests	
		Bennet and Drury (1931)	Ostern and Parnas (1932)		
Blood (rabbit)		30[a]	700[b]	85	⎫
(dog)		—	—	94	⎪
(cat)		—	—	95	⎬ Buell and Perkins
(man)		—	—	178	⎪ (1921)
(pigeon)		—	—	700	⎭
Skeletal muscle	⎫	600	1250	3800	⎱ Buell, Strauss and
Heart	⎪	600	930	2000	⎰ Andrus (1932)
Stomach	⎪	200	500	—	
Skin	⎪	50	—	—	
Brain	⎬ Rabbit	100	230	—	
Lung	⎪	80	230	—	
Liver	⎪	500	1160	—	
Kidney	⎪	500	1100	—	
Suprarenal	⎭	50	—	—	
Pancreas	⎱ Sheep	270	—	—	
Spleen	⎰	330	—	—	
Testis	⎱ Dog	300	—	—	
Ovary	⎰	300	—	—	

[a] Blood cells.
[b] Erythrocytes.

generally depended on the preparation of insoluble salts with silver, barium, calcium, copper, zinc or lead.

Table 11 presents selected data for the concentration of adenosine compounds in a number of tissues. Here the differences in the values reported by the different authors are probably mainly due to their choice of different methods for preparing their extracts. Bennet and Drury (1931) state that the tissues they used were as fresh as possible, but these were minced before being placed in trichloroacetic acid, and no special precautions were taken to minimize autolytic changes. Buell, Strauss and Andrus (1932), on the other hand, plunged the tissue into liquid air within 4 or 5 seconds of its excision. They showed that a large proportion of the adenosine triphosphate might be destroyed during the first few minutes of autolysis. It is not surprising, therefore, that their estimates of the level of adenosine compounds in tissues are higher than those of the other authors.

D. The metabolism of adenosine compounds

The purine derivatives in the body are derived partly from purine derivatives absorbed as such from the food and partly from synthesis of the purine ring, but the details of the processes through which compounds like adenosine triphosphate are formed are not yet known.

The fate of these substances in the body has been more extensively studied, and a number of enzymes have been discovered that can act upon them. Nucleotidase and nucleosidase are widely distributed in animal tissues. The former has the ability to split off phosphoric acid from nucleotides; the latter hydrolyses nucleosides to the sugar and the base (Levene, 1931; Jacobsen, 1931).

Extracts of skeletal muscle also contain deaminases, which remove the amino group from muscle adenylic acid or from adenosine to form inosinic acid or inosine (Schmidt, 1928). This reaction is the source of the ammonia liberated when muscle contracts, or is injured (Parnas and Mozolowski, 1927; Parnas, 1929; Embden and Schmidt, 1935). The reaction is apparently a reversible one and probably plays an important role in muscular contraction, but the nature of the role is still obscure.

The importance of adenosine compounds in muscle metabolism has also been emphasized by the work of Lohmann (1931a, b). He has shown that adenosine triphosphate is the coenzyme of zymase, which converts glycogen to lactic acid in extracts of skeletal muscle. Magnesium must also be present for this reaction to take place. The coenzyme of yeast fermentation, which has been isolated by von Euler and his co-workers, is also an adenylic acid, though one that is not identical with any of the adenylic acids hitherto isolated from tissues.

In the light of these facts it is not surprising that the pharmacological activity attributable to the adenosine compounds rapidly disappears during tissue autolysis, or that adenosine is rapidly destroyed by the juices expressed from tissues. Bennet and Drury (1931) found that the juices expressed from the lungs were especially active in this respect.

In spite of the presence in the body of these mechanisms for destroying adenosine compounds, a certain amount of free adenosine is excreted in the urine (Calvery, 1930).

E. The physiological significance of adenosine compounds[3]

The role of the adenosine compounds in the metabolism of skeletal muscle has already been mentioned, and it remains only to discuss the significance of the fact that these substances have pharmacological effects. Awareness of these effects has suggested to various authors that they may play a part in reactive hyperaemia, in the hyperaemia of muscular activity, and in the hyperaemia that accompanies injury or shock.

Bennet and Drury (1931) found that when an isolated perfused heart was damaged by heat it released at least two substances, which acted pharmacologically like histamine and adenosine. Rigler (1932) found that when a frog's skeletal muscles were perfused and stimulated, the observed vasodilatation was often accompanied by the appearance of a vasodilator substance in the perfusate. The substance could not have been lactic acid, because it also appeared when the muscles were poisoned with iodoacetic acid; and it could not have been acetylcholine, because its action was not abolished by atropine. Rigler suggested that it might be an adenosine compound.

The possibility that adenosine compounds may account for the vaso-dilatation that occurs in active muscle has also been discussed by Zipf (1931a), but the evidence is still inconclusive. There is no proof as yet that adenosine compounds are liberated, and apart from the vasodilatation itself there is no particular reason for supposing that they would be liberated. The possible relationship of these substances to reactive hyper-aemia, burns, shock, etc. will be discussed at greater length later on (see pages 170 *et seq.*).

VI

Unidentified substances present in tissue extracts and body fluids

A. Substances from mammalian blood

1. General

It has sometimes been suggested that the blood circulating in the body normally contains substances, such as adrenaline or the active principle of the posterior pituitary, whose concentration is high enough to maintain the tone of the blood vessels. No convincing support for that suggestion has as yet been brought forward. Adrenaline can indeed be detected in the venous blood from the suprarenals, but arterial blood does not ordinarily contain enough to be detected by any of the available methods (Schlossmann, 1927).

Extracts from freshly drawn blood contain choline, and also an extremely minute proportion of an ester that resembles acetylcholine and may well be identical with it.[1] The latter comes from the corpuscles (Chang and Gaddum, 1933; Gollwitzer-Meier, 1934) and probably the choline does also. Nothing is known about how these substances are held within the corpuscles, but it must be in a way that ensures that they will exert no physiological activity; and it seems that their presence indicates no immediate physiological significance.

Many attempts have been made to detect histamine in normal blood. The earlier ones were uniformly unsuccessful (Hanke and Koessler, 1920; Guttentag, 1931; Zipf and Hülsmeyer, 1933; MacGregor and Thorpe, 1933a, b); but recently Barsoum and Gaddum (1935a) have been able to demonstrate its presence in extracts from normal blood after these were boiled with HCl. Such treatment destroys other pharmacologically active substances and the remaining activity appears to be due almost entirely to histamine.[2] Its concentration in rabbit blood is about 10 µg per cc and in cat blood about 10 µg per l (see page 47). There is some evidence that histamine is liberated and escapes into the blood in skin injuries (see page 155), reactive hyperaemia (see page 170), burns (see page 158) and anaphylaxis (see pages 162 *et seq.*).

The presence of adenylic acid in blood has been mentioned on page 92.

Its concentration is higher in the corpuscles than in the plasma. When blood is whipped or allowed to clot its total content of pharmacologically measurable adenosine is increased.

The evidence for the view that kallikrein, which was first recognized in the urine, is also present in blood though in an inactive form, will be discussed in the section on urine (see pages 113 *et seq.*).

Although circulating blood is usually almost without pharmacological activity, it can acquire a high degree of such activity when it is removed from the body. The following account deals with the substances that are responsible for that activity. They have been studied extensively, and many contradictory statements have been made about their properties. The discrepancies are no doubt due in part to the fact that different investigators have used different pharmacological tests to identify their active materials, and different methods of processing the blood, and in consequence may have been studying different substances.[3]

Until the active substances have been obtained in pure form, it will be impossible to be quite sure that any activity that an extract is found to exert on a series of different tissues is actually due to a single unknown substance. The use of several different pharmacological reactions for the recognition of an unknown substance has been a source of error in the past, because it has been assumed without justification that all the effects can be attributed to the same substance. To avoid error from this source we shall deal separately with the experimental findings obtained with each type of test.

2. *Effects on perfused frog vessels*

The perfused vessels of the frog provide a very sensitive test for adrenaline. Läwen (1904) and Trendelenburg (1910) used this preparation to measure the adrenaline content of blood. Kahn (1912) found that hirudin plasma was less active than serum, and O'Connor (1912) showed that the vasoconstrictor substance in serum was not adrenaline, since it was not destroyed by bubbling oxygen through the serum; adrenaline added to serum was destroyed by such treatment. O'Connor found that when sodium citrate was used as anticoagulant the plasma had no vasoconstrictor effect and he suggested that the effect of serum was due to an unknown substance liberated by the platelets. He discussed the possiblity that the substance might have a significant role in the response to injury: the local vasoconstriction it produced could diminish the loss of blood.

Trendelenburg (1916) confirmed the observation that during clotting a substance is liberated that constricts the perfused vessels of the frog and is not adrenaline. He found that its action was unaffected by doses of ergotamine that inhibited the action of adrenaline.

Heymans, Bouckaert and Moraes (1932a, b), on the other hand, found that ergotamine reversed the vasoconstrictor effects of both adrenaline and defibrinated blood on the perfused vessels of the frog, so that both materials became vasodilator. They obtained the same result on the rabbit ear perfused with salt solution.

Freund (1920, 1921) used differential centrifugation to prepare a suspension of cat platelets in citrated plasma; he showed that the suspension had no activity of its own, but on shaking with glass beads it released pharmacologically active materials which Freund classified, according to the time at which they appeared, into two groups, which he named *Frühgifte* and *Spätgifte*:

(1) *Frühgift*: this was present as soon as the platelets had been broken up, had mostly disappeared within 30 minutes, and had the following effects:

(a) vasodilatation: shown by an increase in flow in the Ringer-perfused frog preparation and in a fall of blood pressure on intravenous injection into frogs or mammals;

(b) inhibition of the isolated mammalian intestine;

(c) contraction of the cat uterus (inhibition when horse platelets were used).

Later Zipf (1931) came to the conclusion that the effects on the mammalian blood pressure and the intestine were due to adenylic acid. The results obtained with other test-objects are discussed below in the appropriate sections.

(2) *Spätgift*: this developed gradually over several minutes, and had the following effects:

(a) vasoconstriction: shown by a decrease in flow in the Ringer-perfused frog preparation and in mammals by a rise in blood pressure;

(b) contraction of the isolated mammalian intestine;

(c) contraction of the cat uterus (inhibition when horse platelets were used).

This classification has some advantages, but the different actions that Freund attributed to *Frühgift* and *Spätgift* are probably due to more than two substances. Thus one of the actions of *Frühgift*, according to Zipf and Wagenfeld (1930a), is the production of urticarial wheals. Now it has been mentioned that Zipf attributes many of the effects of *Frühgift* to adenylic acid. But adenylic acid does not give rise to wheals. Whealing is a characteristic action of histamine.

Some observations by Stewart and Zucker (1913) indicate that there may also be more than one *Spätgift*. They found that citrated plasma caused a strong contraction of the rabbit intestine and uterus, although it had only a slight effect on the isolated carotid strip and on perfused frog vessels. When this plasma was clotted by the addition of calcium it acquired the power of constricting blood vessels, but its action on the intestine and the uterus was not increased.

In the experiments of Haake (1930) the effect of defibrinated blood on the perfused frog, preparation was invariably vasoconstrictor.

Freund (1921) treated serum with 9 volumes of alcohol and found that a vasoconstrictor substance could be recovered from the alcoholic solution. When he made a *Frischblutextrakt* by letting blood flow directly into 9 volumes of alcohol the extract was normally inactive.

According to Borgert and Keitel (1926) the vasoconstrictor material from serum was dialysable and insoluble in ether and acetone; it was not precipitated by phosphotungstic acid. If the serum was allowed to stand for 24 hours it lost its vasoconstrictor property. Berger and Keitel offered further evidence for the view that the material originates from the platelets. In contrast, other authors found that it is neither dialysable nor heat-stable (Handovsky and Pick, 1912a, b; Heymann, 1927).

3. Effects on isolated mammalian arterial strips

A substance is present in serum that contracts an ox carotid artery strip isolated by Meyer's (1906) method (Stewart and Zucker, 1913; Loening, 1913). This substance is almost completely absent from citrate or hirudin plasma, but appears when blood clots or when a platelet suspension is otherwise damaged. Erythrocytes and leucocytes appear not to form this substance (Janeway, Richardson and Park, 1918). The active substance is dialysable and heat stable, and soluble in alcohol but not in ether or chloroform. The correspondence between its properties and those of the *Spätgift* which constricts the perfused vessels of the frog suggests that the two may be the same. However, the possibility has not yet been excluded that the substance that acts on the ox carotid may be histamine.

4. Effects on the perfused rabbit ear

Serum produces vasoconstriction in this preparation. The sensitivity of the vessels to serum is diminished if sodium citrate is added to the perfusion medium. An ear perfused with a saline solution containing serum and sodium citrate is apparently a very sensitive test-object for adrenaline in blood (Schlossmann, 1927).

It has already been mentioned that Heymans, Bouckaert and Moraes (1932a, b) have found that the vasoconstrictor action of serum on the rabbit ear is reversed by ergotamine.

5. Effects on mammalian vessels with normal circulation

It has long been known that an injection of serum may cause a profound fall of blood pressure. The earlier papers have been reviewed by Weiss (1897). Brodie (1900) found that an injection of 3–10 cc gives rise to respiratory and cardiac arrest and vasodilatation. These effects were mostly due to stimulation of sensory nerves in the lungs and were much reduced by severing the pulmonary branches of the vagus. The active material was formed as a result of blood clotting, but originated from the cells since it could not be obtained from the clotting of plasma. It was unstable to heat. After several injections the animals became immune to its action.

Freund (1920), whose experiments on perfused frog vessels have already been mentioned, observed the depressor effect of damaged platelets in mammals and attributed it, as well as the depressor effect that survived vagotomy, to *Frühgift*. These effects, which were very marked in the rabbit, and were also seen in the cat and the frog, were almost completely absent in the dog. After about half an hour the solution acquired pressor activity, which was particularly striking in the cat; this was attributed to *Spätgift*.

Phemister and Handy (1927) also observed that freshly drawn blood that has been damaged at first excites a vasodilator, and later a vasoconstrictor response. They allowed a dog's blood to flow from the femoral artery into a special vessel provided with stopcocks, by means of which the blood could be returned to the artery after varying time intervals. The effect on the limb volume was recorded with a plethysmograph. After several injections the vessels became immune.

The vasodilator effect of serum, plasma, platelets and defibrinated blood has been investigated by Feldberg, Flatow and Schilf (1929) and by Zipf and Wagenfeld (1930). In a later report Zipf (1931b) concluded that the substance responsible for the vasodilator effect is adenylic acid. He isolated adenylic acid from blood and showed that some of its pharmacological properties correspond to those that had been ascribed to *Frühgift*.

Although the vasoconstrictor action of defibrinated blood can be demonstrated in the mammal with normal circulation, large doses are required, and for that reason relatively few studies of this sort have been made on mammals. Cats under urethane anaesthesia have, however, been employed to detect the abnormal presence of pressor substances in the blood of certain patients (Bohn, 1931). Page (1935) has recently discussed these trials. He was able himself to detect a pressor substance in the blood of patients before and during recovery.

6. Effects on the perfused mammalian kidney

Sollmann (1905) perfused dog kidneys with Ringer's solution and found that in this preparation the injection of serum produced vasodilatation. The active principle was destroyed by boiling. Stewart and Harvey (1912) confirmed these findings in experiments on dog, cat and rabbit kidneys, and obtained the same result with hirudin plasma. The active principle was precipitated by alcohol. After removal of the vasodilator principle the serum caused vasoconstriction, and the substance responsible for that effect was heat-stable and soluble in alcohol. Its action was not affected by apocodeine given in doses sufficient to abolish the action of adrenaline.

When the mammalian kidney was perfused with defibrinated blood, its vessels constricted so powerfully as to jeopardize the experiment. Verney and Starling (1922) showed, however, that a satisfactory perfusion could be maintained if a heart–lung preparation were used as the perfusion pump. Eichholtz and Verney (1924) found that replacement of the heart by a mechanical pump while keeping the lung and kidney in the circulation caused a diminution of the blood-flow. Hemingway (1931) found a similar reduction of flow when he substituted an artificial oxy-genator for the lung in a pump–lung–kidney circuit. Such observations can be explained on the basis of a release of 'vasotonin' by mechanical damage to the blood cells. Defibrinated blood loses its vasoconstrictor action on the kidney if it is made to flow through the lungs or other tissues, but reacquires it on being removed from the circulation and allowed to stand for an hour or two (Eichholtz and Verney, 1924; Bayliss and Ogden, 1933).

7. Effects on the heart and coronary circulation

The action of serum on a rabbit heart perfused with Ringer's solution is biphasic. Frequency and force are increased at first and then diminish. The later phase is accompanied by coronary vasoconstriction and is probably due to this (Cushny and Gunn, 1913). According to Manwaring, Mein-hard and Denhart (1916), the two sets of effect are due to two different substances: the one responsible for the first phase dialyses through col-lodion while the other does not. Yanagawa (1916) concluded that the depression was due to proteins, and showed that one could get a similar effect with egg-white, though not with other proteins. When blood is applied to the perfused heart before it clots (see Henking and Szent-Györgyi, 1923, who used the cat heart), or when heparin is used to delay clotting (Herrick and Markowitz, 1929), it has little or no effect.

The isolated coronary artery ring also responds with constriction to serum, defibrinated blood or a suspension of damaged platelets (Löning, 1913; Janeway, Richardson and Park, 1918).

Serum strengthens the beat of the isolated frog heart (see Clark, 1913, who cites the earlier reports). Clark ascribed the effect to lecithin, although he showed that it was not produced by purified lecithin (Eggleton, 1926).

The heart of a heart–lung preparation continues to beat and perform work, although its usual perfusion medium is defibrinated blood. This is probably due to the ability of the lungs to remove or destroy the substance with coronary vasoconstrictor properties. If the lungs are replaced by a mechanical oxygenator after the preparation has been running for some time, the heart readily becomes hypodynamic (Daly, 1927) – a phenomenon that can be prevented if the blood is continuously filtered through fine muslin (Daly and Thorpe, 1932). The effect is thus probably due to the presence of fine clots or clumped platelets, rather than to toxic substances. It cannot be ascribed to the substance that contracts an isolated strip of coronary artery.

8. Effects on the intestine

Defibrinated blood may cause either inhibition or contraction of the isolated intestine, or the two effects in succession (Cannon and La Paz, 1911; O'Connor, 1912; Stewart and Zucker, 1913; Dittler, 1914, 1918; Freund, 1920; Haake, 1930).

According to Dittler (1918), who worked with the rabbit intestine, the stimulating substance dialyses through parchment and is heat-stable. It was not precipitated by phosphotungstic acid (a single experiment). It did not appear if clotting were prevented with hirudin or oxalate. In contrast, the inhibitory substance did not dialyse through parchment, as Heymann (1927) confirmed. Freund (1920) obtained both effects with platelet suspensions, and found that the appearance of the inhibitory substance corresponded in time to that of *Frühgift*, with a similar correspondence for the stimulant substance and *Spätgift*.

Zipf (1931b) concluded that the inhibition is due to adenylic acid. If so it is surprising that it failed to dialyse through parchment in the experiments of Dittler. Zipf's view is however supported by Haake (1930), according to whom the active principle does pass through an ultrafilter. Guttentag (1931) and Jürgensohn (1931) used the isolated intestine of the guinea pig to compare the effects of blood with those of histamine. They found that alcoholic extracts of fresh blood stimulated the gut, the effect roughly matching that of 0.1 µg of histamine per cc of blood. This effect, however, was not due to histamine, for such an extract did not lower the cat blood pressure as did a histamine solution with equivalent activity on the intestine; furthermore it produced contraction of the rat uterus. This conclusion that the action is not due to histamine has been confirmed in a great variety of experiments by Zipf and Hülsmeyer (1933).

Barsoum and Gaddum (1935a) found nevertheless that after destruction of the other pharmacologically active substances by acid treatment, the residual activity of the extracts on the intestine must be due mainly to histamine. Their conclusion was that blood contains, in addition to the known substances histamine, adenylic acid, choline and a choline ester, at least one more active substance, this being one that contracts the intestine and is destroyed by either acid or alkali.

9. Effects on the uterus

Blood that has been shaken with glass beads will contract the isolated uterus of the rabbit (O'Connor, 1912; Stewart and Zucker, 1913) or the guinea pig (Haake, 1930). Freund (1920) found that horse platelets destroyed by whipping inhibited the cat uterus, while cat platelets similarly treated contracted it. These effects were obtained whether the suspension was tested immediately after destruction of the platelets or after standing.

An alcoholic extract of fresh blood stimulates both the rat and the guinea pig uterus (Guttentag, 1931; Zipf and Hülsmeyer, 1933).

10. Discussion of the substances released during clotting

The findings so far reported present a confusing picture of the number and the nature of the substances that appear in blood after it has been removed from the body.[4] We do know that such blood contains adenylic acid, and that this is responsible, at least in part, for some of the blood's pharmacological effects, for example vasodilatation and inhibition of the intestine. The contraction of isolated carotid, renal and coronary vessels, and of the uterus and the intestine, might be due in some cases to histamine, although there is evidence discussed above that they cannot be due entirely to that base.

However that may be, it is clear that at least one vasoconstrictor substance, *Spätgift*, is present. Differences of opinion as to whether this substance is dialysable or not, and whether or not it is destroyed by heating, may be due to the use of different membranes and different tissues in the experiments, and to failure to define conditions of tempera-

ture and pH. On the other hand it is possible that there is more than
one vasoconstrictor substance. The observation that some of the effects
are antagonized by specific ergot alkaloids suggests that the vasoconstrictor substance acts on the same material as sympathetic nerve stimulation. The material may be indentical with the pressor substance of Collip
(see page 122).

B. Substances from other sources

1. Heart

It is clear from the information now available that extracts of heart may
contain a number of known substances that have pharmacological
activity. Both histamine (Thorpe, 1930) and adenylic acid (Pohle, 1929a;
Drury and Szent-Györgyi, 1929) have been isolated from such extracts.
Pharmacological evidence for the presence of an active choline ester has
been offered by several investigators (Witanowski, 1925; Plattner, 1926;
Engelhart, 1930; Chang and Gaddum, 1933); its concentration in the
auricles is particularly high. Heart extracts also contain cholinesterase,
but apparently no histaminase.

There is reason to suppose that the heart contains some additional
active substances. It is known, for example, that stimulation of its sympathetic nerve supply leads under some conditions to the release of an
Acceleransstoff with an adrenaline-like action : see page 146 *et seq.* for
discussion of the evidence. Loewi and Navratil (1926) have shown that
such a substance is at present in heart extracts, and Bacq (1933a) has
presented chemical evidence that heart tissue contains a catechol
derivative which resembles adrenaline.

Demoor (1922) applied watery extracts of sinoauricular node and right
auricle to the isolated left auricle of the rabbit, and found that the latter's
previously irregular rhythm was returned to regularity. Demoor took this
to mean that there is a substance in the auricle that provides the normal
stimulus for regularity of the rhythm. Haberlandt (1924) found that
extracts of frog heart had a similar effect; he called his active principle the
'*Sinushormon*', and later re-named it the '*Herzhormon*' or the '*Automatie-
Stoff*' (Haberlandt, 1927).

The results of these experiments have been confirmed, but the conclusions have been much criticized (Rigler and Singer, 1927; Rigler and
Tiemann, 1928; Weichardt, 1927; Kisch, 1927; Oppenheimer, 1929). It
could in fact be shown that actions of the sort described are produced not
only by extracts of every part of the heart, but also by extracts of skeletal
and smooth muscle, liver, and lung. Such observations destroy the main

argument for supposing the 'Herzhormon' to be a specific principle. One suggestion has been that the observed effects were due to histamine, and indeed it was shown that the extracts did act like histamine in lowering the cat blood pressure, contracting the guinea pig uterus, and producing the typical picture of intoxication in the intact guinea pig. There can be no doubt that extracts of the kind tested by Haberlandt would contain histamine, and that this would contribute to their action on the rabbit heart. On the other hand, histamine by itself would not produce all the reported effects. Haberlandt (1928) claims to have produced the typical action on the rabbit heart with a histamine-free extract, but he fails to state how he removed the histamine and how he demonstrated its absence.

2. Liver

The vasodilator action of liver extracts was first observed by Vincent and Sheen (1903). Since then evidence has been obtained for the presence of several known substances that may have contributed to the effect. Choline was isolated by Kinoshita (1910), and histamine by Best, Dale, Dudley and Thorpe (1927). According to Strack, Neubaur and Geissendörfer (1933) there is no free choline in fresh liver, but it is liberated in the first few hours after death. There is pharmacological evidence for the occurrence of adenosine compounds (Bennet and Drury, 1931) and of a choline ester (Chang and Gaddum, 1933). It has been shown that stimulation of the hepatic nerves leads to the release of 'sympathin E' (see page 145), but the presence of this material in extracts has not yet been clearly demonstrated.

Nevertheless it is probable that extracts of liver contain active substances other than those mentioned above. James, Laughton and Macallum (1925) found two such substances: one a pressor substance, not precipitated by phosphotungstic acid; and one an 'antipressor substance', which inhibited the pressor effect of adrenaline and other drugs and was soluble in ether. They imply that this substance may have been responsible for the clinical effects of liver extracts employed for the treatment of high blood pressure.[5]

Zuelzer (1928) made liver extracts to which he gave the name of 'Eutonon'. He found that they acted on the isolated heart, and attributed the action to a specific hormone. The effects of 'Eutonon' on the heart, on the cat and rabbit blood pressure, and on the guinea pig uterus are similar to those of histamine; and there is no reason to doubt that 'Eutonon' contains histamine and that some of its effects are due to that base. It also has some other effects, which cannot be attributed to histamine; thus it causes a fall of blood pressure in the atropinized rabbit under ether.

According to Krayer (1933) a number of commercial liver extracts used for the treatment of pernicious anaemia will produce cardiac acceleration in the heart–lung preparation. This effect is obtained also with extracts of heart and spleen, though not with extracts of lung, and is only obtained when the organs are stored for 24 to 48 hours after death before extraction.

Tyramine and isoamylamine are known to be formed by bacterial activity, and perhaps are responsible for the positive chronotropic activity seen in these experiments. Krayer (1933) and his co-workers (Grabe, Krayer and Seelkopf, 1934) have recently obtained tyramine from such extracts as the dibenzoate, and it is legitimate to conclude that the effects just described are due in the main to that substance.

3. Spleen

The vasodilator action of spleen extracts was already mentioned by Oliver and Schaefer (1895); they found that aqueous extracts caused a fall of blood pressure followed by a rise.

The first report of the presence of choline in spleen extracts was by Schwarz and Lederer (1908). Later histamine and acetylcholine were isolated from spleen by Dale and Dudley (1929). Bennet and Drury (1931) obtained pharmacological evidence for the presence of adenosine compounds in this tissue.

Besides these known substances other pharmacologically active materials may be present, but they have not been identified with certainty. Zuelzer (1908) asserted that he had extracted a hormone from the gastric mucosa which was present during digestion and evoked intestinal peristalsis. The preparation 'Hormonal', made according to his directions by unpublished procedures, seems however to have been made from spleen under the impression that the spleen is a depot for hormone made by the stomach. 'Hormonal' promotes peristalsis of the small intestine when it is administered to a rabbit (Zuelzer, 1911). Nothing speaks against the idea that this action is due to the known active substances that have been found in the spleen; their presence there was, however, unknown when Zuelzer published his original results.

Stern and Rothlin (1919) obtained partially purified preparations from spleen, to which they gave the name 'lienin'. It is probable that histamine, which has since been isolated from spleen, was the active principle of lienin (Rothlin, 1920b).

4. *Pancreas*

Several groups of authors have independently observed the presence in pancreatic extracts of unknown vasoactive substances. It is not easy to say how these substances are related to one another, or even how many of them exist.

Of the known vasoactive materials, the following probably occur in pancreas:

Choline: Kinoshita (1910)

Acetylcholine: Chang and Gaddum (1933)

Adenylic acid: Bennet and Drury (1931).

The early reports on the vasodilator effects of pancreatic extracts have been reviewed by Santenoise and his co-workers (1932).

(1) *Vagotonin.* Garrelon and Santenoise (1924) found that a number of crude insulin preparations had the following effects:

(a) Increase of the 'oculocardiac reflex' (slowing of the heart in response to pressure on the eyeball).

(b) Slowing of the pulse and respiration.

(c) A slow fall of blood pressure.

These effects were attributed to an increased activity of the parasympathetic nervous system. Subsequent work showed that they were not due to insulin itself but to an unknown substance, which was named vagotonin.

Santenoise and Penau (1932) have described methods for making active preparations of vagotonin. The substance is extracted by acid alcohol (55–60 per cent) and is precipitated at pH 5.6 when the alcohol concentration is raised. It is insoluble in chloroform and in saturated salt solution. It does not dialyse through cellophane.

Santenoise and others have published many reports on these phenomena in recent years, and have confirmed the original observations with purified preparations. Vagotonin has been found to strengthen not only the oculocardiac reflex, but also the reflex fall of blood pressure caused by stimulation of the depressor nerve (nerve of Cyon) or of the carotid sinus. Vagotonin also lowers the blood sugar (Santenoise, Brieu, Fuchs and Vidacovitch, 1932) but is said to differ from insulin in causing a rise in liver glycogen.

Vagotonin differs fundamentally in its action from the vasodilator substances known as kallikrein and angioxyl, which are also present in pancreatic extracts. The fall of blood pressure that it produces is very slow in onset and is prevented by atropine; whereas kallikrein and angioxyl cause a rapid fall of blood pressure which is unaffected by atropine, and unlike vagotonin they do not increase the sensitivity of the depressor reflex.[6]

Santenoise (1932) has put forward the view that vagotonin is a specific hormone, secreted by the pancreas and controlling vagal tone. In support of this view it was found that removal of the pancreas diminishes the response to the depressor reflex (Garrelon, Santenoise, Verdier and Vidacovitch, 1930). This effect may however be a secondary consequence of hyperglycaemia. It seems not to have been proven that an animal deprived of its pancreas and maintained on pure insulin differs from a normal animal in respect of these reflexes.

The fact that pancreatic extracts evoke alterations in the activity of certain nerves does not, of course, prove that the pancreas secretes a hormone that normally controls the tonic activity of those nerves. Thus Oswald (1916) observed similar alterations of nerve function after the administration of thyroid extract, yet later control experiments showed that extracts of other organs could produce the same changes.

2. *Angioxyl.* Gley and Kisthinios (1928, 1929) obtained vasodilator effects with alcoholic extracts of pancreas in rabbits that had been treated with atropine. The possiblity does not seem to have been excluded that these effects were due to adenylic acid. The name angioxyl was bestowed on the hypothetical unknown substance to which the extracts owed their activity. It should be mentioned that von Euler and Gaddum (1931a) found that alcoholic extracts of pancreas had a much weaker depressor effect than similar extracts of other tissues, intestine and brain in particular.

3. *Kallikrein.* Kraut, Frey and Werle (1930a) found that pancreatic extracts had a powerful action of the type that they had identified in purified extracts of urine, and attributed this to a substance that they named kallikrein. Their studies are more fully described in the section on urine (pages 114 *et seq.*).[7]

In order to test the theory of Kraut, Frey and Werle that angioxyl and kallikrein are identical, Elliot and Nuzum (1931) made extracts of pancreas by the method of Gley and Kisthinios and extracts of urine by the method of Kraut, Frey and Werle. The effects of the two extracts was the same. Nevertheless the fact that the angioxyl preparation of Elliot and Nuzum was soluble in 90 per cent alcohol is not immediately reconcilable with the account given by Kraut, Frey and Werle of the properties of kallikrein prepared from urine (see pages 104 *et seq.*).

5. Intestine

The depressor action of intestinal extracts was noted by Bayliss and Starling (1902) when they were working on secretin. Later researches have established the presence of several well-defined substances which must have contributed to this action.

Choline has been isolated from extracts of intestine by Fürth and Schwarz (1908), histamine from extracts of intestinal mucosa by Barger and Dale (1911). The pharmacological behaviour of the extracts indicates that adenosine compounds (Bennet and Drury, 1931) and acetylcholine (Chang and Gaddum, 1933) are also present.

Substance P. Von Euler and Gaddum (1931a) obtained evidence for the presence of a different vasodilator substance in extracts of intestine, and probably also in extracts of brain. In a later paper (Gaddum and Schild, 1934) it was designated 'substance P'.[8] In the atropinized rabbit under ether it produces a fall of blood pressure which is the result of peripheral vasodilatation, and thus differs from histamine and from choline and its esters, which lack that effect. These extracts also produce a slow increase in the tone of the isolated intestine of the rabbit in the presence of atropine. While the latter action is probably also due to substance P, it is naturally impossible to be certain on that point until the active principle has been isolated in a chemically pure state.

Substance P is very stable at its optimal pH. Solutions that have been made neutral to congo red can be boiled for hours without appreciable loss of activity; in contrast, acid or alkaline solutions rapidly lose their activity. In this respect substance P differs from the adenosine compounds, which are destroyed in acid solution but are very stable in alkaline solution.

Substance P is soluble in alcohol and acetone but not in ether. It can be extracted from the intestine with either alcohol or trichloroacetic acid. It passes readily through parchment, parchment paper or cellophane by dialysis or ultrafiltration. It is carried towards the cathode by an electric current. It is stable during autolysis of minced intestine, and it is not affected by nitrous acid under conditions in which histamine is destroyed.

Substance P thus appears to be a rather simple organic base, though the fact that it dialyses only slowly indicates that its molecular weight is higher than that of histamine. It does not require a free amino group for its activity. Beyond these facts nothing is yet known of its chemistry. It is still possible that it will turn out to be a known substance.

Other substances. Lange (1932) and Felix and Putzer-Reybegg (1932) have made observations which they interpreted as indicating the presence in extracts of kidney and intestine of an unknown vasodilator substance that is unstable in alkaline solution but remarkably stable in acid solution. They heated an extract to 100 °C for 18 hours in 30 per cent sulphuric acid and found no loss of activity; in some of their experiments there was even a significant increase. Histamine is known to be formed from histidine under such conditions (Ewins and Pyman, 1911), and the possibility that this reaction accounted for the gain in activity was not excluded. The physical and chemical properties that the investigators attributed to their substance are in many ways like those of histamine, and it is likely that many of their extracts did contain histamine. Gaddum and Schild (1934) could find no evidence for the presence in intestinal extracts of any acid-stable substance, other than histamine, that lowered the blood pressure after atropine.

Felix and Putzer-Reybegg (1932) described the isolation, as a crystalline flavianate, of a substance that they regarded as the active principle of their extracts. The analyses they presented agree well with those expected for methylguanidine, but the activity they observed led them to suggest a larger molecule; and certainly methylguanidine does not have the sort of intense activity that they found. Contamination with a relatively small proportion of histamine, however, would account for both the material's depressor activity in the cat and its remarkable stability to acid.

At earlier stages of the purification the investigators recognized a depressor effect in the rabbit, and this could not have been due to histamine. Whether they could have assumed that the depressor effect in the cat that they observed later on was an equivalent one is not clear from their account. So far as the available evidence goes, it seems possible to say only that the authors have not adequately excluded the suspicion that the preparation they isolated was methylguanidine whose activity was due to a small contamination with histamine.

6. Brain

The vasodilator action of extracts of brain was first observed by Schaefer and Moore (1896). It was more thoroughly investigated by Osborne and Vincent (1900), who showed that the action was apparent in the rabbit, and after atropine, and was due to dilatation of the arterioles.

Vincent and Sheen (1903) examined some of the chemical properties of their active material. They also obtained extracts that would elevate the blood pressure, and made extracts of tissues other than brain whose effects were rather similar. Vincent and Cramer (1904) found that the depressor materials in brain extracts could be separated into two fractions on the basis of their alcohol solubility. Only a part of the activity of the alcohol-soluble fraction was due to choline.

Since these early studies it has been shown that brain extracts can contain a number of depressor substances.

(1) Choline (Gulewitsch, 1899). According to Kaufmann (1911) most of this is formed post mortem. A choline ester, presumably acetylcholine, is also present (Dikshit, 1933a, b; Chang and Gaddum, 1933).

(2) Adenylic acid (Pohle, 1929b).

(3) A substance that lowers the blood pressure of the atropinized rabbit and is unstable in alkaline solution; this is probably identical with substance P, found in extracts of intestine (von Euler and Gaddum, 1931a).

(4) Major and Weber (1929, 1930) have obtained evidence for the presence of still another substance. It has not been chemically identified, but has the following properties:

(a) It lowers the blood pressure of both the dog and the rabbit, and differs from choline in that it still does so after atropine.

(b) It is not precipitated by silver, even from alkaline solution, and in this respect differs from histamine and adenosine.

(c) It is not precipitated by phosphotungstic acid, and in this respect it differs from all the recognized vasodilator substances. Major, Nanninga and Weber (1932) found that phosphotungstic acid precipitates almost all the depressor activity of lung and liver extracts but only about half of the activity of brain and pancreas extracts.

(d) It is not absorbed by animal charcoal (Norit) in alkaline solution, and in this respect it differs from histamine. Norit removes almost all the depressor activity of lung and liver extracts and half the activity of pancreatic extracts, but hardly any of the activity of brain extracts.

(e) It is precipitated by mercuric salts and alcohol in alkaline, but not in acid, solutions.

(f) It survives boiling for 5 minutes in either strong acid or strong alkali.

(g) That the substance is neither histamine nor a guanidine derivative is shown by the fact that an active solution does not give the Pauly reaction or the Sakaguchi reaction.

The experiments on which these conclusions are based have not been described in full detail, and it is sometimes hard to assess the evidence provided. For example, the authors' conclusion that a substance with depressor activity in the rabbit has the remarkable stability and other properties described above, is apparently based on the following observations.

The alcohol was removed from an alcoholic extract, impurities were precipitated with basic lead acetate, and the active material was adsorbed from acid solution on Lloyd's reagent and eluted with baryta. The solution was next treated with phosphotungstic acid and then with silver; after both treatments the solution still exhibited some vasodilator activity, lowering the blood pressure of either the dog or the rabbit after atropine. The action on the rabbit, however, was a very slight one, as is apparent from the published tracing; and there is no clear proof that it was not due to traces of adenylic acid that had not been completely removed by the treatment described.

Similarly the conclusion that the unidentified substance survives boiling with acid and alkali appears to have been based on the finding that brain extracts still lower blood pressure in the dog after such treatment. But crude extracts of brain so treated must inevitably lose part of their activity; for the adenosine and the substance P they contain will be destroyed by one or other treatment. No information has been supplied as to what proportion of the activity was lost in the tests on which the conclusion was based; and again there is no clear proof that the residual effects were not due to traces of histamine or adenosine that were incompletely removed from the extracts that were used in these actual tests.

The conclusion that histamine was not responsible for the effects observed is supported by the following evidence:

(1) The active principle is not precipitated by phosphotungstic acid or by silver ions and is not adsorbed by norit. Such findings, however, can in some cases lead to erroneous conclusions. Thus Barger and Dale (1910) have found that histamine is completely precipitated by silver ions from neutral extracts of ergot, whereas pure histamine is precipitated by such treatment. The possiblity that such an explanation may apply to the results of Major, Nanninga and Weber has not been definitely excluded.

(2) The depressor activity of an extract was evaluated in terms of histamine in one experiment on a dog. The extract was then found to have no action on the guinea pig uterus, although this responded to a much smaller amount of histamine than was needed to match the depressor effect. This piece of evidence is more compelling than the previous one, but no data are given to exclude the possibility that this particular extract owed its activity to a mixture of histamine, adenylic acid and choline, the last two of these being relatively inactive on the uterus.

These experiments of Major and his co-workers have been discussed in especial detail in order to illustrate the considerable difficulties encountered in the pharmacological analysis of tissue extracts.[9] While the findings of these authors are open to criticism in detail, the totality of their results does argue strongly for the view that brain extracts contain a depressor substance that is different from any such substance already isolated from tissues. Weber, Nanninga and Major (1933) indeed claim to have isolated a crystalline substance with the properties of the material they were looking for. The data they present on its chemical properties do seem to exclude the known depressor substances. On the other hand, the only activity hitherto recorded for it, that on the blood pressure of the dog, is of such low magnitude that its physiological significance must be regarded as doubtful.

7. Male accessory genital glands

The injection of prostate extracts may produce either a rise or a fall of blood pressure. Such extracts have also been found to contract the bladder and inhibit the intestine. These effects are apparently due to adrenaline, adenylic acid and substances that have not been chemically identified. Von Euler (1934a, b) gives references to the earlier reports.

Apart from the suprarenals, the prostate appears to be the only organ in the body that contains adrenaline in high enough concentration for detection in its extracts by the available methods. Collip (1928) found that prostate was exceptional in that the pressor action of extracts made from it was increased by cocaine in the same way as the action of adrenaline. Von Euler (1934a, b) obtained the same results with extracts prepared in a different way. He also found that part of the inhibitory action of his extracts on the intestine was probably due to adrenaline, and that biological and chemical estimates of their adrenaline concentration were in agreement. As confirmation of the identity of the active substance with adrenaline, Von Euler showed that it was unstable in the presence of iodine or of oxygen and alkali. His conclusions were further supported by histological evidence for chromaffin tissue in the prostate.

Goldblatt (1933, 1935) and von Euler (1934a, b, 1935) found that extracts of prostate, seminal vesicles and seminal fluid of man and a number of animals contained at least one substance whose properties were in a number of respects like those of the substance P present in intestinal extracts (see page 108).[10] Thus the extracts lowered the blood pressure of the atropinized rabbit and caused a contraction of the isolated rabbit intestine. The ratio of the effective doses was, however, not always the same in tests in which extracts of sex-gland material were compared with extracts of intestine on the two test-objects mentioned. It follows that the effects were not all due to the same substance. Some of the extracts' chemical properties point in a similar direction.

Von Euler (1935) expressed the opinion that the action of seminal fluid on the blood pressure and the intestine of the atropinized rabbit is due to at least two substances. In addition seminal fluid contains a substance that acts like acetylcholine on the frog rectus abdominis; its action is intensified by eserine (Goldblatt, 1935).

8. *Eye*

Plattner and Hintner (1930b) and Engelhart (1931) obtained evidence for the presence of a choline ester in extracts of iris (see page 147).

Velhagen (1931b) studied the pharmacological effects of extracts of various parts of the eye. He found that they produced a fall of blood pressure, which was in part due to a choline-like substance since the effect was diminished by atropine. Other substances, however, were also present, which were depressor in both cat and rabbit after atropine. The action on the cat might have been due to histamine, though there was no clear evidence that histamine was present; the action on the rabbit might have been due to adenylic acid.

Later Velhagen (1932) studied the properties of the choline-like material more fully. It was unstable in alkaline solution; it was destroyed by blood and this action was inhibited by eserine; its activity was increased 50–100-fold by acetylation; it was dialysable and soluble in alcohol. These findings indicate that it was a mixture of choline with an active ester of choline. Velhagen discusses the significance of his results in relation to earlier work on extracts of eyes.

9. *Urine*
(a) *Pressor substances*

Abelous and Bardier (1908) showed that urine may contain a pressor substance, which they named urohypertensin. It was soluble in alcohol and ether, did not dialyse readily, and was not precipitated by lead acetate or mercury. These findings were mostly confirmed by Bain (1914), who also found that the active substance could be distilled with steam, and was readily absorbed on charcoal from which it could be eluted with HCl. Bain isolated a small amount of the active substance in an almost pure state, and was thus able to identify urohypertension as isoamylamine; in spite of a small discrepancy in the elemental analysis there can be little doubt that his identification was correct. The observation by Abelous and Bardier that their active material did not dialyse readily is probably to be explained on the basis of its adsorption on the large molecules of other materials present in their extracts. Isoamylamine is also present in putrefying muscle and placenta (Rosenheim, 1909; Barger and Walpole, 1909). It is formed by the decarboxylation of leucine.

Bain also obtained some evidence for another pressor substance, which was insoluble in ether, but he did not succeed in identifying it.

(b) Depressor substances
(i) Substances of known composition

The following depressor substances of known composition may be present in urine:

Histamine (after parathyroidectomy)	Koch (1913)
Choline	Hunt (1915b)
	Guggenheim and Löffler (1916b)
(after injection of choline)	Shanks (1923)
Adenosine	Calvery (1930)

Dingemanse and Freud (1933) identified a substance in the urine of men, which had previously been designated 'katatonin', as nicotine; as might have been expected it was obtained from the urine of smokers. It is, however, uncertain that all the activity attributed to katatonin was due to nicotine.

(ii) Substances of unknown composition (kallikrein)

Abelous and Bardier (1909) found that urine contained an unknown depressor substance to which they gave the name of 'urohypotensine'. This substance was insoluble in alcohol, did not dialyse and was precipitated by saturated ammonium sulphate. In dogs and rabbits it evoked a fall of blood pressure which was accompanied by a rise in the volume of the brain and a fall in the volume of the kidney and the intestine. The vessels of the rabbit's ear and conjunctiva were visibly dilated. The preparation produced constriction of the pupil (miosis), which could also be seen in the isolated iris and was therefore presumably of peripheral origin.

The substance studied by Abelous and Bardier under the name of urohypotensine is probably identical with the substance that later authors have designated as 'kallikrein'.[11] The latter has not yet been purified and little is known about its chemistry. Its physical properties, however, have been clearly defined, and can be taken as established that the physiological effects of the purified preparations are due to a substance that is neither histamine, choline nor adenosine.

The name kallikrein is derived from the Greek word καλλικρεας (pancreas). It was chosen because Kraut, Frey and Werle (1930a) were convinced by the results they obtained that their material from urine was a hormone secreted by the pancreas into the circulation.[12]

Pharmacological properties (Frey and Kraut, 1928). The pharmacological response on which almost exclusive reliance has been placed for the purification of kallikrein is a broadening of the amplitude of the pulse in the dog.

The blood pressure is recorded with a Frank–Petter manometer whose low inertia and fast response permit accurate registration of the oscillations due to the pulse. An intravenous injection of either ordinary urine or a purified preparation made from it increases the amplitude of the pulse thus recorded. The effect, which is thought to be due to a direct action of the kallikrein on the heart, lasts about 30 seconds and can be repeated at intervals of 5 minutes. The maximum broadening of amplitude provides an index of the effectiveness of the dose. By alternating the unknown preparation with a standard until the effects match, the potency of an extract can be estimated with an accuracy of about 5 per cent. The amount of kallikrein is measured in units which correspond to the activity of 0.5 mg of a particular powder prepared by Frey from urine and adopted as a standard. The unit is contained in about 5 cc of urine from male subjects. A tenth of a unit is enough to produce a clear effect in a sensitive dog.

The effect on the pulse pressure, which has just been described, forms the basis for the concept of kallikrein. But many other effects have also been found to be exerted by kallikrein preparations of very different degrees of purity. The evidence that all these effects are due to the same substance can of course not be finally convincing until the substance has been isolated in the pure state. But if several different substances are in fact involved, they must either have very similar physical and chemical properties, or else they must exist in a physical association that confers such properties on them.

A record of the blood pressure made with a mercury manometer shows only the mean pressure, since this instrument has too much inertia to inscribe the pulse amplitude. Such a record shows the action of kallikrein only as a fall of blood pressure, which is due in part to pulmonary vasoconstriction and in part to peripheral vasodilatation. The site of this vasodilatation has been investigated with the plethysmograph, which reveals that it involves an increase in the volume of limbs, lungs and brain, and a decrease in the volume of intestine, spleen, liver and kidneys. Active vasodilatation must occur in the limbs, but the effect on the abdominal organs is thought to be a passive consequence of the fall in blood pressure. Kallikrein also produces a fall of blood pressure in the atropinized rabbit.

The action of kallikrein on the heart has been studied in experiments in which it was added to the Locke's solution perfusing a rabbit heart. The beat was immediately accelerated and strengthened and the coronary vessels were dilated. Purified preparations had no effect on the frog heart.

In the dog, Hochrein and Keller (1931), as well as Krayer and Rühl (1931), have found that kallikrein produces vasodilatation.

Elliot and Nuzum (1931), using a preparation made from urine by precipitation with uranyl acetate, confirmed some of these findings, but in their experiments on perfused rabbit hearts there was no increase in rate and the force of the beat was diminished. The preparation was inactive when given subcutaneously or intramuscularly.

In the dog, the infection of kallikrein preparations has been found to produce a transient diminution of urinary flow followed by alkaline diuresis. The blood alkali reserve fell; rectal temperature rose; metabolic rate was increased; and the dog shivered (Szakall, 1932). In the guinea pig, kallikrein preparations caused contraction of the intestine and the uterus; in the cat, they inhibited the uterus.

Identification and assay. The action of kallikrein on the blood pressure of the dog, which is the test used for its detection, was described in the preceding section (see page 115). By itself this response is not a specific reaction for kallikrein, for a similar action is evoked by histamine, 10–20 µg of which will match a unit of kallikrein. It is therefore necessary to remove histamine from extracts as completely as possible before testing them for kallikrein. This can be achieved by dialysis or by extracting the histamine with alcohol.

The physical properties of kallikrein which may be used to distinguish it from other substances are the following:

(1) It is insoluble in alcohol (90 per cent or higher) and in practically every other solvent but water.

(2) It is destroyed when its aqueous solution is boiled for 10 minutes.

(3) Crude preparations do not dialyse through parchment, but a very pure preparation lost about a third of its activity during dialysis for 48 hours, and the loss was attributed to the dialysis.

(4) It is inactivated by blood, and the activity can be recovered by treatment with acid.

(5) It is precipitated from extracts by uranyl acetate. Highly purified preparations are not precipitated by this reagent.

Kallikrein differs from histamine not only in these physical properties, but also in its pharmacological behaviour.

(1) Its activity relative to that of histamine is different in different tests. For example, kallikrein has comparatively little depressor activity in the cat, which is especially sensitive to histamine.

(2) It causes a fall of blood pressure in the rabbit under ether.

Table 12. *Distribution of kallikrein in various tissues*

Organ	Animal	Kallikrein in units per g of tissue
Pancreas	Pig	50
	Dog	10; 27
	Ox	18; 13; 15
	Sheep	9.5
	Man	3.6; 4.5
	Horse	1.6; 2; 0.6
Blood	Dog	2
Spleen	Dog	14
Thymus	Dog	Approx. 1
Thyroid	Dog	Approx. 1
Kidney	Dog	0.3
Liver	Dog	0.2
Skeletal muscle	Dog	0.12
Heart	Dog	< 0.05
Urine	Man	0.2

(3) It lowers liver volume and raises lung volume in the dog under conditions in which histamine has the opposite effects.

(4) In guinea pigs it fails to produce the severe bronchospasm which is a characteristic effect of histamine.

(5) It relaxes the cat uterus.

(6) The action of kallikrein is unaffected by atropine but it is strengthened by cocaine.

Weese (1933) considered the possibility of using animals other than the dog for detecting kallikrein, but he concluded that the dog is particularly suitable for this purpose because of its high sensitivity to kallikrein as compared with the heat-stable depressor substances of urine.

When urine is mixed with 10 volumes of alcohol, kallikrein can be recovered from the precipitate. Similar procedures have been applied to blood; however the presence of kallikrein has also been detected by digesting the proteins with papain in a collodion bag, so that the products of digestion were removed as fast as they were formed.

Distribution. In some of their studies of the distribution of kallikrein in tissues Kraut, Frey and Werle (1930) prepared aqueous extracts with acetic acid or ammonia and removed histamine by dialysis. In other cases they first extracted the tissues with acetone and ether and recovered kallikrein from the solid residues. Table 12 is based on the data of these authors: it states the kallikrein content of a number of tissues as units per g of tissue.

It will be seen that the pancreas of some species is richer in kallikrein than any other organ. The activity of two pancreatic cysts was found to be 1–2 units per cc whereas serous exudates from other parts of the body contained no activity at all.

The active constituents of pancreatic, blood and urine extracts have in common that they are destroyed by boiling, are inactivated by serum and can be regenerated by treatment with acid. These observations led to the view that kallikrein is a specific hormone secreted by the pancreas into the circulation.

Purification (Kraut, Frey, Bauer and Schultz, 1932; Schultz, 1928). Kallikrein can be prepared from the precipitate formed by the addition of alcohol to urine; the precipitate is re-dissolved and crude kallikrein is then precipitated by adding uranyl acetate. It is usual to omit the alcohol step and add the uranyl acetate directly to the urine. The kallikrein, after elution from the precipitate, is then adsorbed on freshly precipitated benzoic acid. Further purification is accomplished by adsorbing impurities on animal charcoal and lead phosphate. Aluminium hydroxide ('Tonerde') adsorbs kallikrein from neutral or acid but not from alkaline solution. Sodium carbonate can in fact be used to elute the active material from the precipitate. By combining these purification procedures preparations have been obtained whose activity corresponds to one unit per 100 µg of the dry material. In all such preparations the Pauly diazo reaction was positive, and the colour produced by one unit was in no case weaker than that produced by 1 µg of histamine. Tests for guanine (0.1 µg), adenine (0.02 µg), reducing sugar (0.42 µg) and tyrosine (0.4 µg) were all negative. The numbers in parentheses are the quantities that would have been detected if they had been present in one unit of kallikrein. The smallest amount of phosphorus found in any preparation was 0.06 µg per unit.

The commercial preparation of kallikrein marketed as 'Padutin' contains procaine as a preservative.

Inactivation (Kraut, Frey and Werle, 1930b). If kallikrein is mixed with blood and tested at intervals, the activity disappears in the course of a few minutes, but is regenerated if the mixture is made acid. The substance in blood that causes the apparent loss of added kallikrein is known as the inactivator. It is destroyed by boiling, passes through an ultrafilter, and is insoluble in alcohol. It is destroyed by trypsin, and may therefore be a polypeptide. The amount of inactivator in a solution is determined by adding kallikrein; a unit of inactivator is the amount required to inactivate one unit of kallikrein. A preparation of the inactivator has been made in which one unit is contained in only 6 µg of solid material.

Inactivation apparently depends on the formation of a loose chemical linkage between the inactivator and the kallikrein. At pH 8 the reaction proceeds at its maximum rate and is nearly complete within half an hour. At pH 5 the reaction is reversed, so that the solution regains its physiological activity. If 1 cc of N HCl is added to 9 cc of an inactivator-kallikrein mixture, and 1 cc of N NaOH is added 2 minutes later, the inactivator is destroyed but the kallikrein is not. The inactivator content of many tissues has been measured; it is especially high in blood, lymph, lymph nodes and parotid gland.

Physiological significance. In support of their theory, Frey, Kraut and Schultz (1930) showed that the ratio of kallikrein to urinary solids falls after removal of the pancreas. It is not clear from their publication whether they have observed an absolute reduction of kallikrein excretion or whether the effect they report was due to an increase in total urinary solids. The fact that insulin given to the pancreatectomized animal restores the normal ratio of kallikrein to solids is not easily reconciled with the concept that kallikrein is a specific hormone of the pancreas.

According to the theory of Frey and Kraut, kallikrein is made in the pancreas and circulates in an inactive complex with the inactivator. It is released in the periphery in response to local acidification of the tissue fluid and there brings about functional vasodilatation.

10. Saliva

Saliva contains a substance, or substances, with a choline-like action (Hunt, 1915b; Guggenheim and Löffler, 1916b). Secker (1934a, b) thought that the vasodilator material he found in saliva might be acetylcholine. More recent work by Gibbs (1935) and by Feldberg and Guimarãis (1935) has shown that this material is not acetylcholine, nor is it identical with any known tissue constituent. It is thermolabile in neutral, acid or alkaline solution. With a pure sympathetic saliva from the cat, the injection of 1 cc of 0.1 per cent solution evoked a sharp transient fall of blood pressure. When chorda saliva is given by vein to a cat, a rise of blood pressure often precedes the fall; given to a dog, the effect is a pure fall. Both sympathetic and chorda saliva cause a secretion of saliva when injected into the artery of the gland (Demoor, 1913; Guimarãis, 1930, 1936). That the secretagogue and depressor effects of saliva are due to the same substance is doubtful, though both effects are lost on heating (see Feldberg and Guimarãis, 1936).

11. Gastric juice

Gastric juice injected intraperitoneally causes gastric juice to be secreted. Komarow (1933) has shown that the substance responsible for this effect can mostly be recovered from the histidine–arginine fraction when the bases are separated by the conventional procedure; he suggested that the active substance was histamine.[13]

12. Sweat

A substance that acts like acetylcholine on the leech can be detected in human sweat, but according to the preliminary results of Dale and Feldberg (1934b) its properties are different from those of acetylcholine.[14] The presence of choline in sweat has long been known. Its concentration is said to be elevated during menstruation (Klaus, 1926).

13. Frog skin

Grant and Jones (1929) have shown that the vessels in the mucosa of the frog tongue react to local irritation in much the same way as those of human skin. In the latter the response is due to the local release of histamine or a closely allied substance (see page 172). In the frog tongue it must be due to some other substance, because frog vessels are insensitive to histamine whether it is applied locally or injected into the blood stream. If an extract of frog skin is applied to the frog tongue by a superficial puncture, it reproduces the vascular effects of mild injury just as faithfully as a human skin extract, or histamine, evokes the analogous response in human skin. Human skin extracts freed from inorganic salts were as inactive as histamine on the frog; but in contrast, extracts of frog skin had a histamine-like action on human skin, on the cat blood pressure or on the guinea pig uterus. It is of course uncertain how many substances could have contributed to these effects, but it is clear that extracts of frog skin possess activity on frog vessels that is due to some unidentified substance.[15] The substance can be extracted with aqueous alcohol, and after removal of impurities with basic lead acetate the active substance can be precipitated with phosphotungstic acid and then with silver in weakly alkaline solution; so that its chemical properties in this respect are like those of histamine.

Grant and Duckett Jones (1929) found that, in addition to this unknown substance, their extracts contained a vasodilator substance, presumably choline, whose effects are abolished by atropine.

14. Peptone

Dale and Laidlaw (1910) drew attention to the similarity between the effects of peptone and those of histamine. Abel and Kubota (1919) believed that the similarity was due to the presence of some free histamine in peptone. This is no doubt the case in some samples of Witte's peptone, but histamine cannot be the only active principle.

It was apparent from early work on peptone that it makes the blood incoagulable. Histamine may have some action of this kind, but it is very feeble compared to that of peptone. It is likely that peptone owes much of its action in the dog, where that action has been chiefly studied, to a direct injury of the liver cells by one of its components. Damage of that sort could readily account for the incoagulability of the blood. Thus the anticoagulant action of peptone is evident only in the whole animal, or when peptone-treated blood is made to flow through the liver, but not when peptone is added to blood *in vitro*. Damage of the liver cells in anaphylactic shock readily leads to the release of histamine in the dog, and many of the consequences observed in the various organs can plausibly be attributed to the histamine thus liberated (cf. Dale, 1929).

An account of the pharmacology of peptone and of the fractions that can be obtained from it, with numerous references, will be found in a paper by Chittenden, Mendel and Henderson (1899).

Tomaszewski (1918) and Popielski (1920b) found that Witte's peptone, unlike histamine, evoked no secretion of gastric juice. Hanke and Koessler (1920) applied their chemical test to measure the histamine content of a sample of Witte's peptone and obtained an estimate of 33 mg per kg. They also made a histamine-free preparation of peptone by digesting fibrin with acid and alkali in the presence of toluene. They subjected this preparation to the procedures that they used for purifying histamine (see page 40); after which it gave no chemical test for histamine and caused no contraction of the guinea pig uterus, though it did lower the blood pressure of the dog.

In contrast, their starting material, the 'histamine-free' peptone of the preceding paragraph, produced a typical profound peptone shock in the dog and contraction of the guinea pig uterus. Thus a large part of the activity was lost during the final purification.

The proof that the characteristic effects of peptone are not due to histamine rests on two facts:

(1) The active principle did not survive purification. This might be due to some special property of the peptone.

(2) The action of the final purified product on the blood pressure of

the dog could not be due to histamine. The evidence for that conclusion might have been more convincing if the action of the extract had been quantitatively compared with that of histamine, and if it had then been shown that the corresponding amount of histamine would have been detected in the chemical test, and would have evoked a contraction of the guinea pig uterus. Nevertheless, one's knowledge of the sensitivity of the various tests does make it probable that the conclusion of Hanke and Koessler was correct.

The evidence has been further discussed by Abel and Geiling (1924). They refer to unpublished work of Vermooten, which appeared to show that histamine could be present in peptone even in the absence of bacterial growth.

Clark (1924) came to the conclusion that peptones contain at least two principles that are dialysable, alcohol-soluble and ether-insoluble; one of them is a general stimulant of smooth muscle, while the other has sympathomimetic activity. The lethal action of peptones on mice, however, was mainly due to alcohol-insoluble substances. Clark's paper contains many references to work on the pharmacology of peptones.

Addendum: Pressor substances

The well-known pressor substances of the suprarenals and the pituitary lie outside the scope of this book. The experiments on adrenergic nerves described in the next chapter show that most tissues are capable of liberating sympathomimetic substances when these nerves are stimulated; and there is also some evidence for the presence of such substances in tissue extracts (Loewi and Navratil, 1926a). The pressor substances in blood, urine and saliva have been discussed on pages 95 *et seq.*, 113, and 119.

References to other work on pressor substances may be found in a paper by Collip (1928).[16] He found that pressor substances could be detected in many tissues, provided that suitable chemical procedures were used to remove the depressor substances that were generally present in relatively high concentration.

Collip used the following method of extraction: the tissue was first extracted with boiling water; the aqueous extract was concentrated and treated with three volumes of absolute acetone, in which the pressor substance is soluble; the process is repeated, so that the pressor substance is now in 98 per cent acetone, and depressor contaminants are removed from this solution with ether; the watery layer now contains these contaminants in high concentration, while the ether layer is relatively rich in the pressor substance.

By this procedure a small amount of active substance is obtained from a large amount of starting material. The active principle is certainly not adrenaline, though the two have some common properties. It does not elevate the blood sugar. Its pressor action is enhanced by small doses of ergotamine, but is reversed by larger doses; it is interesting that the actions of sympathin are similarly influenced by ergotamine (see page 161). The pressor action is unaffected by atropine or nicotine, but it is abolished by cocaine. Small amounts of the substance have no effect on the intestine, the uterus or the heart; large doses are inhibitory. Collip gives some information about the chemical and physical properties of his substance, but its chemical composition is still unknown.

VII

The release of specific active substances by nerve impulses

A. General

Evidence that various nerves produce their effects by liberating chemical substances has been accumulating rapidly in recent years. A number of reviews of the subject have been published (Dale, 1929, 1933a, 1934; Cannon, 1931, 1933; Kroetz, 1931; Hirschberg, 1931; Fredericq, 1927; Parker, 1932; and others).[1]

The possibility that the phenomenon of antidromic vasodilatation might result from chemical changes in muscle spindles was considered by Bayliss as early as 1901. But the concept that the effects of nerve impulses could be transmitted by the peripheral liberation of specific chemical stimulants was first clearly expressed by Elliott (1904, 1905).[2] He suggested that the close correspondence between the effects of sympathetic nerves and those of adrenaline could be explained by supposing that these nerves produce their effects by liberating adrenaline at their endings. Dixon (1906, 1907) extended the theory when he suggested that parasympathetic nerves might similarly act on their effector organs by liberating a muscarine-like substance. His attempts to confirm his theory by experiment are described below (page 162). Howell and Duke (1908) found that stimulation of the vagus caused the release of potassium in the heart, and they supposed that the action of the vagus on the heart might be due to that of potassium.

It will be seen that the idea of the transmission of nervous effects by the liberation of chemical substances emerged from several quarters. Nevertheless the great advances that have been made recently in our knowledge of such mechanisms began with the work of Otto Loewi (1921).[3]

Loewi stimulated the vagus nerve supplying a frog heart suspended from a Straub cannula. The usual effect of this stimulation was inhibition of the heart. When the Ringer's solution from that heart was transferred to a second heart, the second heart too was inhibited. When atropine was present in high enough concentration, the inhibitory effect on both hearts disappeared. Under these conditions the action of the sympathetic fibres, which in the frog run in the vagus trunk (Gaskell, 1884), was revealed.

Stimulation of the nerve now produced an increase in the force and the frequency of the beat, and this action also could be transferred to the second heart. Loewi concluded that the first effect, the inhibition, was due to the release at the nerve endings of a substance that he called *Vagusstoff*; while the second effect, the stimulation, was due to the release of *Acceleransstoff*. Stimulation of the combined vagosympathetic trunk causes the release of both substances; the effect on the heart depends on their relative concentrations and this is affected by such factors as the conditions of stimulation and the season.

The researches to be summarized in this chapter support the concept that most if not all of the body's efferent nerves act by releasing substances that resemble either *Vagusstoff* or *Acceleransstoff*. Recent experiments have shown, further, that the old division of the autonomic pathways into sympathetic and parasympathetic fails to indicate which of the two specific substances is set free in the periphery. For this reason Dale (1933b) has proposed a new classification of nerve fibres based on pharmacodynamic principles, which designates these fibres as *cholinergic* or *adrenergic*.[4] This nomenclature has already come into general use. Its adoption has greatly simplified the description of experimental results, and it will be used in the following presentation. Cholinergic fibres are those whose actions are mediated by the release of an acetylcholine-like substance; adrenergic fibres are those whose actions are mediated by release of an adrenaline-like substance. Adrenergic fibres are found almost exclusively in sympathetic postganglionic nerves, but cholinergic fibres have a much wider distribution. There is evidence that the following fibres are cholinergic:

(a) most, and probably all, preganglionic autonomic fibres;
(b) most, and probably all, postganglionic parasympathetic fibres;
(c) in some animals, postganglionic sympathetic fibres to the sweat glands, certain blood vessels, and the uterus;
(d) somatic motor fibres to striated muscle.

B. Cholinergic nerve fibres

1. Heart

(a) The frog heart

The liberation of *Vagusstoff* in the frog heart can be seen with especial clarity if ergotoxine or ergotamine is added to the medium so as to eliminate the opposing action of *Acceleransstoff* (Loewi, 1924). Ergotamine also has the favourable effect of inhibiting the esterase that destroys the *Vagusstoff*. Under these conditions it can be shown that *Vagusstoff* exerts all the characteristic action of vagal stimulation. It not only diminishes the force of the beat, but also, in higher concentration, it slows the frequency and produces auriculoventricular block.

Several authors failed to confirm this basic observation (Asher, 1921, 1923, 1925, 1931; Bohnenkamp, 1924; Nakayama, 1925; and others), and suggested that Loewi's results must have been due to faulty technique. Loewi (1926) answered these criticisms. Since then his observation has often been confirmed (see for example Plattner, 1925; Hirschberg, 1931), and a great number of experiments of other kinds have shown that Loewi's conclusions were correct.[5].

Thus ten Cate (1924) detected *Vagusstoff* by letting the medium flow from the first heart to the second. Bain (1932a) improved the procedure by ensuring that the flow was independent of the activity of the donor heart; he was able to obtain complete standstill of both hearts by stimulating the vagus centre in the medulla oblongata. The experiment worked best when the pH of the medium was about 6.5. Kahn (1926) used a special double cannula, so that both hearts pumped into the same reservoir; he found that when he stimulated the vagus to one of the hearts he produced inhibition of both.

(b) Mammalian hearts.

Appearance of Vagusstoff *in blood.* In order to test whether *Vagusstoff* is liberated in mammalian hearts, Duschl and Windholz (1923) used the method known as parabiosis. They joined the circulatory systems of two rats in a preliminary operation; 2–4 weeks later they recorded the electrocardiograms of both rats, and found that stimulation of one rat's vagus caused the heart of the other to slow. They attributed the effect to transport of *Vagusstoff* by the blood to the second rat. Their work was criticized by Enderlen and Bohnenkamp (1924), who were unable to repeat the finding in experiments on dogs.

Hansen and Rech (1932), however, obtained positive results in experiments on pregnant guinea pigs. They recorded the electrocardiograms of mother and foetus, and stimulated the maternal vagus. The bradycardia that this produced in the maternal heart was followed, after a delay of about 10 seconds, by a pronounced bradycardia of the foetal heart. Hansen and Rech discuss the mechanism underlying the effect, and conclude that it could only have been due to transport of *Vagusstoff* through the placenta into the foetal circulation.

Many other investigators have employed methods that involved withdrawing heart blood from the circulation and testing it. Thus Duschl (1923) stimulated the vagus in cats and rabbits until the heart stopped.

He then took blood samples by cardiac puncture and found that when injected into a second animal the heart slowed and the blood pressure fell; both effects were abolished by atropine.

Brinkman and van de Velde (1925a, b, c) withdrew carotid blood from a rabbit during vagal stimulation; when injected into a second animal, it produced slowing of the heart and contraction of the stomach. Zunz and Govaerts (1924) did similar experiments on dogs, in which the blood collected during vagal stimulation lowered the blood pressure of a second dog, but raised the frequency of the heart. Popper and Russo (1925) found that blood taken from mammalian hearts during vagal stimulation inhibited the frog heart. Rylant (1927b) demonstrated transmission of the vagus effect in a pair of blood-perfused cats hearts. Viale (1929) found that blood taken from a dog heart during vagal stimulation had a stronger stimulant action on the dog intestine.[6]

Experiments of the sort described above were criticized by Plattner (1926) and by Tournade, Chabrol and Malméjac (1926). Their own attempts to demonstrate the release of *Vagusstoff* in the dog were unsuccessful. Plattner attributed his failure to the fact that *Vagusstoff* is very quickly destroyed by blood.

The fact that some workers were more successful than others is probably to be attributed to the use of different animal species with different levels of cholinesterase in their blood. In the successful experiments of Feldberg and Krayer (1933), that factor was eliminated by the use of eserine. Their paper will be discussed in the section that deals with the effects of *Vagusstoff* on other organs (see page 129).

Perfusion with physiological salt solutions. Jendrassik (1924) perfused rabbit hearts with physiological salt solution and applied the effluent to a preparation of isolated intestine. He found that after vagal stimulation the perfusate contained a substance that stimulated the intestine. His report is discussed in the next section.

Rylant (1927) succeeded in demonstrating the transmission of *Vagusstoff* from one rabbit heart to another when both were perfused with physiological salt solution; atropine abolished the effect.

(c) *The chemical nature of* Vagusstoff *and its effects on other organs*
In Howell's (1906) experiments on the isolated hearts of frogs and terrapins he investigated the effect of altered potassium concentration on the action of the vagus. In the absence of potassium the action was diminished or absent; as the potassium concentration was raised, the vagal effect increased. When the concentration was high enough to affect the

beat by itself, the change corresponded to that produced by vagal stimulation. In later experiments, Howell and Duke (1908) perfused the hearts of dogs, rabbits and cats with Locke's solution, and found that when they stimulated the vagus potassium was released into the perfusate. They suggested that the inhibitory action of the vagus was a direct consequence of this mobilisation of potassium.[7]

Asher (1923), who was unable to confirm Loewi's original observations, did get positive results when he used potassium-free salt solutions. He attributed these results to the liberation of potassium. But it is quite clear from the studies to be described below that the actions of *Vagusstoff* cannot all be due to potassium.

Brinkman and van de Velde (1925b) measured the surface tension of rabbit serum (diluted 1000-fold with water) and found it to be diminished by vagal stimulation. They did not, however, expressly propose that *Vagusstoff* is a surface-active substance.

Atzler and Müller (1925) suggested that the phenomena attributed by Loewi to *Vagusstoff* and *Acceleransstoff* were really due to changes in the acidity of the solutions. They did show that a small increase of acidity weakened the contractions of the heart, while alkalinity had the opposite effect.

Loewi himself (1921) acetylated the salt solution that had been in contact with the heart, and found that choline diffused out even in the absence of nerve stimulation, though more of it appeared as a result of stimulation. This choline, however, was present in too small quantity and had too little intrinsic activity to be the *Vagusstoff* itself. On the other hand much evidence has been secured that is consistent with the view that *Vagusstoff* is either acetylcholine or some closely related choline ester. Thus *Vagusstoff*, like acetylcholine, is dialysable, unstable in alkali (pH 12.1), relatively stable in acid solution (pH 2), and soluble in alcohol though not in ether (Witanowski, 1925).

Watery extracts of heart tissue inactivate both *Vagusstoff* and acetyl-choline. The agent responsible for the inactivation is presumably the cholinesterase discussed on page 81. The enzyme is destroyed by ultra-violet light or by heating to 56 °C, and it is inactivated by physostigmine (eserine) or ergotamine (Loewi and Navratil, 1926a, b). Blood, which contains cholinesterase, likewise destroys the *Vagusstoff* obtained from the frog heart (Plattner, 1926; Hirschberg, 1931).

The probability that *Vagusstoff* is a choline ester has been greatly increased by pharmacological studies. Brinkman and van Dam (1922), ten Cate (1924) and Bain (1932a) used a proceudre by which a frog heart and a frog stomach or intestine were perfused consecutively with physiological solution. When the cardiac vagus was stimulated, the perfusate acquired the power of increasing the tone of the stomach or intestine.

Jendrassik (1924) did similar experiments on rabbits. The coronary vessels were perfused with a physiological salt solution and the vagus was stimulated; in order to concentrate the *Vagusstoff* as much as possible the perfusion was stopped during the stimulation. The fluid was then washed out and tested on an isolated gut preparation. In control experiments, with no vagal stimulation, the solution was inactive, but when the vagus had been stimulated it produced a strong increase in intestinal tonus. The effect could be obtained only once with each heart; a second vagal stimulation produced intestinal inhibition.

Brinkman and van de Velde (1925) found that arterial blood withdrawn from a rabbit during vagal stimulation increased the tone and peristaltic movements of the rabbit stomach. Viale (1929) obtained similar results in experiments on dogs; he tested the blood on the isolated intestine.

Feldberg and Krayer (1983) collected blood directly from the coronary sinus of eserinized cats and dogs. They found that blood taken during and shortly after vagal stimulation showed two properties that are typical of acetylcholine. It caused a contraction of a leech muscle that had been pretreated with physostigmine (eserine), and it caused a fall of blood pressure in a cat. The effect on the blood pressure was abolished by atropine. The substance responsible for these effects was destroyed by blood but was comparatively stable in blood that had been treated with physostigmine.

Krayer and Verney (1934) used the same method to demonstrate the release of a choline ester from the frog heart when the vagus centre was stimulated reflexly by raising the blood pressure.

2. Intestine

The importance of choline and choline esters for the activity of the intestine was recognized early (see Magnus, 1930).[8] Neukirch (1912) had been investigating the diffusion of pilocarpine out of a piece of rabbit intestine that had been treated with that drug. Weiland (1912) carried out the control experiment without pilocarpine, and found that the salt solution in which a piece of normal rabbit intestine had been suspended acquired the ability to stimulate a second piece of normal intestine. The active substance was stable to boiling, soluble in alcohol, and antagonized by atropine. It was identified by Le Heux (1919) as choline. The first clue to its identity came from the observation that its activity was greatly increased by treatment with glacial acetic acid. Pure choline was then isolated in crystalline form and identified by the melting-points of its gold, platinum and mercury salts, and its identity was further confirmed by quantitative comparison of its physiological activity with that of choline. The amount of choline that diffuses out of the whole length of a rabbit intestine in an hour totals 3–4 mg.

This diffusion also occurs when a piece of normal intestine with intact circulation is immersed in physiological salt solution (Sawasaki, 1925). Isolated loops of intestine cannot synthesize choline (Girndt, 1925). The choline content of the intestine has been found to be unaffected by prolonged chloroform anaesthesia, laparotomy, peritonitis or diarrhoea (Arai, 1922; Zunz and György, 1914).[9]

These experiments made it possible to explain certain puzzling and apparently contradictory observations on the action of atropine on the intestine. A small dose of atropine will inhibit the freshly excised intestine, and this is attributed to antagonism of the free choline naturally present in the tissue. But if this choline is washed out by frequent changes of the Ringer's solution in which the intestine is suspended, atropine no longer produces inhibition. Under these conditions small doses of atropine have no effect at all, and larger doses have a direct stimulant effect (Le Heux, 1920).

Evidence suggesting that active choline esters can be synthesized by isolated intestinal loops was obtained soon afterward. Rona and Neukirch (1912) had found that sodium acetate and sodium pyruvate differ from the sodium salts of other organic acids in having a stimulant action on the isolated intestine. Le Heux (1921) confirmed that, and suggested that their action was due to synthesis of the active esters acetylcholine and pyruvylcholine. He showed that both were much more active than choline. Sodium acetate and sodium pyruvate lose their stimulant action on the intestine after its choline has been washed out, and their action on the fresh intestine is antagonized by atropine. Le Heux investigated the sodium salts of other organic acids and found a rough parallelism between their stimulant action on the intestine and the increase in activity that choline undergoes when it is esterified with the acid in question. Sodium succinate has no action on the intestine, and the succinyl ester of choline is no more active than choline itself. The only exception to the rule was sodium butyrate, which had more activity than would have been expected from the effect of butyrylcholine.[10]

These conclusions have been confirmed by later work. The synthesis of active choline esters has been demonstrated *in vitro* (Abderhalden and

Paffrath, 1925; Ammon and Kwiatkowski, 1934). Evidence that acetylcholine is normally present in the intestine was presented by Chang and Gaddum (1933).

The use of eserine has provided new methods for studying the properties of the choline esters in the intestine. Like many other organs, the intestine contains cholinesterase (Loewi and Navratil, 1926a, b; Abderhalden and Paffrath, 1925; Plattner and Hintner, 1930a); and the stimulant action of eserine on this organ is probably due in the main to its effect on that enzyme. It had been shown that after a dose of eserine, stimulation of the chorda led to the appearance of detectable amounts of acetylcholine in the veins of the submaxillary gland. The first attempt at such an experiment on the intestine was a failure (Freeman, Phillips and Cannon, 1931); but Feldberg and Rosenfeld (1933) found later that the blood leaving the intestine of an eserinized dog contains a substance that acts like acetylcholine on the leech and on the cat blood pressure. Both actions were enhanced by eserine, and the one on the blood pressure was abolished by atropine. When the two responses were compared quantitatively, the estimates of acetylcholine concentration agreed. Blood leaving the spleen was inactive. Donnomae (1934) has repeated the experiment on the cat, with the same result. He and Feldberg (1934) used a somewhat different method for detecting the release: in cats whose blood had been made incoagulable they shunted the portal blood into the femoral vein, by-passing the liver; in the absence of eserine the bypass procedure generally had no effect on the blood pressure, but in the presence of eserine the blood pressure fell, and the fall could be relieved with atropine.

Another approach was taken by Feldberg and Kwiatkowski (1934), who perfused the isolated small intestine of the cat with Ringer's solution; the fluid emerging from the vein acted like acetylcholine on both the leech and the cat blood pressure, and was destroyed by alkali. Even more convincing, perhaps, was the identification in the experiments reported by Dale and Feldberg (1934), in which the acetylcholine was found in the effluent from the perfused stomach of the dog. In these experiments the fluid was tested not only on the leech and the blood pressure, but also on the frog heart and the frog rectus abdominis; all four tests were in quantitative agreement. Studies such as these encourage the idea that under normal conditions acetylcholine is continually being released from the stomach and intestine and destroyed by the esterase, so that in the absence of eserine no acetylcholine is to be found in the portal blood.

Dale and Feldberg (1934a) found in the dog with intact circulation, and also in the dog stomach during perfusion with Locke's solution, that vagal stimulation causes a two- to four-fold increase in the amount of acetyl-choline released by the stomach. The purpose of these experiments was to show if possible that nerves whose action is not abolished by atropine, like the vagal supply to the gut muscle, can nevertheless be cholinergic. Unfortunately Dale and Feldberg did not exclude the possibility that the acetylcholine they found to be released by nerve stimulation came from the glands in the mucosa.

3. Lungs

One would expect the pulmonary branches of the vagus to be cholinergic, because their effects on the bronchi and the pulmonary vessels are abolished by atropine and enhanced by eserine (Dixon and Ransom, 1912; von Euler, 1912). Thornton (1934) supplied direct evidence for this. He perfused the lungs of a guinea pig with eserine-containing physiological salt solution, and found that vagal stimulation led to the appearance in the perfusate of a substance that acted like acetylcholine on the leech and the cat blood pressure (see also Saalfeld, 1934).

4. Oculomotor nerve.

Velhagen (1930) and Plattner and Hintner (1930b) sought for evidence that excitation might be chemically transmitted from the oculomotor nerve to the iris, but their results were inconclusive.

The first convincing evidence that the oculomotor nerve is cholinergic was provided by Engelhart (1931). When the eye of a rabbit was treated with eserine and then exposed to light, a substance appeared in the aqueous humour that acted like acetylcholine on the toad heart. The action matched that of a solution that contained 50 µg or more of acetylcholine per litre. In the absence of either eserine or illumination the aqueous humour was almost or quite inactive. The active material was destroyed when it was left in contact with the heart, and its effect was abolished by atropine. It was soluble in 90 per cent alcohol. No such material could be found in cats under urethane; this was attributed to the anaesthetic (see also pages 113 and 163).

5. Salivary glands

Demoor (1911, 1912, 1913) found that the injection of a dog's saliva into a dog caused secretion of saliva. The observation suggested that some chemical mechanism might play a part in the activity of the salivary glands, but for many years no further findings pointed in that direction.

Guimarãis (1930) confirmed Demoor's observation. He also perfused the gland with blood diluted with Lock's solution, and showed that the injection of saliva into the perfusion fluid induced secretion.[11]

Somewhat similar experiments have recently been done by Secker (1934a, b). He did not confirm the above results, but he found that saliva secreted in response to stimulation of the chorda tympani or the cervical sympathetic, or after the injection of pilocarpine or adrenaline, contained a substance that acted like acetylcholine on the cat blood pressure and the rabbit intestine. Still more recent experiments by Feldberg and Guimarãis (1935) and by Gibbs (1935) have shown that this substance is not acetylcholine (see page 119); and it seems increasingly doubtful that the pharmacological activity of saliva has anything to do with the transmission of nervous excitation to the gland. The following results, however, do seem to be significant in that context.

Von Beznák (1932) carried out experiments in which blood from the submaxillary vein of one dog was transferred with a syringe to the maxillary artery of another dog. He also tested the blood on the frog heart. From these tests he came to the conclusion that stimulation of the chorda releases an acetylcholine-like substance. These observations were suggestive, but could not be regarded as conclusive.

Later in the same year the accounts were published of three independent investigations which showed convincingly that the chorda supplies the submaxillary gland with cholinergic fibres. In each case the results were made possible by the use of eserine to inhibit cholinesterase (Babkin, Stavraky and Alley, 1932; Henderson and Roepke, 1932; Gibbs and Szeloczey, 1932a, b).

Experiments with intact circulation. Stimulation of the chorda produces a slight fall of blood pressure in the cat; this fall is greatly increased by the injection of eserine beforehand. This observation, which suggests a cholinergic mechanism, had been made some years earlier by Babkin and Gibbs; it was first published by Babkin, Gibbs and Wolff (1932) after they had examined the problem more thoroughly. They considered that the effect was due partly to local vasodilatation, but partly to the release into the general circulation of a substance like acetylcholine.

Feldberg (1933b) has found that the strong depressor effect of chorda stimulation in the eserinized cat is abolished by atropine. Since the local vasodilatation is unaffected by atropine, this observation shows that the fall of blood pressure is largely due to the release of some substance into the general circulation.

Babkin, Alley and Stavraky (1932) showed that after 0.5 mg of eserine had been injected into a cat, stimulation of the chorda on one side evoked secretion from the submaxillary gland of the opposite side, even when that gland had been denervated. This effect was not a result of the fall of blood pressure, for it did not occur when the blood pressure was lowered in other ways; nor did it depend on the carotid-sinus reflex or on the release of adrenaline from the suprarenals. Rather, it depended, at least for the most part, on the release of a substance from the salivary gland into the circulation, for the secretion was very much less when the vein from the stimulated gland was clamped off. The authors suggested that the very small effect that was sometimes seen after occlusion of the vein was due to a little of the same substance making its way into the blood via the lymphatics.

Perfusion experiments.[12] Reference has already been made to the reports of Demoor and Guimarais, who investigated the action of saliva on the perfused submaxillary gland.

Further perfusion experiments were carried out by Henderson and Roepke (1932, 1933b) on the submaxillary glands of dogs. They used Ringer's solution, with or without the addition of gum-acacia, as their medium, and they found that on chorda stimulation a substance appeared in the perfusate that acted like acetylcholine on the frog heart. This substance was soluble in 95 per cent alcohol, but insoluble in ether or chloroform, and it was destroyed when it was heated for 5 minutes to 60 °C in a weakly alkaline solution. Its action was abolished by atropine. In Henderson and Roepke's first experiment the substance appeared only in the presence of eserine, but later they had positive results without eserine. The release of this active substance persisted for 2–3 hours after the stimulation. No activity was detected in blood or saliva. Atropine did not prevent the release of the active material into the perfusate, although its concentration was high enough to stop secretion. The amount of active substance released in 2 minutes corresponded to some 0.25–0.5 µg of acetylcholine.

Gibbs and Szelöczey (1932) used a perfusion arrangement that allowed the glands to be perfused only temporarily with Ringer's solution, normal blood flow being restored in the intervals. A similar method had been used by Guimarais (1930). It appears to be very useful for studying the local release of chemical transmitters. Gibbs and Szelöczey were thus able to demonstrate that stimulation of the chorda in the dog or the cat caused the release of a substance that acted like acetylcholine on the blood pressure and salivary glands of the cat as well as on the isolated heart of the frog and the intestine of the rabbit. These actions were diminished by atropine, but the perfusates also contained a second substance, which inhibited the frog heart after atropine.[13] The substance responsible for the major part of the effect was destroyed by blood; and once more, eserine inhibited the destruction, while atropine in dosage sufficient to block salivary secretion did not interfere with the appearance of the active substance in the perfusate. According to the data of Gibbs and Szelöczey about 0.02 μg of acetylcholine were released in 30 seconds. When the perfusate was re-injected it resulted in the secretion of saliva at about half the original rate.

6. Bladder

Henderson and Roepke (1933b, 1934) perfused the bladder, and found that stimulation of the pelvic nerve led to the appearance of a substance that acted on the frog heart like acetylcholine in a concentration of about 12 μg per l.

7. Suprarenals

A number of facts indicate that the splanchnic fibres that elicit the discharge of adrenaline from the suprarenal medulla are cholinergic: in other words, the release of adrenaline by splanchnic stimulation is secondary to the release of acetylcholine.

The action of choline esters in causing a discharge of adrenaline was discussed in the section on the pharmacology of acetylcholine (page 57). Feldberg and Minz (1933) have examined the suprarenal venous blood of dogs and cats, and have found that splanchnic stimulation liberates a choline ester. In untreated animals the only active substance in their blood samples was adrenaline; but if the animals had been given eserine just before, the nerve stimulation led to the appearance of a material that was identified as a choline ester by the following tests: it contracted the eserinized leech muscle but had no effect on the uneserinized preparation; it was destroyed when the blood was allowed to stand, and the destruction could be prevented by adding additional eserine; and when an active blood sample was tested on the cat blood pressure, its effect was indistinguishable from that of a mixture of adrenaline and acetylcholine.[14]

Feldberg, Minz and Tsudzimura (1934) confirmed these conclusions. In their experiments they allowed the suprarenal venous blood to flow back into the cat in the normal way, but the animal was eviscerated, so as to confine the effect of splanchnic stimulation to the suprarenals. An injection of eserine then modified the blood pressure response to such stimulation in several ways, which indicated that a choline ester was entering the circulation along with adrenaline. Thus the pressor response was delayed, or was preceded by a fall of blood pressure. When the cat was made insensitive to adrenaline by the injection of adrenaline in large amounts, the splanchnic stimulation in the presence of eserine produced a pure fall of blood pressure, which was prevented by atropine. In other experiments a record of salivary secretion was made in animals that had been made insensitive to the sialogogue action of adrenaline by pretreatment with ergotoxine. Under these conditions splanchnic stimulation in the presence of eserine evoked a secretion of saliva that could be abolished with atropine. This effect could only have been due to a choline ester liberated from the suprarenals and circulating in the blood.

Although it was thus demonstrated that splanchnic stimulation releases a choline ester from the suprarenals, the causal relationship between the release and the discharge of adrenaline had still to be proved. The fact that extracts of the suprarenal medulla have a higher acetylcholine equivalent value than extracts of the cortex (Feldberg and Schild, 1934) suggested that the medulla was the source of the released ester; more direct evidence was supplied by Feldberg, Minz and Tsudzimura (1934) when they found that eserine increased the quantity of adrenaline released by splanchnic stimulation, as judged by the size of the pressor response – an effect that was particularly striking when atropine had been given. These experiments thus provided evidence that the cholinergic mechanism in the suprarenals is directly concerned with the secretion of adrenaline. They also provided the first evidence for a cholinergic mechanism in preganglionic sympathetic nerves.

8. Sympathetic ganglia

The first evidence for the concept that impulses are transmitted by a chemical mechanism across the synapses in a sympathetic ganglion was obtained by Kibjakow in 1933. He perfused the superior cervical ganglion of the cat with Locke's solution and recorded the withdrawal of the nictitating membrane produced by stimulation of the cervical sympathetic nerve trunk. He collected the effluent and tested it by perfusing it through a second ganglion. Kibjakow's finding was that the fluid collected during stimulation of the cervical sympathetic caused stimulation of the second ganglion, as shown by contraction of the nictitating membrane.[15]

Some years before this, Witanowski (1925) had detected the presence of something in sympathetic nerves and ganglia that acted like acetylcholine. Chang and Gaddum (1933) came across it again, using tests that gave clearer evidence of its identity.[16] They supposed that the substance Kibjakow had discovered might be acetylcholine.

Feldberg and Gaddum (1933, 1934) then used Kibjakow's technique for perfusing the superior cervical ganglion, but they were unable to confirm his results. When there was no eserine in the perfusion medium they could detect no activity in the effluent; but when eserine was present stimulation of the sympathetic trunk led to the appearance of a substance that was shown by pharmacological tests to be acetylcholine. The identification was based on the following findings: the active material was destroyed by alkali but was relatively stable in weak acid; it produced effects indistinguishable from those of acetylcholine on the eserinized leech, the frog heart, the cat blood pressure, the normal or eserinized rectus abdominis of the frog, and the rabbit auricle; and estimates of the activity of the solution in terms of acetylcholine gave quantitatively concordant values in all six tests.

The tissues perfused in the experiments of Feldberg and Gaddum included the nodose ganglion of the vagus, the carotid sinus and the carotid body, in addition to the superior cervical ganglion itself. It might be suggested that the acetylcholine was released by one of these extraneous tissues. Evidence against such a suggestion was secured by Barsoum, Gaddum and Khayyal (1934), who perfused the inferior mesenteric ganglion of the dog and showed that stimulation of the preganglionic fibres led to the appearance of acetylcholine-like activity in the effluent. In this case practically no tissue other than the ganglion itself was perfused.

The work on the perfused superior cervical ganglion of the cat was continued by Feldberg and Vartiainen (1934). They found that neither stimulation of the vagus nor antidromic stimulation of the postganglionic fibres of the sympathetic, backfiring impulses into the ganglion cells, caused the discharge of any detectable amount of acetylcholine. Their experiments thus confirmed the idea that preganglionic impulses release acetylcholine at the synapses.

The quantity of acetylcholine set free by each maximal shock applied to the cervical sympathetic trunk was about 0.001 µg. Presumably the impulses arrived at all the synapses. Since there are about 100 000 nerve cells in the superior cervical ganglion of the cat (Billingsley and Ranson, 1918), the amount of acetylcholine released per impulse and per synapse corresponds to about 10^{-9} µg, or some 3 000 000 molecules. In further experiments Feldberg and Vartiainen studied the release of acetylcholine from ganglia whose circulation was intact, and were able to detect it in the ganglion's venous blood after they had administered eserine to the cat by vein. This observation increases the probability that the release of acetylcholine is a normal physiological process, and not a consequence of the artificial conditions of a perfusion.[17]

The perfused ganglion is a convenient preparation for studying the pharmacology of a neuronal synapse. Feldberg and Vartiainen found that the ganglion was excited not only by acetylcholine, but also by choline, hordenine methyl iodide, nicotine or potassium ions applied in sufficiently high concentration. Eserine (10^{-6} g per cc) only slightly potentiated the effects of these drugs, whereas it increased the effect of acetylcholine 10- or 20-fold. (After eserine the same submaximal effect could be obtained with a tenth or a twentieth of the former dose.) Eserine in this low concentration also sensitized the ganglion to the action of preganglionic impulses generated by a series of submaximal shocks. In high concentration, however, eserine paralysed the ganglion, making it insensitive both to drugs and to nerve impulses. Potassium in concentrations too low to evoke a discharge from the ganglion could sensitize it to the action of either acetylcholine or preganglionic impulses.

Since acetylcholine is an effective stimulant of sympathetic ganglia, these results support the theory that the mechanism by which nerve excitation normally crosses the synapse depend on the liberation of a small quantity of acetylcholine. If that is indeed the case, the cholinergic mechanism in a ganglion must be completely different from the other cholinergic mechanisms that have been studied. For the delay when a single impulse is transmitted through the ganglion is very short, as is the refractory period; there is no summation of the kind seen when postganglionic nerves are stimulated, though under some conditions the effect of a submaximal stimulus can be increased by a preceding stimulus (facilitation), while under other conditions the first stimulus can depress the response to the second. The properties of ganglia thus have many points in common with those of the synapses of the central nervous systems. On the basis of these and other facts, the chemical transmission of excitation in ganglia has been disputed (Eccles, 1933a, b, 1934a, b; Brown, 1934).[18]

9. *Sweat glands*

The sweat glands of the cat's paw are supplied by nerve fibres that belong anatomically to the sympathetic system but have an anomalous response to drugs. They are stimulated by alkaloids like pilocarpine that belong to the muscarine group and have actions that are abolished by atropine (Ott and Field, 1878); they do not respond to adrenaline (Langley, 1922).

These facts led Dale (1933b) to suggest that the nerves to these glands differ from most postganglionic sympathetic nerves in being cholinergic. The suggestion was confirmed experimentally by Dale and Feldberg (1934b). They perfused a cat's paw with eserine-containing Locke's solution and applied pharmacological tests to the perfusate. Stimulation of the sympathetic chain led to the appearance of a substance that they identified as acetylcholine on the basis of its action on the eserinized leech and the cat blood pressure. The two tests agreed quantitatively in their estimates of the acetylcholine concentration. The identity of the active material was confirmed by its lability in alkali, and by the fact that its depressor action in the cat was enhanced by eserine and abolished by atropine.

The sweat glands of the cat's paw are confined to the hairless pads of the foot. Exclusion of these pads from the perfusion in a control experiment resulted in failure of acetylcholine to appear in the effluent in response to stimulation.

These results show that in the cat the secretory fibres to the sweat glands of the paw are cholinergic. This is probably true also of the sweat glands in man, for they are affected by pilocarpine and atropine but not by adrenaline (Elliott, 1905). In other animals, for instance the horse or the sheep, the sympathetic fibres to the sweat glands seem to be adrenergic, like most other postganglionic sympathetic fibres (for details see Dale and Feldberg, 1934b).

10. *Vasodilator nerves and nerves that cause pseudomotor contractions*

When a muscle is made to contract by stimulation of its motor nerve, its blood vessels dilate. The phenomenon will be discussed on page 169. In this chapter we shall concern ourselves only with those nerve fibres whose main, and perhaps exclusive, function is to cause vasodilatation. The possibility that these may act by releasing vasodilator substances has been discussed by Bayliss (1901a), Gaskell (1920), and Langley (1923). In recent years evidence has been obtained that vasodilator fibres of several different types may perhaps act in this way.

(a) *Axonal branches of sensory nerves, responsible for 'axon reflex' and antidromic vasodilatation*

Direct evidence for the existence of this mechanism is at present limited to the vessels of the skin. Bayliss (1901a) found that antidromic vasodilatation in the dog's limb was trivial after removal of the skin.

Lewis and Marvin (1927) furnished evidence for the chemical transmission of the antidromic vasodilator impulses to the skin of the cat's paw. Their results seemed to show that the transmitter is a substance that is relatively stable at the site of its liberation; in their view it resembled histamine rather than acetylcholine. On the other hand the axon reflex dilatation, which presumably represents the normal function of this innervation, is mainly concerned with larger arterioles than those that are dilated by histamine; it has more similarity with actions that occur at other sites and are recognized as cholinergic.

It was formerly thought that indirect evidence for the cholinergic nature of these sensory vasodilator branches had been obtained through study of the 'pseudomotor' response of Sherrington, exhibited by muscles deprived of their motor innervation – a response that was supposed to be elicited by antidromic stimulation of sensory fibres. That response, however, had all the features ascribed to the pseudomotor effects evoked by cholinergic vasodilator fibres on voluntary muscle that had lost its motor nerve supply (Hinsey and Gasser, 1930; Dale and Gaddum, 1930). According to more recent studies by Hinsey and Cutting (1933), the Sherrington contracture is in no way due to antidromic stimulation of sensory fibres; rather, it results from the stimulation of sympathetic fibres originating in the sacral sympathetic plexus and joining the sciatic plexus. The results of Hinsey and Cutting have recently been confirmed by Bülbring and Burn (1935; see also page 142). The evidence for the cholinergic nature of the fibres responsible for the effect of course remains valid; and what the newer work indicated is that the skeletal musculature receives cholinergic fibres, presumably vasodilator in function, from the classical sympathetic system. This leaves open the question of the nature of the chemical transmitter of those vasodilator actions that are certainly due to antidromic impulses or axon reflexes; it is a question that can only be answered by further investigation.

(b) *Other vasodilator nerves to voluntary muscle.*

The literature indicates that most mammalian voluntary muscles are supplied by cholinergic fibres other than those that run to the muscle fibres themselves (see page 144). If the conclusions of Hinsey and Cutting are correct, most of these fibres must run in sympathetic nerves. They were mentioned incidentally in the preceding section, and must now be considered in greater detail. In this category, besides those already mentioned, are the vasodilator fibres that run in the chorda tympani to the tongue musculature (Heidenhain, 1883), and those that run in the cervical sympathetic to the facial muscles of the dog (Rogowicz, 1885; von Euler and Gaddum, 1931b).

The first evidence for the theory that these nerves have effects that depend on local chemical transmission was based on the following observation. If the vasodilator nerves are excited a week or more after the motor nerves to the muscles concerned have been transected, they produce a paradoxical contracture of the striated muscle fibres, which resembles the response evoked in those fibres under the same conditions by an injection of acetylcholine. The theory has been proposed that the reason for the resemblance is that the vasodilator nerves normally liberate acetylcholine or a similar substance; some of this substance reaches the muscle fibres, which have been made sensitive to acetylcholine by the denervation, and accordingly respond by contracting (Bremer and Rylant, 1924; Dale, 1929; Hinsey and Gasser, 1930).

Dale and Gaddum (1930) have discussed the phenomenon in detail. They observed that the nerve-induced contracture was intensified by eserine, and pointed out that such behaviour might provide evidence for a cholinergic mechanism. They studied the apparent difficulties raised by the effects of atropine and adrenaline on these phenomena (see pages 157, 160), and concluded that these effects were compatible with the idea that both vasodilatation and contracture were due to acetylcholine. The same phenomena have been studied in the lip musculature of the dog by von Euler and Gaddum (1931b), who considered the fibres responsible to be an example of cholinergic fibres of true sympathetic origin. The later reports by Feldberg (1933a) and by Hinsey and Cutting (1933) have confirmed this view.

The reasoning here involves the assumption that the vasodilatation and the contracture are due to the same nerve fibres. This is hard to prove. The fibres that subserve the two effects have the same anatomical distribution, but Boeke (1927) has described another set of fibres that is distributed similarly. As far back as 1883 Heidenhain offered evidence for the view that the tongue contracture was produced by impulses in vasodilator nerves, when he showed that the two effects responded similarly to changes in the strength and frequency of the stimulation. Bremer and Rylant (1924) have also obtained indirect evidence that the tongue contracture is evoked by vasodilator fibres and not by the fibres of Boeke; however, the question cannot be regarded as finally settled.

In the meantime other, and more direct, evidence has been obtained that indicates the cholinergic nature of certain vasodilator nerves. Bain (1932b, 1933) perfused the dog tongue with physiological salt solution and tested the perfusate on a piece of rabbit intestine. Stimulation of the chorda tympani led to the appearance of a substance that raised the tone and motility of the intestine. When the nerve supply of the tongue was intact, the Traube–Hering waves of arterial pressure were in synchrony with periodic waves of intestinal activity, which might well have been due to the periodic liberation of a choline ester.

Feldberg (1932) collected tongue blood from the lingual vein and tested it on the cat blood pressure and the eserinized leech. If the dog had been given eserine, stimulation of the chorda tympani led to the appearance in the blood of a substance that could be identified pharmacologically as acetylcholine. The effects of the blood on the two test-objects matched those of acetylcholine in the same concentration; and again, the substance was destroyed by blood and the destruction was prevented by eserine; finally, its effects were increased by eserine and abolished by atropine.

Bain and Feldberg both supposed that the acetylcholine they identified had been liberated by vasodilator fibres. This conclusion was probably correct, but it should be noted that there are scattered mucous glands on the surface of the tongue of the dog, and that the transmission of secretory impulses to these might have been responsible for the appearance of acetylcholine.

Bülbring and Burn perfused the hind limbs of dogs and cats with blood; under these conditions suitable stimulation of the sympathetic nerve supply caused vasodilatation in dogs but not in cats; eserine intensified the effect and atropine abolished it. The results are surprising because they offer the first clear example of a vasodilator innervation whose action is blocked by atropine. The observation that atropine antagonizes the action of some, but not all, cholinergic nerves is discussed on page 144.

In recent experiments by Bülbring and Burn (1935) the muscles of the dog's hind limb were perfused with eserine-containing Locke's solution. The effluent obtained during sympathetic stimulation contained a substance that behaved pharmacologically like acetylcholine.

(c) Penis

Lewis and Marvin (1927) found that the vasodilator supply of the penis in the nervi erigentes behaved differently from the vasodilator supply to a limb: in the case of the penis, the vasodilator effect could not be held in abeyance during a temporary interruption of the circulation to appear when the flow was restored, as it could be in the case of the limb. Lewis and Marvin concluded that if the vasodilatation is chemically mediated the mediator must be of low stability compared to the substance that mediates the antidromic vasodilatation.

Henderson and Roepke (1933a) investigated the possibility that the fibres involved are cholinergic. They found that the vasodilator effect is greatly increased by eserine, an observation that speaks in favour of a cholinergic mechanism. They also stimulated the nerve supply of a perfused penis and tested the perfusate on the frog heart, but they did not succeed in getting evidence for the release of a choline ester; nor could they antagonize the vasodilator effect with atropine.

According to Bacq (1935) cholinergic fibres to the penis appear to be present also in the hypogastric nerve.

(d) Other vessels

Gollwitzer-Meier and Otte (1933) perfused portions of the femoral artery, the mesenteric artery and vein, and the splenic artery and vein of cats and dogs with eserine-containing physiological salt solution. The effluent was deproteinized when necessary and tested pharmacologically; suitable control experiments were done to ensure that the deproteinization itself had no effect. Vasodilatation was produced either by stimulating the vasodilator nerves directly, or reflexly by stimulating the nerve from the carotid sinus. This led to the appearance in the effluent of a substance that acted like acetylcholine on the frog heart and the cat blood pressure. Though some experiments gave clear results, unfortunately this was not the case in those in which the dorsal roots of the spinal nerves were stimulated. The question as to whether these roots contain cholinergic fibres is therefore still undecided. Gollwitzer-Meier and Otte also made extracts of blood vessels, and apparently found the acetylcholine content of the vascular wall roughly doubled as a result of the vasodilator stimulation.

11. Uterus

The fact that acetylcholine and other drugs with a muscarine-like action evoke a contraction of the uterus that is antagonized by atropine gave Sherif (1935) the idea that cholinergic fibres might participate in the normal regulation of the organ. He did find that stimulation of the hypogastric nerves in eserine-treated dogs led to the appearance in the uterus of a substance that was identified pharmacologically as acetylcholine. The finding that eserine increased the contraction of the uterus induced by hypogastric stimulation confirmed his conclusion that the contraction was produced by a cholinergic mechanism. Nicotine did not affect the response, which indicates that the fibres responsible for it are cholinergic. The experiments thus represent another example of a cholinergic postganglionic sympathetic innervation.

12. Motor nerves to voluntary muscle

Several observations pointed to the idea that the motor nerves to voluntary muscles might also be cholinergic.[19] Zucker (1923) found that treatment of frog muscle with eserine lowered its threshold for electrical stimulation.

Samojloff (1925) found that temperature changes had a greater effect on neuromuscular transmission than on conduction in the nerve itself. He concluded that a chemical mechanism must be responsible for the transmission of excitation from nerve to muscle.

A number of authors (Hess, 1923; Shimidzu, 1926; Brinkman and Ruiter, 1924, 1925) have observed the release of a substance like acetylcholine when they stimulated the mixed nerve to a frog's skeletal muscle. In their experiments the possibility was not excluded that the active material might have been released by the stimulation of sympathetic or parasympathetic fibres, and there is no evidence that the authors themselves related their experiments to the transmission of impulses from nerve to voluntary muscle.

Dale and Feldberg (1934) perfused the cat tongue with Locke's solution and stimulated the hypoglossal nerve after its sympathetic fibres had degenerated. The stimulation caused the appearance in the effluent of a substance pharmacologically identified as acetylcholine. Except for its sympathetic component the hypoglossal nerve is composed almost entirely of somatic motor fibres. It does, however, contain some sensory fibres that are apparently connected with muscle spindles, and come from the nodose ganglion, from the second cervical ganglion, and from a variable and rudimentary ganglion on the hypoglossal itself (cf. Langworthy, 1924). Thus the hypoglossal cannot be freed with assurance from its small sensory component. Dale, Feldberg and Vogt (1935) have, however, extended their experiments to the muscles of the dog's limb, which they have perfused with Locke's solution to which eserine had been added, and in this preparation they have observed a similar discharge of acetylcholine on stimulation of the ventral spinal roots after extirpation of the abdominal sympathetic chain. Under these conditions, apparently, only motor fibres were involved in the stimulation. Similar observations by Feldberg and Schriever (1935) on the skinned frog limb gave the same result.

Such evidence strongly suggests that the ordinary motor impulses to voluntary muscle fibres are transmitted from the nerve endings to the receptive end-plate of the muscle fibre by the release of acetylcholine. There are many pharmacological and other analogies between the junction of the motor nerve fibre and the muscle end-plate on the one hand, and a synapse in an autonomic ganglion on the other hand, which suggest the possibility of a similar method of chemical transmission at both sites.

Investigations of the effect of eserine and related synthetic compounds in the therapy of myasthenia gravis have brought to light some interesting facts that are probably relevant to the problems that have just been discussed. In this disease the mechanism for transmission of impulses from the motor axons to the striated muscle is abnormally susceptible to fatigue. This state of affairs can be relieved temporarily with eserine (Walker, 1934; Pritchard, 1935; Laurent, 1935). It is thought, naturally, that the improvement is due to inhibition of the esterase, which destroys the acetylcholine that transmits the excitation to the muscle. The details of the mechanism, and the exact nature of the defect in myasthenia gravis, are still unknown.

C. Adrenergic nerves

1. *General*

Adrenergic nerve fibres are those that liberate a local chemical transmitter whose immediate action resembles that of adrenaline. This statement does not necessarily mean that the substance is adrenaline itself, though there is no present reason to doubt that this is so. The available evidence shows at least that the substance released by these nerve fibres must be closely related to adrenaline in its chemical structure.

In the opinion of Cannon and Rosenblueth (1933), however, the matter is not so simple. According the these authors the substance released by the adrenergic nerves, designated 'sympathin' when it is carried by the circulation to distant organs where its action is detected, may have only excitatory effects or only inhibitory effects, depending on the nature of the nerve fibres that release it. Cannon supposes that the material originally released is a substance like adrenaline that is capable of producing both sorts of effect. But this substance is converted, according to Cannon, by the tissue on which it acts, into either sympathin E (excitatory) or sympathin I (inhibitory). It is these two substances, with excitatory and inhibitory actions respectively, that escape into the circulation and have effects on distant organs. For example, stimulation of the nerves to the cat intestine produces both augmentor and inhibitor actions on distant organs, which match those of an appropriate dose of adrenaline. But when the nerves to the liver – or to the heart, the limbs or the tail – are stimulated, the distant inhibitor effects are feeble compared with the augmentor effects. Experiments on the hepatic nerves provided an especially clear indication that sympathetic nerve stimulation can liberate sympathomimetic substances that are different from adrenaline.

Cannon's interpretation has been criticized by Bacq (1934), and Cannon and Rosenblueth (1935) have replied to Bacq's arguments.[20]

The different types of experiment will be discussed below in the appropriate sections. So far as we know all the adrenergic fibres of mammals are postganglionic sympathetic fibres; but as already pointed out, there are some postganglionic sympathetic fibres that are cholinergic rather than adrenergic.

Methods for detecting the adrenergic transmitter. The first evidence for an adrenergic mechanism was obtained by Loewi (1921) on the isolated frog heart. Thereafter many other investigators have studied the properties of the substance set free in that preparation. Finkleman (1930) has demonstrated an adrenergic mechanism in the isolated intestine.

In the year in which Loewi's first paper appeared, Cannon and Uridil (1921) found that stimulation of the hepatic nerves caused the liberation into the blood of a substance that accelerated the denervated heart.[21] In more recent years Cannon and others have investigated, in animals with intact circulation, the mechanism by which various other sympathetic nerves exert their effects. They employed the heart and many other tissues as test-objects. Cocaine was used in many of the experiments, because of its action in sensitizing tissues both to adrenaline and to the substance released by sympathetic nerves (Rosenblueth and Schlossberg, 1931). (See page 158 for the mechanism of sensitization by cocaine.) The released substances were able to accelerate the denervated heart, raise the blood sugar and evoke a flow of saliva from the denervated submaxillary gland (Canon and Bacq, 1931). They could also cause retraction of the denervated nictitating membrane (Rosenblueth and Cannon, 1932), diminish limb and spleen volume, and inhibit the non-pregnant uterus of the cat (Cannon and Rosenblueth, 1933) as well as the intestine (Bacq, 1933b). Details of the methods employed will be found in the papers cited.

2. Heart

Frog heart. Mention has already been made (see page 124 *et seq.*) of the fact that Loewi (1921) was able to detect the release by the isolated frog heart of an *Acceleransstoff* that appeared under conditions in which stimulation of the vagosympathetic increased the heart's force or frequency, and that could itself increase the force and frequency of a second heart. The fibres in the vagus trunk that were responsible for this action are true sympathetic fibres, which join the vagus near its exit from the skull (Gaskell, 1884; Langley and Orbeli, 1910).

The effect of *Acceleransstoff* on the heart can be demonstrated by the method that Loewi used for *Vagusstoff*, namely to transfer the fluid from a heart isolated by Straub's method to another heart prepared in the same way. The effect is clearest if atropine is used to abolish the opposing action of *Vagusstoff* (Loewi, 1921). Kahn's double cannula, already mentioned in connection with the cholinergic action of the frog vagus, can also serve for detecting the transference of the augmentor effects (Kahn, 1926). Lanz (1928), instead of stimulating the whole vagosympathetic trunk, followed a recommendation of ten Cate (1922) and stimulated the sympathetic chain at a point where its fibres had not yet joined the vagus.

The known physical properties of *Acceleransstoff* are identical with those of adrenaline. It is destroyed by ashing, and by ultra-violet irradiation (Loewi, 1926). Extracts of frog heart have an augmentor action on another heart if atropine is present, and it is likely that this action is mainly due to the substance that is liberated by sympathetic stimulation. The active material in extracts is destroyed by ultra-violet light, but not by heating to 56 °C for 20 minutes. Simple extracts of heart also contain some agent that slowly destroys the augmentor substance during autolysis. This destructive agent is itself destroyed by heat (Loewi and Navratil, 1926a).

The substance released by sympathetic nerves in the frog heart also has inhibitory actions like those of adrenaline. Thus Brinkman and van Dam (1922) tested the fluid from a frog heart on a perfused frog stomach, and observed inhibition when the cardiac effect of vagosympathetic stimulation had been augmentor. Lanz (1928b) obtained the same effect when he stimulated the sympathetic nerve before it joined the vagus; in his tests the active substance was destroyed by boiling. Lanz also showed that his active solutions shared with adrenaline the property of producing vaso-constriction in the perfused vessels of the frog (cf. also Tschannen, 1933).

Külz (1928) found that frog hearts isolated by Straub's method con-tinuously liberated some substance that inhibited the stomach and rectum of the frog; but he was unable to detect any increase in its rate of liberation when the sympathetic was stimulated.

The action on the heart of the substance liberated by the sympathetic nerves is weakened by ergotamine (Loewi, 1924a), but like that of adrenaline it cannot be entirely eliminated by that drug. Ergotamine does not prevent the release of the *Acceleransstoff*, it only opposes its action (Navratil, 1927; see also page 161).

Mammalian heart. The sympathetic nerves of the mammalian heart are likewise adrenergic. Rylant (1927) perfused two rabbit hearts in series with the same saline solution, and found that stimulation of the sym-pathetic nerves to the first heart increased the frequency and amplitude of the beat of both hearts; both effects were abolished by ergotamine. Later he and Demoor (1927) obtained similar effects on blood-perfused cat hearts. Blood apparently contained an agent that rapidly destroyed the sympathetic transmitter.

Jendrassik (1924) showed that stimulation of the sympathetic nerve supply of an isolated rabbit heart released a material that is inhibitory to the rabbit intestine.

Cannon and Rosenblueth (1933) found that when they stimulated the cardioaccelerator fibres arising in the cat stellate ganglion, a substance or substances entered the circulation whose effect was to cause withdrawal of the denervated nictitating membrane and relaxation of the denervated non-pregnant uterus. A brief stimulation produced only the stimulant action on the nictitating membrane, with no inhibition of the uterus. If the cat was then given a dose of adrenaline which had about the same effect on the nictitating membrane as the brief nerve stimulation, the uterus was now inhibited. Cannon and Rosenblueth considered that this experiment was evidence that the sympathetic nerves release relatively more sympathin E than sympathin I.

3. Liver

Stewart and Rogoff (1920) had found that stimulation of the splanchnic nerves accelerated the denervated heart even after removal of the supra-renals. Cannon and Uridil (1921) showed that this effect, and the accompanying rise in blood pressure, were both due to the release of some substance from the liver. The effect could be produced by stimulation of the hepatic nerves alone, and it was not prevented by occluding the portal vein and the vena cava below the liver. Cannon and Griffith (1922) also investigated the effect and found that they could duplicate it by re-inject-ing into the inferior vena cava blood that they had collected from the hepatic veins during stimulation. Later papers from Cannon's laboratory have emphasized that the substance from the liver can reproduce a number of the augmentor actions of adrenaline but cannot reproduce its inhibitory actions. Thus it caused a withdrawal of the nictitating mem-brane (Rosenblueth and Cannon, 1932), and a fall in the volume of the leg and the spleen; but it failed to inhibit the uterus of the non-pregnant cat; and after ergotoxine, when the actions of adrenaline were reversed, so that blood pressure fell and limb volume rose, the action of the substance from the liver was diminished or abolished, but never reversed (Cannon and Rosenblueth, 1933). These phenomena were explained on the basis that the hepatic nerves liberate sympathin E, but not sympathin I.

The fact that stimulation of the hepatic nerves liberates a substance whose augmentor effects are especially pronounced was shown with particular clarity in one of Cannon and Rosenblueth's (1933) experiments. They recorded simultaneously the movements of the nictitating membrane and of the non-pregnant uterus of a cat, and compared the effects of injecting adrenaline, stimulating the splanchnic nerves, and stimulating the hepatic nerves. Stimulation of the splanchnic nerves (the suprarenals having been removed) caused contraction of the nictitating membrane and inhibition of the uterus. The injection of adrenaline also produced these two effects; but stimulation of the hepatic nerves produced only contraction of the nictitating membrane without any inhibition of the uterus.

In another experiment the substance released into the circulation on stimulation of the hepatic nerves caused a fall in limb volume, while adrenaline caused a rise; the nictitating membrane responded with a small contraction to both treatments.

4. *Splanchnic nerves*

When the suprarenals have been removed, the principal effects of splanchnic stimulation are the augmentor effect on the blood vessels and the inhibitory effect on the smooth muscle of the gut. It can therefore be understood that under these conditions both augmentor and inhibitory effects can be produced on distant organs by splanchnic stimulation.

Gley and Quinquaud (1923) did a cross-circulation experiment on two dogs, and found that even after removal of the suprarenals a stimulation of the splanchnic nerves in one dog was followed by a rise in the blood pressure of the other dog. They did not, however, attribute the effect to the release of an active substance.

Later studies have shown that in the cat, splanchnic stimulation after removal of the suprarenals can lead under suitable conditions to quickening of the denervated heart (Cannon and Bacq, 1931), retraction of the denervated nictitating membrane and inhibition of the virgin uterus (Cannon and Rosenblueth, 1933).

Finkleman (1930) used isolated loops of cat intestine to demonstrate chemical transmission of inhibitory effects. He found that stimulation of the mesenteric nerves could produce either contraction or relaxation of the intestine; but after the loop had been in use for some time he observed only the inhibitory effect. He then suspended two intestinal loops in a warm chamber, where they were washed with Ringer's solution which ran first over the upper piece and then over the lower piece. When stimulation of the mesenteric nerves to the upper piece caused inhibition, that effect was transmitted by the Ringer's solution to the lower piece. The effect could be obtained no more often than once or twice with the same loop. Finkleman did not succeed in transmitting augmentor effects in this way.[22]

5. *Abdominal sympathetic chain*

When the abdominal sympathetic chains of the cat are transected at the level of the third lumbar ganglion and their peripheral ends are stimulated, substances with a pressor action might be liberated from the pilomotor and vasomotor supply to the tail, from the vasomotor supply to the hind limbs, and from a small area of the rump which receives pilomotor fibres. These different areas have usually been studied separately, but in some experiments all the fibres have been stimulated together. The latter was the case in the experiments of Rosenblueth and Schlossberg (1931), who showed that the rise of blood pressure that followed such stimulation was magnified by cocaine. Cannon and Rosenblueth (1933) found that after ergotoxine, stimulation of the lower abdominal sympathetic caused a rise followed by a fall of blood pressure, even when the injection of adrenaline caused a pure fall. They used this observation as an argument for their theory that stimulation of the lower abdominal sympathetic liberates two distinct substances, sympathin E and sympathin I.

Hind limb. Bacq and Brouha (1932) used a different approach. They showed that stimulation of the peripheral end of the severed sciatic nerve caused acceleration of the heart, which must have been due to sympathin carried to it by the circulation. This was because the effect was not produced by occluding the circulation of the limb, and so could not be ascribed to ischaemia. It was not due to stimulation of somatic motor fibres, because it could be elicited after these had degenerated; nor was it affected by curare. It was absent, however, when the sympathetic postganglionic fibres had degenerated. The sympathin must therefore have come from these fibres. The results are particularly interesting because the area from which the sympathin was liberated contains no chromaffin tissue.

Bacq and Brouha (1932) also showed that stimulation of the sciatic or of the lower abdominal sympathetic leads to inhibition of the intestine, dilatation of the pupil, and hyperglycaemia; and they attributed all these effects to the release of sympathin.

Sympathetic stimulation reduces the fatigue of the mechanism by which motor excitation is transmitted to striated muscle (Orbeli phenomenon). Corkill and Tiegs (1937) have examined this effect of sympathetic stimulation; they conclude that it is mediated by an adrenergic mechanism, because no nerves by which it might be directly transmitted can be detected histologically.

Tail. The sympathetic nerves to the cat tail contain many pilomotor fibres, and probably vasomotor fibres as well. Under suitable conditions stimulation of these nerves leads to the release into the blood of a substance or substances that accelerate the denervated heart, raise blood pressure, evoke secretion of saliva (Cannon and Bacq, 1931) and retract the nictitating membrane (Rosenblueth, 1932). Cannon and Bacq failed to see any inhibition of the intestine, a finding that Cannon and Rosenblueth explained by postulating that there was no release of sympathin I.

Bacq (1933a) scarified and cupped the shaven tail of a cat and tested the serous exudate. Normally this fluid acted like adrenaline on the denervated heart and pilomotor muscles of the cat; but in a cat whose sympathetic tail nerves had degenerated no such actions could be observed.
Bladder, uterus and colon. Cannon and Bacq (1931) found that stimulation of the sympathetic fibres to the bladder and the uterus or to the colon caused an increase in heart rate.

6. Cervical sympathetic
Bacq and Brouha (1932) and Bacq (1933a, b) found that stimulation of the cervical sympathetic in a rabbit or dog, but not in a cat, caused the liberation into the aqueous humour of a substance that acted like adrenaline on the toad heart and on the cat pilomotor musculature; there was no observable effect on the isolated intestine. The chemical evidence for the release of an adrenaline-like substance into the aqueous humour will be discussed below (see page 152).

These workers also found that stimulation of the cervical sympathetic causes release into the blood of a substance, presumably sympathin, that inhibits the virgin uterus and contracts the retractor penis; there was a suggestion also of contraction of the spleen.

Cattell, Wolff and Clark (1934) stimulated the cervical sympathetic after destruction of all its fibres except those that supply the salivary glands, and found that the stimulation led to the release of a substance that was probably sympathin. It contracted the nictitating membrane and caused generalized vasoconstriction. These effects, as well as the immediate secretory effect of the stimulation, were potentiated by cocaine.

7. The chemical nature of sympathin
The striking resemblance between the pharmacological effects of sympathin and those of adrenaline makes it likely that the two are closely related in chemical structure. This conclusion is reinforced by the fact that the degenerative section of sympathetic nerves sensitizes the tissues they supply to both substances. Nor is the conclusion weakened by the evidence

that there are two kinds of sympathin, E and I, since it is unlikely that these are very different in chemical composition.

The case based on the pharmacological evidence is supported by the chemical evidence. It has been shown that stimulation of adrenergic nerves leads to the release of a substance that gives a reddish-brown substance on treatment with oxidizing agents, just as adrenaline does.

Thus Lanz (1928) used Russman's reagent (mercuric iodide, potassium iodide and sulphanilic acid) to test the fluid obtained from an isolated frog heart during sympathetic stimulation. In some of his experiments a faint colour developed when he had concentrated the solution before testing it.

Bacq (1933a) used a similar reagent with certain modifications (Viale's test). With this test he could detect adrenaline in a concentration of 1:30000000, and his results were more convincing than Lanz's. He collected and examined the following kinds of fluid:

(1) He stimulated the cervical sympathetic on one side and then withdrew aqueous humour from both eyes. In his experiments on dogs the fluid from the stimulated side gave a stronger colour reaction than the fluid from the control side. In his experiments on rabbits the fluids were tested with ferric chloride and sulphanilic acid (von Baeyer's test); in 6 out of 10 experiments he obtained a stronger reaction on the stimulated side. Similar experiments on cats showed no difference in the fluids taken from the two eyes, as was to be expected from the physiological results.

(9) Blood samples from three cats were subjected to Viale's test: one cat was normal, one was suprarenalectomized, and one was deprived of all its sympathetic nerves. The results suggested that the colour reaction was weaker after loss of the suprarenals, and weaker still after sympathectomy; but the blood of all three cats gave a clear positive reaction.

(3) Extracts of endocardium gave a stronger reaction with Viale's reagent than extracts of myocardium. Degeneration of the sympathetic supply reduced the colour by about 50 per cent.

(4) Ringer's solution from the perfused frog heart was examined spectrographically with ultra-violet light. Stimulation of the vagosympathetic trunk increased the absorption in the general region of the spectrum where the absorption bands of aromatic compounds are located.

These chemical results suggest that sympathin is a catechol derivative with a side-chain. The side-chain is probably similar to that of adrenaline, but there is no chemical evidence on this point. It is possible that the solutions to which these tests were applied contained adrenaline, but adrenaline could not have been the only substance responsible for the reactions, because the physiological activity was much lower than the chemical tests had indicated. It should be kept in mind, however, that if the transmitter were indeed adrenaline, the fluid collected would probably contain the early products of its inactivation as well. It is also possible that in solutions of such low concentration there might be changes subsequent to collection.[23]

D. Chemical transmission in the central nervous system

Some of the phenomena of reflex activity in the spinal cord have led those who studied them to consider the possibility of chemical mechanisms, by which states of excitation or inhibition might persist beyond the duration of the impulses that gave rise to them (Sherrington, 1925; Fulton, 1926; Brücke, 1929). Dikshit (1934), for example, thinks that vagal impulses arriving in the medulla oblongata might be transmitted by acetylcholine. At this time, however, no evidence is as yet available that would justify speculations about chemical transmission at central synapses.

E. The action of drugs on chemical transmission

1. *General features of drug action*

The drugs that act on chemical transmission or simulate its effects may be classified as follows.[24]

(a) *Drugs with excitatory effects*

(i) *Drugs that have actions like those of specific chemical transmitters.* Acetylcholine, muscarine, pilocarpine and arecoline are examples of drugs whose actions resemble those of the substance liberated by the vagus. Adrenaline acts like the substance or substances liberated by the sympathetic. Such drugs do not potentiate the action of the nerve, but drug action and nerve action may be additive under some conditions; in some cases degeneration of the nerve makes the tissue more sensitive to the drug.

(ii) *Drugs that potentiate the action of the nerve.* These may act either by inhibiting the destruction of the transmitter, or by increasing the sensitivity of the effector organ to excitation. Eserine acts on cholinergically innervated tissues in the first of these ways, though possibly also in the second way to some extent. Cocaine has a potentiating effect on some adrenergic mechanisms, but its mode of action is unknown. Under special conditions adrenaline increases the sensitivity of mammalian voluntary muscle, both to drugs (after the muscle has been denervated) and to nerve stimulation.

(iii) *Drugs that act by liberating a specific transmitter.* It has not yet been conclusively proved that drugs can act in this manner but it is possible that tyramine has this effect on adrenergic nerves. On the other hand the findings of von Beznák (1934) on the heart, Brown and Feldberg (1935a) on the superior cervical ganglion of the cat, and Feldberg and Guimarãis (1935b) on the salivary gland all indicate that potassium can liberate acetylcholine. As to the significance of this release for the action of potassium, nothing can be said until the matter has been explored in greater detail.

(b) Drugs with inhibitory effects

(i) *Drugs that make the effector organ insensitive to the transmitter.* Examples of this type of action are the actions of atropine and ergotoxine on the vagosympathetic innervation of the frog heart (Loewi and Navratil, 1924; Navratil, 1927), of nicotine on the superior cervical ganglion (Feldberg and Vartiainen, 1934), and of curare on the junctions of motor nerves with voluntary muscles (Dale, Feldberg and Vogt, 1935). In the cases that have been studied the quantity of the active substance that is released by nerve stimulation is not affected.

(ii) *Drugs that prevent the transmitter from reaching the structures that are sensitive to it.* Either adrenaline or pituitrin can act in this way to modify the contracture produced in denervated voluntary muscles by the stimulation of vasodilator nerves (see page 140). This action may be due to circulatory changes (Dale and Gaddum, 1930).

(iii) *Drugs that prevent the release of the transmitter by nerve stimulation.* It is conceivable that a substance like cocaine may paralyse nerve fibres and their endings, so as to block the impulses that would normally liberate the transmitter and thus prevent its release, without necessarily affecting the sensitivity of the effector organ. There is also some evidence that the release of a chemical transmitter may be suppressed by fatigue (Dale and Gaddum, 1930; Rosenblueth, 1932; Orias, 1932), or by overheating (Fryer and Gellhorn, 1933), without impairing the responsiveness of the tissue to the transmitter when it is artificially applied.

2. The mode of action of individual drugs

(a) Eserine

The influence of eserine on the effects of other drugs. The actions of acetylcholine are practically always greatly increased after the injection of eserine. This phenomenon has been discussed in the section on the metabolism of acetylcholine (page 83 *et seq.*); it is presumably due to the inhibition of cholinesterase.

Eserine also increases the effects of other unstable esters of choline, but not the effects of other drugs (Loewi, 1912; Dale and Gaddum, 1930). There are, however, cases where the action of drugs that are not choline esters is somewhat strengthened by eserine. Thus it slightly strengthens the effect of barium on the leech (Fühner, 1918b, c; Vartiainen, 1933) and the effects of choline, potassium and hordenine methiodide on the superior cervical ganglion (Feldberg and Vartiainen, 1934). But in these cases the potentiation is small compared to that of choline esters.

The influence of eserine on the effects of nerve stimulation. Eserine increases the actions of cholinergic nerves wherever it has been tested (see Table 13). The increase is not always easy to demonstrate. In the case of the superior cervical ganglion, for example, it seemed at first to be absent; but later it could be shown that under suitable conditions the transmission of impulses through the ganglion is potentiated (see page 138).[25]

This effect of eserine is presumably due for the most part to its inhibition of the destruction of the choline ester released by the nerve; but it is not safe to assume that every nerve action that is increased by eserine is necessarily cholinergic. Nevertheless a very pronounced potentiation does speak strongly for a cholinergic mechanism. It has been found to have no effect on the action of adrenergic nerves (Heinekamp, 1925; Granberg, 1925), and there is no evidence that it can increase the action of any but cholinergic nerves.

When one does experiments on the effect of eserine on the response to nerve stimulation, it is advisable to make control injections of other drugs, in order to show that the observed potentiation is not due to an unspecific increase in the excitability of the innervated tissue. For example when Dale and Gaddum (1930) found that the contractures of denervated muscle elicited by the stimulation of vasodilator nerves were potentiated by eserine, they showed that the effect was a specific one, because in the same experiment the contracture produced by injecting tetramethylammonium salts was not potentiated.

One difficulty in experiments of this kind is that it is sometimes impossible to inject enough eserine to produce the desired increase of sensitivity without at the same time producing such strong general effects – circulatory disturbances for example – that the interpretation of the result is compromised. It is generally possible to inject 0.1–0.3 mg per kg without killing the animal, but much larger doses cannot be tolerated. This difficulty can readily be surmounted in cases where the response under study is not blocked by atropine: when an animal is first given atropine it is possible to give much greater amounts of eserine without producing serious general effects.

It is probable that some, if not all, of the consequences of administering eserine are due to inhibition of the ongoing destruction of acetylcholine

Table 13. *Nerves whose action has been shown to be potentiated by eserine*

Nerve	Tissue	Authors
Parasympathetic nerves		
Vagus	Heart	Winterberg (1907)
		Heinekamp (1925)
		Gibbs (1926)
		Eccles (1933b)
	Bronchi	Prevost and Saloz (1909)
		Dixon and Ransom (1912)
	Bronchi and lung vessels	von Euler (1932)
	Stomach	McSwiney (1933)
Oculomotor	Iris	Anderson (1905)
Pelvic	Bladder	
Chorda	Salivary gland	Loewi and Mansfeld (1910) and others
Nervi erigentes	Penis	Henderson and Roepke (1933a)
Sympathetic preganglionic nerves		
Splanchnic	Suprarenals	Feldberg, Minz and Tzudzimura (1934)
Cervical sympathetic	Superior cervical ganglion	Feldberg and Vartiainen (1934)
Sympathetic postganglionic nerves		
Secretory fibres	Sweat glands	Burn (1934)
Vasodilator fibres	Limb muscles	Bülbring and Burn (1934)
Fibres evoking contracture	Denervated striated muscle	Dale and Gaddum (1930)
		Hinsey and Cutting (1933)
Fibres evoking a contraction	Stomach	McSwiney (1933)
Other nerves		
Somatic motor	Striated muscle	Zucker (1913)

that is always occurring in the body. Anderson (1905) found that the constrictor action of eserine on the pupil disappeared after degeneration of the postganglionic ciliary nerves. This finding indicates that the drug's usual action depends on inhibition of the destruction of acetylcholine liberated by impulses spontaneously generated in the ciliary nerves, for section of the nerves in an acute experiment did prevent the eserine miosis. The explanation was offered that the trauma of the operation generated impulses that passed continuously down the ciliary nerves. The well-known direct action of eserine on organs such as the heart and the intestine is probably due to a similar phenomenon, and is related to the activity of the local parasympathetic ganglia.

(b) Adrenaline
(i) *Action on the contractures of denervated mammalian muscles*
Adrenaline has two opposed actions on the contractures of denervated mammalian voluntary muscle (Dale and Gaddum, 1930). If one injects acetylcholine or some similar drug at a time when there is adrenaline in the circulation, no contracture occurs. A similar inhibition of the contracture is produced by the pressor principle of the pituitary; its effect is antagonized by histamine. This effect is not caused by stoppage of the blood flow through the muscle; it is probably due to an action on the capillaries, which are made impermeable to acetylcholine.

This inhibitory action of adrenaline contrasts with another action, which is to increase the sensitivity of denervated muscle fibres to acetylcholine and other drugs. This action can be demonstrated by using an isolated piece of mammalian muscle denervated 10 days earlier, but it can occasionally be seen as a second phase of the action of adrenaline on a muscle with intact circulation. Perhaps this effect of adrenaline is somehow related to its ability to diminish fatigue of the mechanism responsible for transmitting motor nerve impulses to the muscle fibres (for discussion see Corkhill and Tiegs, 1933).

At one time it was thought that the action of adrenaline on contractures elicited by stimulating vasodilator nerves was different from its action on drug-induced contractures. The evidence on this point was re-examined by Dale and Gaddum (1930), who concluded that the action was essentially the same in both cases.

(ii) *Action on adrenergic nerves*
Burn (1932) found that stimulation of the abdominal sympathetic chain lost its effect on the vessels of the perfused hind limb of a dog, unless adrenaline was added to the perfusing blood. He suggested that the added

adrenaline was absorbed from the blood by a process associated with the endings of the adrenergic fibres, and that it was stored there and made available for release when the nerve was stimulated. A rather similar view was expressed by Cannon and Rosenblueth (1931).

(c) Cocaine

Cocaine increases some of the actions of adrenaline (Fröhlich and Loewi, 1910: Burn and Tainter, 1931; and others). This action is not seen when isolated smooth muscle is exposed to a cocaine solution; it seems to have more to do with the washing-out of cocaine than with its actual presence. Cocaine also sensitizes the smooth muscle of the nictitating membrane both to adrenaline released from the suprarenals and to sympathin present in the circulation (Rosenblueth and Schlossberg, 1931). This effect of cocaine is seen even when the muscle has already been sensitized by degenerative section of its sympathetic nerves. The effects of the two kinds of sensitization are additive.

It might be expected from these facts that cocaine would also increase the responses to the stimulation of adrenergic nerves. This expectation has been confirmed in the case of the sympathetic nerve supply to the cat nictitating membrane (Rosenblueth and Rioch, 1933).

The mechanism of this action of cocaine is still unknown.

(d) Tyramine

Tyramine has a number of effects that resemble those of sympathetic nerve stimulation. It loses its effectiveness when these nerves degenerate. In the perfused hind limb the effect of tyramine and the effect of nerve stimulation are both increased by the addition of adrenaline to the perfusion fluid. On the basis of these observations it has been proposed that tyramine acts on the nerve endings themselves, in a way that promotes their regular function of releasing sympathin in response to nerve impulses (Burn, 1930; Burn and Tainter, 1931).

(e) Ephedrine

Ephedrine resembles tyramine chemically, and in cats its mydriatic effect is lost when the sympathetic supply to the eye degenerates. It may act in the same manner as tyramine (Pak and Tang, 1933).

(f) Potassium

The only substance so far definitely known to release acetylcholine from tissues is potassium. The observation by von Beznák (1934) on the frog heart, that raising the hydrogen ion concentration of the medium also leads to the appearance of free acetylcholine, could not be confirmed in experiments on the perfused superior cervical ganglion of the cat (Brown and Feldberg, 1935b). Experiments with potassium have now been carried out on a number of organs, with eserine present to prevent the destruction of acetylcholine.

Von Beznák (1934) found that the frog heart released acetylcholine into the medium when this was calcium-free and its potassium concentration was raised. Omission of calcium from the Ringer's solution was by itself an effective stimulus for the release. Brown and Feldberg (1935a) have made a more detailed study of acetylcholine release by potassium on the perfused superior cervical ganglion of the cat. When they tripled the potassium concentration of the perfusion fluid the change was sufficient to produce a detectable release; and when they injected 2–5 mg of KCl in 0.5 cc of medium into the inflow tubing, the first cc of effluent contained 0.02–0.1 µg of acetylcholine. No acetylcholine was released by the injection of a high-calcium medium; calcium in fact diminished the release of acetylcholine by potassium.[26]

Little or no acetylcholine was released by potassium from a ganglion whose preganglionic fibres had been severed and allowed to degenerate.[27]

Feldberg and Guimarãis (1935b) have detected the release of acetylcholine from the sweat glands of the pads of the feet and from the tongue in perfusion experiments on cats, and from the submaxillary glands of cats and dogs with maintenance of normal blood flow. In their experiments on the salivary glands of eserine-treated dogs, they injected 4–7 mg of KCl into the artery of the gland (the NaCl of the solution was reduced to preserve isotonicity), and collected the venous effluent and tested it for acetylcholine. It produced a contraction of the leech muscle (having been diluted with potassium-free salt solution to avoid potassium effects on the leech) and it lowered the blood pressure of the cat; the latter effect was abolished by atropine. The active substance was destroyed by alkali. In the experiments on the salivary glands of cats, it was observed that the injection of 4–5 mg of KCl into the artery of the gland normally caused no change of blood pressure, but after eserine caused a sharp fall of pressure which was eliminated by atropine.

The release of acetylcholine by potassium is of theoretical interest, because it is widely accepted that the conduction of impulses by nerve is associated with a mobilization of potassium, or even that this is the basic physico-chemical process underlying excitation.

It should be mentioned that the pharmacological actions of potassium may not depend, or may depend only in part, on the acetylcholine it liberates (see Brown and Feldberg, 1935). This matter requires further study before it can be stated how much of the action of potassium is due to acetylcholine.[28]

(g) Atropine

Small doses of atropine completely inhibit all the muscarine actions of acetylcholine. In many cases they also inhibit the actions of cholinergic nerves. The simplest explanation of the latter action is that atropine prevents the choline ester released by the nerve from exerting its effect. That this is its only mode of action has been shown by experiments on the heart and on the salivary glands.

That the release of the active choline ester by nerve stimulation is unaffected by small doses of atropine was first shown by Loewi and Navratil (1924). They set up two frog heart preparations in the usual way, and when they added atropine to the first heart in a concentration that just sufficed to abolish vagal inhibition, they found that they could still demonstrate the release of *Vagusstoff* when they stimulated the vagus to the first heart and transferred its medium to the second heart. Under these conditions some of the atropine disappeared from the fluid in contact with the first heart, so that its concentration was not high enough to inhibit the action of the *Vagusstoff* on the second heart. Similar results were obtained by Rylant and Demoor (1927) in experiments on the cat heart and by Henderson and Roepke (1933b) in experiments on the salivary glands.

The dosage of atropine necessary to abolish the effect of nerve stimulation varies greatly from tissue to tissue. The vagus fibres to the bronchi are paralysed by very small doses, and the actions of the nerves to the heart, the iris and most glands are also easily paralysed. The action of the vagal motor fibres to the stomach, on the other hand, though it is weakened by atropine, is not completely antagonized even by very large doses. The motor effects of the vagus on the intestine, and the effects of the pelvic nerves on the bladder, the uterus and the retractor penis are even more resistant to atropine. As to vasodilator fibres, those in the chorda, the nervi erigentes and the dorsal roots are little if at all affected by atropine, while those in the sympathetic nerves to the limbs are put out of action (see page 142). In spite of these variations, however, the actions of muscarine-like drugs on all these tissues are readily abolished by small doses of atropine (Heffter, 1924).

It is unlikely that fundamentally different mechanisms are involved in these different cases. Since atropine does not paralyse any aspect of the vagal action on intestinal smooth muscle, its effect on the acetylcholine released in that tissue must be that it prevents it from reaching its site of action. The other anti-acetylcholine actions of atropine probably involve a similar mechanism. The effect of atropine on the cardiac vagal response is not due to interference with the formation of the *Vagusstoff*, but to interference with its reaching its site of action. The variety of its effects on the different nerve actions becomes comprehensible if it be assumed that the route by which the chemical transmitter reaches its site of action is in some cases one that is blocked by atropine, while in other cases it is more direct and allows no opportunity for atropine to interfere.[29]

Unless one adopts some such theory, one is obliged to suppose that chemical transmission occurs only at the terminations of nerves whose action is blocked by atropine, or else to postulate a whole series of chemical transmitters, whose actions are liable in different degrees to the antagonism of atropine.[30]

The action of eserine speaks so clearly for the identity of the parasympathetic transmitter with an unstable choline ester that the resistance of some of the transmitter's effects to atropine cannot throw doubt on the concept. It should, further, be stressed that if the atropine evidence is regarded as critical, every other choline ester will have to be eliminated as a possible transmitter.

(h) *Hypnotics*

Moraeus (1928) found that paraldehyde, chloralose and amylenechloral increased the sensitivity of the frog heart to vagal stimulation and to acetylcholine. According to Plattner and Hou (1930) this effect is partly due to the fact that the drugs depress the heart and make it more sensitive to inhibitory influences. In the case of paraldehyde another factor is involved, for paraldehyde inhibits cholinesterase.

(i) *Ergotoxine and ergotamine*

Navratil (1927) has shown that ergotamine, in sufficient concentration, abolishes the augmentor effect of the sympathetic on the frog heart without preventing the liberation of *Acceleransstoff*.

Cannon and Bacq (1931) confirmed this finding in experiments on cats, in which the release of blood-borne sympathin was detected. They found that the first effect of a small dose of ergotoxine or ergotamine was actually to increase the observed effect of sympathin on the heart. Similar observations were made in experiments in which other organs, for example the retractor penis, were used as indicators of released sympathin (Bacq, 1933a). These observations are still unexplained.

(j) Dioxane derivatives

A number of authors have reported that certain dioxane derivatives antagonize the actions of adrenaline but not those of the adrenergic nerves (for references see Vleeschhouwer, 1935 and Bacq and Bovet, 1935). These findings seem to offer a kind of analogy to those made with atropine and certain cholinergic nerves, whose actions are unaffected by atropine though atropine abolishes the corresponding actions of acetylcholine. (The problem has been discussed in detail above: see page 160.) The experiments with the dioxane derivatives have also been used to support the argument that adrenergic nerves do not act exclusively through the release of adrenaline.

The dioxane derivatives also have a depressor action on the vasomotor centres in the central nervous system.

F. Active substances in extracts

1. *The effect of nerve stimulation*

It is not yet known whether the active substances are set free from a preformed store, or whether they are synthesized from inactive precursors as a direct consequence of the arrival of nerve impulses. In the hope of answering this question, a number of investigators have made extracts of tissues whose nerves had been stimulated, and have compared them pharmacologically with extracts of unstimulated tissues. The results have been conflicting. It is at least clear that in such experiments nerve stimulation has no marked or consistent effect.[31]

(a) Vagusstoff *in heart extracts*

The first attempt to study the effect of vagal stimulation on the amount of a *Vagusstoff* in heart extracts was made by Dixon (1906, 1907). He removed the heart of a dog while it was under vagal inhibition, and made a concentrated and partly purified extract. This had an inhibitory action when it was applied to a beating frog heart, and the action was abolished by atropine. Dixon's extract could not have contained the labile substance that we now recognize as the transmitter of vagal actions; its effects were probably due to choline.

Witanowski (1925) examined the effect of alcoholic extracts of the frog heart when tested on other frog hearts. His finding was that vagal stimulation increased the level of extractable *Vagusstoff* by 100 to 500 per cent. Plattner (1926) made alcoholic extracts of small pieces of ventricular muscle from the hearts of dogs, rabbits and cats and tested them on the frog heart; his results agreed with Witanowski's. Engelhart (1930) did similar experiments with extracts of rabbit auricle and ventricle. He found that vagal stimulation increased the activity of his auricular extracts, but at first he generally found that his ventricular extracts were inactive. This negative finding turned out to be due to small differences between his and Plattner's technique. When the two workers did experiments together they found that extracts of auricle were indeed much more active than extracts of ventricle, but that vagal stimulation did increase the *Vagusstoff* content of both chambers by 100 to 500 per cent.

These findings are open to the criticism discussed above in connection with the detection of acetylcholine in tissue extracts, namely that the frog heart is a very unspecific pharmacological reagent. During vagal stimulation the isolated frog heart releases both *Vagusstoff* and *Acceleransstoff* into its fluid medium, and the possibility seems not to have been excluded that the extracts too will contain both substances, and that the observed effect might be due to a decrease of *Acceleransstoff* rather than to an increase of *Vagusstoff*.

The basic experiment has been repeated by Vartiainen (1934); he used trichloroacetic acid for extraction and the leech muscle as his test-object. He could find no support for the idea that vagal stimulation raises the acetylcholine content of either the frog heart or the rabbit auricle. There can be no doubt that his methods would give a truer reflection of the tissue's total acetylcholine level than the methods of Plattner and Engelhart. The significance of the effect obtained by the latter authors is still unknown; Vartiainen failed to confirm their findings even when he used their methods, and von Beznák (1934) was equally unsuccessful.

(b) Eye

Engelhart (1931) confirmed the observation of Plattner and Hintner (1930b) that extracts of iris contain a material that resembles acetylcholine (cf. also Velhagen, 1932). The amount of the material present in the iris and the ciliary body of the cat was unchanged by illumination of the eye, but the results of two experiments suggested that the activity of the extract was increased by stimulation of the oculomotor nerve. In these experiments the result was complicated by the fact that the two eyes were not removed simultaneously. In order to exclude reflex effects the control eye was taken out before the nerve to the other eye was stimulated. When both eyes were removed simultaneously no difference could be found between the two extracts, even when the nerve had been divided centrally before the stimulation was applied.

(c) Salivary glands

Henderson and Roepke (1932, 1933b) found that extracts of a dog's salivary glands contained a substance that acted on the frog heart like acetylcholine. Chang and Gaddum (1933) used the frog rectus abdominis as test-object and obtained the same result. Neither team found that stimulation of the chorda made any difference to the activity of their extracts.

(d) Skeletal muscle

Plattner (1932, 1933) observed an increase in the acetylcholine equivalent of extracts of frog skeletal muscles when he stimulated their mixed nerves.

(e) Suprarenals

Plattner (1934) found no alteration of the acetylcholine equivalent of the suprarenals as a result of stimulating the splanchnic nerves.

(f) Blood vessels

Gollwitzer-Meier and Otte (1933) found that a vasodilatation elicited by nerve stimulation apparently doubled the acetylcholine equivalent of extracts of the blood vessels.

2. The effect of degenerative section of nerves

If the acetylcholine or sympathin detected in extracts of tissues were derived entirely from the nerves that liberate these substances, it would be expected that degeneration of these nerves would lead to the complete disappearance of the substances from the extracts. The information available on this point is not very extensive. It indicates that there is usually some loss of activity but that the active substance does not disappear completely.

Thus Engelhart (1931) found that two weeks after section of the postganglionic sympathetic nerves the acetylcholine content of the rabbit iris was markedly reduced, while that of the cat iris was at an undetectable level.

Chang and Gaddum (1933) severed the chorda in dogs and found that the acetylcholine content of the submaxillary gland fell by about 50 per cent within 24 hours, a change that was too rapid to have been due to degeneration of nerve fibres. Von Beznák (1934) found no evidence of such an effect. It must be remembered that in experiments of this sort the submaxillary ganglion was not removed, so that it is most unlikely that the parasympathetic fibres would all have degenerated.

Bacq (1933a) found that after degenerative section of the sympathetic supply to the heart the substance responsible for the Viale colour reaction for adrenaline was reduced by about half. The experiments of Bacq quoted on page 151 require a similar interpretation: the fluid that oozed from the scarified tail of the cat exhibited adrenaline-like activity only when the sympathetic fibres to the tail were intact.

Quite recently Brown and Feldberg (1935a) have shown that 3–5 weeks after section of the cervical sympathetic there is a loss of 80 to 90 per cent in the acetylcholine content of the superior cervical ganglion.

G. The release of active substances from nerve trunks

In the experiments of Gaddum and Khayyal (1935), various nerves of the cat were removed and immersed in Locke's solution. Stimulation of cholinergic nerves such as the cervical vagus or the cervical sympathetic released acetylcholine or a substance with similar activity. Nerves that contained sympathetic fibres released a substance that acted on the frog heart like adrenaline. The amounts of the active substances detected in these experiments were minute; nevertheless the results demonstrate that the axon shares some of the special properties of the nerve ending where the chemical transmitter is set free. The same conclusion follows from the experiments on the regeneration of transected nerves described in the next section.

Binet and Minz (1934) obtained results that are somewhat reminiscent of those just described, but they ascribed them to the release of a substance that sensitizes the leech to acetylcholine. Cowan (1934) found that nerves from *Maia* release potassium when they are stimulated. The results of Binet and Minz could be ascribed to liberation of potassium, which does make the leech more sensitive to acetylcholine, except for one discordant observation that they made: their material was destroyed by bubbling air through a solution containing it. The findings of Gaddum and Khayyal, on the other hand, cannot be accounted for as solely due to the release of potassium.

H. Nerve regeneration experiments

Between 1897 and 1904 Langley, alone or with Anderson, carried out a series of experiments on cats, in which the central end of a transected nerve was united with the peripheral end of a second transected nerve, and an opportunity was given for the fibres to regenerate. In some of the cases there was development of a functional connection, so that stimulating the central end of one nerve evoked a response from the effector organ of the other nerve. In other cases no such connection could be formed. Anderson and Langley found that nerves could be classified on that basis, and that all the nerves of one class could replace one another reciprocally, while nerves of different classes could not do so. Later it became apparent that their classification corresponded to our present classification of nerves as cholinergic and adrenergic. Cholinergic nerves can substitute only for other cholinergic nerves, adrenergic nerves only for adrenergic ones (Dale, 1935a, b). The following paragraphs summarize the results of Langley and Anderson in terms of our current understanding.

(1) *Nerves that contain cholinergic fibres can assume new cholinergic functions*

After regeneration, fibres from the vagus, from the chorda-lingual, from the phrenic, or from the fifth cervical nerve can all take over the functions of the cervical sympathetic trunk and stimulate the superior cervical ganglion (Langley, 1898; Langley and Anderson, 1904a). Fibres from the cervical sympathetic can regenerate and act on voluntary muscles (diaphragm, sternomastoid, vocal cords: Langley and Anderson, 1904b). These findings can be explained on the basis of the theory discussed above (see page 143), according to which the motor fibres to voluntary muscles are cholinergic. It might be expected that after extirpation of the ciliary ganglion followed by regeneration, stimulation of the preganglionic fibres would exert a directly stimulant effect on the iris. This apparently does not happen; but Anderson (1905) found that while the pupil failed to respond to eserine for some time after removal of the ganglion, the reaction did reappear later on, after the preganglionic fibres had regenerated into the postganglionic trunk. This phenomenon can be explained on the basis that the preganglionic fibres were taking over the function of the postganglionic fibres, to the extent that the iris was made responsive to eserine. It appears, though, that while the regenerated preganglionic fibres are able to release acetylcholine in the neighbourhood of the iris's smooth muscle cells, they never develop the close contact with the contractile mechanism that is apparently required for full physiological control of the muscle.

(2) *Nerves that contain few or no cholinergic fibres cannot assume cholinergic functions*

When they regenerate, postganglionic fibres from the superior cervical ganglion do not acquire an action on the superior cervical ganglion of the opposite side or on the voluntary muscles of the tongue (Langley and Anderson, 1904b; two experiments only).

(3) *Nerves that contain adrenergic fibres can assume new adrenergic functions*

After section of the postganglionic fibres originating in the superior cervical ganglion, the fibres regenerate, but the individual fibres do not necessarily perform the same functions that they had before the operation. This was demonstrated by examining the spinal origin of the fibres responsible for such effects as dilatation of the pupils, erection of the hair, etc. After regeneration the distribution of these fibres failed to correspond to the original pattern (Langley, 1897). Postganglionic fibres from the superior cervical ganglion can sprout into the hypoglossal and induce vasoconstriction in the tongue (Langley and Anderson, 1904b).

(4) *Nerves that contain no adrenergic fibres cannot assume adrenergic functions*

After removal of the superior cervical ganglion, the preganglionic fibres appear to regenerate into the postganglionic trunk, but they never elicit typical adrenergic effects (Langley and Anderson, 1904b).

(5) *Sensory fibres cannot assume cholinergic functions*

The following fibres failed to form a functional union with the superior cervical ganglion:

(a) fibres that run centrally from the nodose ganglion of the vagus;

(b) fibres in the proximal stump of the severed great auricular nerve, as tested by antidromic stimulation (Langley and Anderson, 1904b).

These regeneration experiments therefore demonstrate that the release of an active substance is a function of the neuron itself, and not that of some special tissue that is related to the nerve ending. This idea also finds confirmation in the absence of any histological evidence for the existence of such an organ.

VIII

The release of active substances in other tissues

A. Introduction

This chapter offers a brief survey of the present state of our knowledge concerning the local regulation of the circulation through the release of active substances by tissue elements other than nerves. In recent years this field has been explored by a number of monographs and articles, most of whose conclusions are still valid (see especially Lewis, 1927; Krogh, 1929: Feldberg and Schilf, 1930). It seems unnecessary, therefore, to discuss in great detail the questions of current interest; however repeated reference will be made to reports that offer a fuller account of them.[1]

B. The local chemical regulation of the circulation

In every small blood vessel the flow is precisely adapted to the needs of the tissue it supplies. The vessels of an active tissue dilate; when the activity ceases the vessels constrict, so that there is no wasteful flow through tissues which do not require it. These facts, long familiar, have been vividly illustrated in the book by Krogh (1929). Krogh kept the smallest vessels of skeletal muscles and other tissues under observation for long periods, and described how their patterns of flow are continually changing. In a resting tissue at any moment many of the capillaries are completely closed, but some of them keep opening and closing, giving an observer the impression that each is individually regulated. When a tissue becomes active each capillary stays open much longer, so that more capillaries are open at once.

The phenomena of reactive hyperaemia show the same mechanism under rather different conditions. If the circulation to a limb is shut off for seconds or minutes, and then restored, it is found that the anaemia has provoked a local dilatation of the vessels, which is evident after release of the circulation as an increase in limb volume. The longer the arrest of the flow, the greater and more enduring is the compensatory dilatation. If the limb is warmed the effect is larger, but not longer-lasting. The dilatation appears to involve both arterioles and capillaries. Some dilatation has been detected after only 5 seconds' stoppage of flow (Roy and Graham Brown, 1879; Dale and Richards, 1918; Lewis, 1927; Krogh, 1929).

The conditions of these experiments cannot really be thought of as artificial; what the observations reveal is a physiological mechanism. There is no reason to doubt that the mechanism that causes an individual capillary to open after it has been closed for some time is the same as the mechanism that causes all the vessels to dilate after they have all been constricted. With increasing activity of the tissue the blood flow increases; but nothing warrants the supposition that the mechanism that adjusts flow to need in an active tissue is different from the one that controls flow in a resting tissue.

When the motor nerve to a skeletal muscle is stimulated, the vasodilatation is not seen until the contraction is over. It seems that the contraction itself makes the muscle anaemic by its direct mechanical effect. The blood that is in the muscle at the outset is partly squeezed out into the veins and partly squeezed back toward the heart (Anrep, Blalock and Samaan, 1934a; Anrep, Cerqua and Samaan, 1934b).

The regulation of blood flow is partly nervous and partly chemical. Nervous influences such as the Loven reflexes (see Bayliss, 1908) and axon reflexes (Bruce, 1910; Lewis, 1927) produce effects over fairly wide areas, and probably control the tone of small arteries. The most direct and important action of the chemical factors, on the other hand, is probably on the capillaries, because these are in more intimate contact with the tissues than vessels of any other type.

The theory that vasodilatation in active tissues results from metabolites formed by the activity remains an attractive one. First clearly stated by Gaskell (1880), it provides a simple explanation of Krogh's microscopic observations and also of reactive hyperaemia. For one can suppose that even in resting tissues metabolites slowly accumulate around closed capillaries, until finally these vessels open and the metabolites are washed away.

Since acid metabolites such as carbon dioxide are the most important end-products of metabolism, it has often been suggested that it is really the rise in hydrogen ion concentration that dilates the vessels. Gaskell showed that lactic acid is a vasodilator, and Bayliss (1901b) showed that carbon dioxide had the same effect. Their observations have been confirmed and extended by Schwarz and Lemberger (1911) and by Hooker (1912). Vasodilator effects have been seen with carbon dioxide and with stronger acids, but not with very weak acids like glycine and alanine. That means that only acids strong enough to liberate carbon dioxide will be effective, but of course it does not prove that the dilatation is due to carbon dioxide.

The rise of hydrogen ion concentration that develops during local anaemia has been directly observed by Rous and Drury (1929) with the aid of indicators, so there is every reason to believe that under certain conditions blood flow may be regulated by changes in hydrogen ion concentration. There are also findings that indicate that oxygen lack, rather than the accumulation of acid products, can sometimes be the factor responsible for vasodilatation (Krogh, 1929).

The discovery in recent years of a number of potent vasodilator substances in the tissues, however, makes it likely that the local chemical control of the circulation is more complex than was at first supposed. It is difficult to believe that these potent substances play no part in the local regulation, and there has been much discussion about which of them plays the most important part. Whenever a new substance has been discovered there has been a tendency to assert that its release is entirely responsible for the vasodilatation of activity or injury, and for many other responses as well.[2] What is more likely is that several substances are released and that their relative contributions vary with the type of response.

In the reactive hyperaemia of the cat both arterioles and capillaries dilate (Dale and Richards, 1918). Histamine might be the substance whose accumulation opens the capillaries in such experiments, and acetylcholine or adenosine the substance that opens the arterioles. But no single known tissue component could be responsible for both effects.

Barsoum and Gaddum (1935b) have found that temporary occlusion of the circulation in the limb of a dog caused the release of a substance that they could identify pharmacologically as histamine. Anrep and Barsoum (1935) showed that stimulation of the mixed nerve to a dog's gastrocnemius also released histamine. Such observations do suggest that histamine acts as a capillary dilator in reactive hyperaemia and plays a part in the regulation of vessel calibre. But histamine is probably not the only active substance to be liberated.[3]

Adenylic acid is intimately involved in muscle metabolism. Muscular activity can lead to the formation of adenylic acid from adenosine triphosphate, but this conversion should involve a loss rather than a gain of vasodilator activity. But the details of the process are still far from clear; perhaps further research will show that the adenylic acid of resting muscle forms a complex with some even larger molecule that is inactive or indiffusible, and that muscular activity can liberate diffusible active groups from such a complex. Rigler's (1932) experiments show that a vasodilator substance does become free, but there is no experimental proof that it is adenylic acid.

Certain facts suggest that kallikrein circulates in blood as part of an inactive complex, from which it is liberated when the reaction becomes acid (see pages 118 and 119). The view has been put forward that this chemical conversion plays a part under physiological conditions in the control of blood flow, but the concept has not received convincing support and is unlikely to be valid.[4]

From the above discussion it will be clear that the evidence for the liberation of specific vasodilator substances under physiological conditions is still incomplete. The theory that these specific substances adjust blood flow to the metabolic needs of the tissues seems to be based mainly on the teleological argument that it is hard to imagine that such active substances should exist in the tissues without having a physiological function.

The mechanism through which these substances are set free is still unknown. Lewis (1927) has pointed out that the phenomena of reactive hyperaemia suggest that some vasodilator substance is normally always being produced by tissue metabolism and washed away by the blood, so that it does not accumulate until the circulation is stopped. Yet there is no evidence that any one of the specific vasodilators is actually being continually produced in this way.

Although the discovery of new vasodilator substances seems at first sight to have confused the original simple picture of regulation by acid metabolites, the two sets of facts can be reconciled if we suppose that the liberation of vasodilator bases is a secondary consequence of the accumulation of acid. Some observations by Barsoum and Gaddum (1935b) may indicate that the release of histamine is regulated in this manner. According to Zipf (1927), tissues generally take up bases from the alkaline solution and set them free when the solution is made acid. There is therefore experimental support for the notion that accumulation of acid metabolites should lead to the release of bases, including the vasodilator bases. Since in most tissues vasodilator bases predominate over pressor bases, this reaction might provide a link between the older theories that emphasize acid metabolites and the newer ones that are based on the discovery of specific vasodilator substances.

C. The effect of mild skin injury

Only a brief summary will be given here of the observations that have been made on the responses of human skin to mild injury. A full account, with references to the original work, has been given by Lewis (1927). The following brief description has been taken from his book.

1. The 'white reaction'

If a blunt point is drawn with slight pressure over the warm skin of the forearm or the back, the skin blanches as the point travels over it. The blanching is caused by a simple and transitory displacement of blood from the superficial vessels of the skin, and the natural skin colour returns almost immediately. This first white reaction is followed by a second white reaction, which reaches a maximum in one-half to one minute. The second white reaction also occurs in an arm whose circulation has been stopped, so it is due to an active contraction of the small blood vessels that give the skin its colour, the subpapillary venous plexus; a contraction of small arterioles and capillaries has also been directly observed. This white reaction is apparently a direct response of the small blood vessels to stretching. There is no reason to suppose that any vasoactive substance is liberated.

2. The 'triple response'

When the pressure applied to the blunt point is heavier, the second white reaction does not occur, but is replaced by a local *red reaction*; and if the pressure is still further increased, a diffuse red *flare* appears, which is due to reflex dilatation of arterioles, in addition, local *oedema* develops. These three reactions constitute the 'triple response', which has already been described in detail as a pharmacological effect of histamine.

The triple response is the typical reaction of the skin to almost any form of injury. It has been produced by pricking, scratching, freezing or burning; by electric currents; or by the introduction of substances so varied as acids or alkalis, formaldehyde, silver nitrate, copper sulphate, morphine, atropine, chloroform, mustard oil, and many more. It appears after flea bites and nettle stings. The triple response can be shown especially clearly when the veins are occluded by a cuff placed on the upper arm. Under these conditions the whole forearm surface is cyanosed and forms an effective background for the reaction.

Histamine produces the triple response in lower concentration than any other pure substance that has been tested. From pharmacological experiments on the cat blood pressure there is evidence that extracts of skin, and especially of epidermis, contain histamine (Harris, 1927); the effect of such an extract in eliciting a triple response matches that of a histamine solution of equivalent strength.

The fact that such a complex response can be produced by so many different stimuli strongly suggests a common mode of action. There are many reasons for believing that all these stimuli liberate the substance that Lewis has called 'H-substance', which is almost certainly identical with histamine or with a compound of histamine. This histamine apparently exists preformed in the tissues, since it can be extracted by such mild processes as electrodialysis or treatment with alcohol.

Since in the absence of irritation or injury the histamine is present but inactive, it is to be supposed that it is normally bound in some way, about which we know nothing, within the protoplasm of the cells, especially those of the epidermis.

Lewis mentions a number of further observations that are readily explained by the theory of a released active substance, but would be hard to explain in terms of any other theory.

(1) The local red reaction, due to the introduction of histamine, first appears in about a minute and a half. During the next few minutes its area increases, which can be accounted for by the diffusion of the histamine.[5] The local red reaction due to mechanical, thermal or electrical irritation spreads in a precisely similar way, and the only ready explanation is that released H-substance is diffusing from the site of its liberation.

(2) When the circulation is interrupted ('occlusion test') by the application of a pneumatic cuff to the upper arm, the H-substance (or histamine) cannot be washed away by the blood as it otherwise would be, so its effects are prolonged. The following three observations may then be made:

(a) When the skin is irritated during circulatory arrest, the local red reaction persists at full intensity for at least 25 minutes instead of fading away completely during this period as it ordinarily does.

(b) The flare does not appear during circulatory arrest, so the occlusion test must be used in a modified form to show that the flare is due to a substance that is washed away by the blood. The circulation is arrested for 10 minutes, and two similar mechanical (or thermal or electrical) injuries are inflicted, one at the beginning and one at the end of the period. The circulation is now restored, and the two flares can be seen to develop and fade with the same time-course; but if the experiment is done without stopping the circulation, the first flare will have almost vanished before the second injury is inflicted. Another way of doing the experiment is to injure one spot and then apply a tourniquet in such a way that its upper edge presses firmly on the injured spot. One can then observe an abnormally persistent flare just above the tourniquet, in skin whose circulation is normal.[6]

(c) Wheal formation also does not occur during circulatory arrest. The wheal appears when the circulation is restored, but if the circulation is kept occluded for more than 5 minutes after the injury (or the injection of histamine) there is no whealing at all. If the arm is kept hot during the period of circulatory arrest, the whealing is diminished and if it is kept cold, the whealing is increased. The fact that whealing does not occur when the flow is stopped for many minutes cannot be attributed to removal of histamine by diffusion; for the observations just mentioned on the persistence of the local red reaction and the flare show that histamine is only carried away when the blood is actually flowing. The explanation here is that when histamine is present throughout the period of circulatory arrest, the capillaries lose their sensitivity to it, so that histamine no longer makes them permeable to the plasma of the blood. The insensitivity can be demonstrated by injecting a second dose of histamine into the same spot, and observing that it has practically no effect. This lack of responsiveness to histamine develops faster in a warm arm than in a cool one.

3. A general reaction to histamine

If the skin of a patient with urticaria factitia is subjected to even mild injury, so much histamine can enter the circulation that the cheeks become flushed, the skin temperature rises (Lewis and Harmer, 1926, 1927), and gastric juice is secreted (Kalk, 1929).[7]

D. The effect of irradiation on the skin

It is well known that when the skin is exposed to sunlight its vessels dilate. This response is elicited by radiation of wave-length about 300 mμ. It does not appear immediately, but develops after a latent period of about an hour. The reaction shows all the features of the triple response, which as already pointed out (see page 172) is the general reaction of the skin to irritation or injury, and which is apparently due to the release of H-substance. According to Lewis the skin's reaction to X-irradiation or exposure to radium is the same, except that the latent period is longer. Lewis believes that all these reactions to radiation involve the release of preformed H-substance, and differ from the reactions to mechanical injury only in the slower time-course of the release. The same sort of slow reaction can be produced by other agents, including bacterial poisons and dichloroethyl sulphide (mustard gas), and it too can be attributed to the same mechanism (Lewis, 1927).

Krogh (1929) has pointed out that it is hard to understand how the H-substance (histamine) can exist so long in such high concentration in the skin, since experiments with various kinds of stimulation show that it is readily diffusible. He proposes that these long-lasting responses involve the release not only of H-substance but also of indiffusible vasodilator materials (H-colloids).

Ellinger (1928, 1929, 1930), who had found that histidine can be converted by ultra-violet light into histamine (see page 19), holds the view that this reaction can occur in the skin and can account for the above-described effects on the vessels. This theory does not account for the latent period of the response, for the chemical reaction is a rapid one. Moreover the conversion of histidine to histamine is due to radiation of very short wave-length which could not penetrate the horny outer layers of the skin, and so should not have any physiological consequences. Irradiation of the skin with light whose wave-length is about 300 mμ does elicit the typical skin reaction; but such light does not convert histidine into histamine *in vitro* (see page 20). It is therefore probable that Ellinger's theory is incorrect, and that the reaction of the skin to irradiation depends, like its reaction to other irritants, on the release of preformed H-substance from damaged cells.

E. The effect of burns and scalds

Mild burns of the skin elicit the above-described 'triple response'. With more severe burns there is a considerable loss of fluid from the burnt surface. Harris (1927) found that scalding had no immediate effect on the histamine content of cat skin; however, with the development of oedema there was a significant fall in the amount of histamine that could be extracted from an area of skin whose boundaries had been marked out beforehand. These observations support the view that a scald does not cause the formation of new histamine; rather, it releases preformed histamine. Presumably this released histamine is washed away by the blood and lymph when oedema develops.

Bennet and Drury (1931) found that when the isolated heart of a cat was subjected to burn injury, substances with the pharmacological properties of both histamine and adenosine compounds were liberated. The conditions of these experiments were grossly unnatural, but the results do show that in a burn more than one substance can be set free and may well contribute to the shock that ensues.

There is thus no doubt that toxic substances may be released from burned tissues. But it remains doubtful that these are produced in sufficient quantity to account for the shock. Underhill and Fisk (1930) are of the opinion that the loss of fluid and chloride from the damaged surface is enough to explain the shock; Simonart (1930) is more inclined to support the toxaemic theory.

Burns are associated with congestion, and eventually with necrotic changes, in the suprarenals. Hartman, Rose and Smith (1926) have also observed increased secretion of adrenaline and a fall in the adrenaline content of the suprarenals. This might be a direct response to circulating histamine, but it could equally well be the consequence of the reflexes that are evoked by a fall of blood pressure however produced. Hyperchlorhydria and duodenal ulcers have both been observed after burn injuries, and they also might be due to histamine carried by the circulation.

Barsoum and Gaddum (1935c) have recently provided direct evidence for a rise in the blood histamine level of patients suffering from burns. The normal value is about 35 µg per litre, and about the fifth day after the burn the value rose to 100–200 µg per litre. This rise of blood histamine appears at a time when a 'toxic shock' is particularly liable to develop, and suggests some sort of relationship between the shock and the release of histamine. Nevertheless there is no direct evidence that the histamine levels reached in these patients would be able to cause shock.[8]

The treatment of burns with tannic acid (Davidson, 1925; Wilson, 1929) was introduced with the object of preventing the absorption of toxic substances into the circulation. But it is also possible that it acts by preventing fluid loss, or the painful excitation of sensory nerves.

F. Traumatic shock

The term 'traumatic shock' is applied to the profound collapse that sometimes appears several hours after extensive injuries, for example those from multiple gunshot wounds. It is primarily due to a reduction of the circulating blood volume (Keith, 1919). The blood pressure is low, the heart is rapid, and the stroke volume of the heart is diminished.

The evidence presented earlier leaves no doubt that histamine, or a substance with very similar properties, is released when skin is injured, and that when the damage is severe enough, this histamine can enter the general circulation and can contribute to the general symptomatology. But traumatic shock cannot be fully accounted for by a reaction of this sort. Three main theories have been proposed to explain the condition.

(1) *The neurogenic theory.* On the basis of studies begun in 1899, Crile (1921) came to the conclusion that shock is due to prolonged stimulation of sensory nerves, leading to exhaustion of the vasomotor centre. Crile showed experimentally that a similar state could be produced by direct stimulation of nerves. A similar theory has been proposed in recent years by Hoet (1929) and by Simonart (1930), who found that crushing of the limbs of a dog led to shock only when the nerves to the limbs were intact, whereas clamping the vessels to the limb did not influence the development of shock. These authors emphasize the difference between traumatic shock and shock due to burns. Their conclusion is that the former is mainly due to nervous stimulation, while the latter is due to the entry of toxic materials into the blood stream (cf. also the experiments of O'Shaughnessy and Slome, discussed on page 178).

The neurogenic theory may account for some of the manifestations of shock, but not for all. Shock has been produced by damaging a denervated limb (Cannon, 1923; Freedlander and Lenhart, 1932; Holt and Macdonald, 1934).

(2) *The toxaemic theory.* During the World War of 1914–1918 when traumatic shock was common, it was generally thought that the condition was due to the absorption of toxic substances. The problem is discussed in the papers and books by Dale (1919a) and Cannon (1919, 1923). The possibility that histamine was one of the toxic substances was widely considered, but there was nothing to suggest that it was the only substance involved. The similarity between histamine shock and traumatic shock was emphasized by Dale and Laidlaw (1919). It is probable that the primary cause of both conditions is a reduction of circulating blood volume. The erythrocyte concentration does not usually rise as high in traumatic shock as it does in histamine shock, but that could be attributed to the blood loss that is inseparable from severe trauma. Ether intensifies both traumatic shock and histamine shock, while nitrous oxide does not. Rats are susceptible to neither traumatic nor histamine shock (Voegtlin and Dyer, 1924).

There was experimental and clinical support for the toxaemic theory. It was observed that if the blood supply to the injured tissues of a badly wounded man were occluded for some time and then suddenly restored, the restoration gave rise to shock, which was attributed to the presence of toxic substances. Cannon and Bayliss (1919) repeated this as an experiment, using cats whose limbs were crushed. At the end of the experiment they removed and weighed both limbs to exclude the possibility that the shock could be due to the loss of blood into the injured tissues. They decided that it could not be, and concluded that the presence of toxic substances in the circulation must be a factor.

The toxaemic theory was severly criticized by Smith (1928). He collected the blood returning from a severely injured limb and was unable to find toxic substances in it. It had no depressor action, but raised the blood pressure to its level before the blood was removed from the animal. However when histamine was introduced into the wounded limb, its presence in the blood leaving the limb could easily be detected.

(3) *The blood-loss theory.* Having convinced himself that the toxaemic theory was inadequate, Smith was driven to conclude that shock of the type he had been examining must be due to the local loss of fluid into the injured limb. This view was supported by Blalock (1930), who repeated the experiments of Cannon and Bayliss. His guess was that they had amputated the limbs at too low a level before weighing them, so he adopted the technique of cutting across the body at the mid-abdominal level and splitting the lower part of the vertebral column. With this approach he found that whenever the trauma was severe enough to cause a substantial fall of blood pressure, the weight gain of the injured limb always corresponded to at least half the calculated blood volume. Holt and Macdonald (1934) have confirmed Blalock's results, and their paper gives references to the literature on this topic.

O'Shaughnessy and Slome (1935) have confirmed many of the findings discussed above, and have added some further ones. They discard the toxaemic theory for several reasons. Their most impressive piece of evidence came from experiments in which the circulating blood of a traumatized dog was continuously dialysed against the blood of another dog by means of a special device; the second dog failed to go into shock. O'Shaughnessy and Slome do stress the importance of nervous influences, for they found that animals under spinal anaesthesia did not develop typical shock. They also pointed out that interrupting the circulation of a traumatized limb by clamping its vessels not only prevented blood loss into the limb, but also anaesthetized its nerves.

There are clearly several factors that can contribute to the production of shock, and it is likely that the relative importance of each factor depends on the conditions of the experiment. Nervous influences are probably important for the immediate reaction, which is sometimes referred to as primary shock. Secondary shock, which takes some hours to develop, is probably due in the main to the local loss of fluid into the injured tissues. But there is also clear evidence from the experiments on skin injuries that trauma may involve the release of toxic substances; and the possibility that their entry into the circulation may be a contributing factor in some types of shock ought not to be forgotten.[9]

G. The anaphylactic symptom complex[10]

The detailed consideration of anaphylaxis as a problem in immunology is outside the scope of this book. The phenomenon concerns us here, because the different syndromes characterizing the anaphylactic shock, in the different species in which this reaction has been studied, show in each case a remarkable similarity to the symptoms produced, in the same species, by an intravenous injection of histamine; though the anaphylactic shock also includes certain symptoms which histamine does not produce. It is relevant to our subject to examine any evidence which can throw light on the meaning of these resemblances and these differences.

Originally, when Richet (Portier and Richet 1902) first described it in the dog, anaphylaxis was regarded as a paradoxical increase of the normal sensitiveness of an animal to a naturally toxic protein, such as can be obtained from sea anemones or mussels, produced by a previous, sub-lethal injection of the same poison. It was supposed to be the direct opposite of immunity. Later, after Theobald Smith (1906) had noticed the sensitiveness of guinea pigs to horse serum, produced by a previous small injection of that normally harmless product, the phenomenon was studied in great detail by a large number of observers (Otto, 1906; Rosenau and Anderson, 1905, 1907; and others); it became clear that the condition depended on the formation of an antibody, probably the ordinary precipitin, but that this was present in the blood in such small amount that it frequently could not be detected. (For earlier literature, see Doerr (1913), or, in much less detail, Dale (1919b, 1920b).) The experiments of Weil (1913) showed clearly, in fact, that when the antibody was artificially introduced, the anaphylactic condition developed as the antibody passed out of the blood, and reached its maximum intensity when no more remained in the circulation. A little earlier it was shown by Schultz (1910a, b, 1912) and then in much greater detail by Dale (1913a) that the isolated plain muscle of the anaphylactic guinea pig, freed as far as possible by perfusion from traces of blood and serum, had a specific and intense sensitiveness of reaction to the sensitizing antigen.

Meanwhile Biedl and Kraus (1910) had drawn attention to the fact that the widely different syndromes constituting the anaphylactic shock in the guinea pig and the dog were accurately reproduced by intravenous injection of Witte's peptone. The guinea pig, in either case, died of the acute pulmonary distension, due to spastic constriction of the bronchioles, which Auer and Lewis (1910) had first described in the anaphylactic shock of that species; the dog, on the other hand, responded to the sensitizing antigen, if anaphylactic, or to Witte's peptone, with an acute engorgement of the liver and the splanchnic area, general vasodilator collapse, and loss of coagulability of the blood. When the action of histamine was discovered, Dale and Laidlaw (1910) found that it too reproduced, with a striking fidelity, the main features of these contrasted syndromes in the two species; it failed, however to render the blood incoagulable, or even seriously to lengthen the clotting-time. For some years the anaphylactic reaction in the rabbit – a less regular phenomenon than in the other species – seemed to provide an exception to this general similarity to the effects of histamine. Dale and Laidlaw had described death from histamine in this species as due to failure of the right side of the heart, in consequence of an intense constriction of pulmonary vessels, whereas the heart of a rabbit dying of anaphylactic shock had been described as resembling one poisoned by digitalis. Coca (1914), however, who later investigated the phenomenon in the rabbit, attributed anaphylactic death to pulmonary constriction and right-sided heart failure; so the parallel was again restored. It might be thought to be completed by the further observation, that attempts to render the rat and mouse anaphylactic were unsuccessful, and that these same species were unaffected by doses of histamine which would cause fatal or urgent symptoms in guinea pig, dog, etc. This negative correspondence, however, must be considered also in connection with Kellaway's (1930) more recent observations, mentioned below.

When the action of histamine was first discovered, the prevalent theories of the anaphylactic reaction postulated the production, by some kind of interaction between antigen and antibody in the blood, of a new, non-specific toxic agent, the so-called 'anaphylatoxin'. It was easy, then, to suggest that the anaphylatoxin was histamine, or something very much like it. When, in due course, the evidence made it necessary to locate the reaction between antigen and antibody in the tissue cells such as those of the plain muscle coats in the guinea pig, or those of the liver in the dog, it was again easy to suggest that the formation of the anaphylatoxin or histamine took place in the cells, owing to some enzymatic cleavage following upon the immune reaction in the protoplasm. Such a conception, as Dale (1920) pointed out, while not impossible, went far beyond the evidence available. When allowance was made for the colloidal nature of the antigen, its action on the sensitized muscle seemed to be as immediate

as that of an ordinary drug, and only slightly slower than that of histamine itself. There was, further, no experimental warrant for the supposition that the aggregating reaction of antigen with antibody in the
cell protoplasm would initiate an enzymatic production of histamine. On
the evidence then available, as Dale suggested, we were entitled to postulate only the occurrence of aggregation, due to reaction of antigen with
antibody; but this might, by itself, be the effective stimulus, and the
similarity of the effect to that of histamine might mean that the action of
histamine was to cause an analogous physical change in the protoplasm.

The correspondence between the anaphylactic symptoms and those
produced by histamine remained in this position of uncertainty, and
without convincing explanation, until 1927. Then Thomas Lewis published the convincing demonstration, by the experiments of himself and
his co-workers, that a wide range of types of injury or irritation caused
the release from the epidermal cells of something acting like histamine. In
the same year Best, Dale, Dudley and Thorpe published the first complete
evidence of the occurrence of histamine, preformed, as a general constituent of the living cells of various organs. The suggestion was ready to
hand, that the substance released from cells when they were injured or
irritated was the histamine already present, and held inactive within
them. The truth of this view was confirmed by an experiment of Harris in
Lewis's laboratory, showing that, when skin was scalded, its content of
extractable histamine did not increase, but diminished, as the resulting
oedema developed. The suggestion that an anaphylactic reaction of the
skin represented only a special case of such histamine-releasing injury to
the sensitized cells was first formulated by Lewis. It was then an obvious
step to an explanation of the central symptom-complex of the anaphylactic
shock in different species, with its hitherto puzzling identity with the
symptoms produced in the same species by histamine, in terms of a general
release of preformed histamine, from cells injured by the occurrence, in
their protoplasm or their surface layers, of the aggregating reaction
between antigen and antibody (see Dale, 1929). It was now not difficult
to understand why in the guinea pig, with the extreme and fatal
sensitiveness of its bronchiolar plain muscle to histamine, an injection of
the sensitizing antigen, leading to immediate release of histamine in the
tissues, should cause, like histamine, a rapid death from asphyxia; and
why in the dog, in which Manwaring (1911) had early shown that the
immediate effect of the re-injected antigen was on the liver, the liver
became immediately turgescent with blood, through restriction of the
histamine-sensitive venous outlet, and the resulting symptoms presented

combined effects of back-pressure on the vessels draining into the portal vein, with histamine effects on other organs.

This release of histamine, in the liver of the dog and the lungs of the guinea pig during anaphylactic shock, was no merely theoretical assumption. It was supported by direct evidence. Manwaring and his co-workers (1925) had shown that the blood leaving the liver of a dog during the anaphylactic shock, if transfused into a normal dog, would produce effects like those of histamine on the plain muscle of such organs as the bladder of this second animal. More recently Gebauer-Fuelnegg, Dragstedt and Mullenix (1932), by collecting the lymph draining from the liver into the thoracic duct of an anaphylactic dog, were able to show that, during the shock, a substance appeared in the lymph which showed, not only the physiological actions, but chemical properties of solubility, which were entirely compatible with the view that it was histamine itself. In the guinea pig an even fuller demonstration was given by Bartosch, Feldberg and Nagel (1932a, b, 1933), who perfused the lungs from an anaphylactic guinea pig with Locke's solution.[11] Dale (1913a) had shown that, under these conditions, addition of a small dose of the sensitizing antigen to the perfusion fluid caused the typical obstructive constriction of the bronchioles. Bartosch, Feldberg and Nagel found that, when this occurred, a substance appeared in the fluid leaving the lungs by the pulmonary veins which, in the following properties, was indistinguishable from histamine. It was soluble in alcohol and stable to boiling. It produced effects like those of histamine on the guinea pig's uterus, intestine, and lungs, the cat's blood pressure, and the dog's suprarenals. Like histamine it had no action on the rat's uterus, or the eserinized leech. When estimates of the histamine content of the fluid were made by two different methods, these estimates agreed quantitatively with one another. There can be little doubt that this substance actually was histamine. Successive portions of the effluent were tested both for histamine, and for the antigen which had been injected. It was thus found that the liberation of histamine reached a maximum some time after most of the antigen had been washed out of the tissue. This seems to indicate that the histamine is liberated from the damaged cells of the lung and is not formed by cleavage of the antigen, Bartosch (1935) has recently shown that the perfused guinea pig lung has a somewhat reduced histamine content after induction of the anaphylactic bronchial spasm. This is further evidence that the histamine detected in the perfusate does come from the lung tissues; it is set free, and not newly formed, as a consequence of the antigen–antibody reaction.

Daly and Schild (1934, 1935) have repeated the perfusion experiments of Bartosch, Feldberg and Nagel on the guinea pig lung, and have confirmed their findings. They were able to strengthen the identification of the substance detected in the perfusate when they showed that it was inactivated by histaminase.

The concept of anaphylaxis presented here not only gives a satisfactory interpretation of the remarkable similarity between histamine action and anaphylactic shock in different species; it also renders intelligible the fact that anaphylactic shock is accompanied by additional effects, which histamine does not produce. One of the effects of the cell injury, produced by intracellular reaction of antigen and antibody, is the release of histamine; but we have no right to assume that histamine is the only substance so released, or that the direct injury to the cells is, of itself, without effect. The additional effects of the shock, such as loss of coagulability of the blood and the appearance of capillary haemorrhages in various organs, are seen to some extent in all species in which the anaphylactic reaction has been studied in detail. They are most conspicuous in the dog, as might be expected from the fact that, in that species, the primary anaphylactic effect is a massive injury of the liver cells. For this reason the resemblance of the anaphylactic shock in the dog to the effects of an intravenous injection of peptone is much closer than that to histamine. The interpretation is, in that instance, somewhat complicated by the fact that the indefinite products of peptic digestion, known commercially as 'peptone', regularly contain a certain amount of histamine. Apart, however, from the direct effects due to this, there can be no doubt that constituents of these 'peptones' have a directly injurious effect on the liver of the fasting dog, and thereby produce a syndrome resembling in all details, including the failure of the blood to clot, that seen in the anaphylactic shock.[12]

There is no evidence yet available to justify speculation as to whether any of the other vasodilator substances, with which this book is concerned, are released into the body fluids by the anaphylactic reaction, and play any part in the resulting symptoms. It would be no matter for surprise if one or another of them were found to contribute some detail of the total syndrome, though the effects due to histamine would still be predominant. There is one special phenomenon which may ultimately find its explanation in terms of the release of some other substance. As mentioned above, the rat and mouse are peculiarly insusceptible both to the anaphylactic shock and to histamine. The lack of reaction to histamine, however, is not the only explanation of the resistance to sensitization. Longcope (1922) showed that a precipitin to an antigen could be produced in a rat, but that this would not passively sensitize a guinea pig to the antigen, as the precipitin from a rabbit so readily does. The rat's antibody is apparently, for some reason, unsuitable for cellular fixation. On the other hand Kellaway (1930) was able, by injecting a strong rabbit precipitin into a rat, to confer on the rat some sensitiveness to the antigen, with which the rabbit had been immunized. The isolated uterus of such a rat responded to the antigen, not by relaxation, which histamine causes in the rat's uterus, but by contraction. It is possible that this reaction may be related to the release of some other, cellular constituent, which stimulates the plain muscle of the rat's uterus.[13] On the other hand, it seems possible that here we have a genuine example of the precipitating reaction acting as a direct stimulus to the plain muscle.

Notes

I. Introduction

1 Dale's introductory chapter for the Gaddum monograph is one of about thirty commentaries, comparable in length to this one, that he published in the decade before World War II. (Many of these were printed versions of invited lectures, and their titles and references can be found in the bibliography attached to Feldberg's account of Dale's life, in *Biog. Mem. Fell. R. Soc.* **16**, 77–174, 1970.) These commentaries touched on a wide range of scientific topics, but most of them were concerned, as was Gaddum's book, with the field of Dale's own most important researches – the field that he helped to open, and that he liked to call 'autopharmacology'. (I'm sorry that this word, which I think felicitous, hasn't been as widely adopted as some of his other coinages have been.)

In his *Adventures in Physiology* (Pergamon, Oxford, 1953; Wellcome Trust, London, 1965) Dale reprinted five of these lectures along with twenty-five of his research papers, adding his own retrospective comments to each. The lectures he chose to omit from that collection are sometimes hard to locate, but I find them equally rewarding to read, and not only for the novel ideas they expressed. Dale was a stylist; his lucid and sonorous prose is particularly enjoyable when one reads it aloud. That was a quality he aimed at, I think. In his later years he used to dictate his first drafts, and eavesdropping juniors were impressed by his ability to conceive each paragraph whole, so that the sentences and subordinate clauses fell accurately into place as he spoke. He did of course revise and polish those drafts later on. But it didn't seem to his associates that his other activities – and he had a great many at the time I'm recalling – slackened noticeably when he had a review or lecture to prepare.

As to the content of these surveys, Dale always found some new features to emphasize, even when he had discussed the topic repeatedly before. In this Introduction, for example, we note his concern with the 'form of combination or physical association' that allows an active tissue component like histamine or acetylcholine to be stored inside cells that can both respond to it and destroy it, as soon as it becomes free.

185

2 As we can now see, Dale exaggerated the contrast between the vascular actions of the two hormones and those of the tissue principles. He was well aware of the vasodilator potency of adrenaline, which his own ergotoxine experiments had long ago dramatically revealed. But those had been acute experiments on animals, and physiologists in 1936 were not yet aware that in man, when adrenaline is given by slow intravenous infusion, its constrictor action on the cutaneous and visceral vessels is roughly balanced by its dilator action on the vessels of striated muscle, so that the mean arterial pressure changes only a little. Nor could Dale and his contemporaries have guessed that the bottled 'Adrenaline' on their shelves, or the adrenomedullary hormone that was released *in vivo* by their experimental procedures, always contained a significant proportion of noradrenaline, which causes almost pure vasoconstriction. These were facts that came to light only in the following decade.

As to vasopressin, it was already appreciated by renal physiologists like Verney that the vascular action to which that hormone owes its names is of pharmacological rather than physiological significance. (Vasopressin has its primary action on the renal tubules; to get a significant pressor response one must give it in much higher dosage.) Two comments seem appropriate here. The first is that Dale was not, perhaps, very sensitive to merely quantitative differences in the responses to the agents he studied. ('Come on, let's get an effect!' I can hear him saying, 'Let's give ten times the dose'.) The second comment is that when Dale was writing his chapter it wasn't yet clear how many active principles there might be in posterior pituitary extracts. It was natural for him to suspect that the antidiuretic principle, whose importance he recognized, might be different from vasopressin, just as oxytocin (whose effect on the uterus he had discovered) was already known to be, from the work of his colleague Dudley.

3 I think we can be reasonably sure that as far back as 1906 Reid Hunt suspected, though he did not come close to demonstrating, the presence of acetylcholine in the suprarenals, and probably also in the brain. (It was in 1906 that Hunt and Taveau first reported the compound's intense pharmacological activity: see Bibliography for reference. Some years earlier, Hunt and others had supplied chemical and pharmacological evidence for the presence of free choline in the brain and other tissues: for references see Halliburton, *J. Physiol.* **26**, 229–43, 1901.)

4 My guess is that the candidates at the head of Dale's list were substance P and kallikrein.

5 In the case of acetylcholine, the first explicit attempt to get information about the nature of its intracellular binding seems to have been made by Beznák (*J. Physiol.* **82**, 129–53, 1934). Further studies were in progress when this monograph came out, and were published soon afterwards: in

particular by Wen, Chang and Wong in Peking and by Corteggiani, Gautrelet and their colleagues in Paris. References to these studies, which helped to popularize the concept of 'free' and 'bound' acetylcholine, can be found in the reviews by Feldberg (*Physiol. Rev.* **25**, 596–642, 1945) and by Burgen and MacIntosh (in *Neurochemistry*, ed. I. H. Page, J. H. Quastel and K. A. C. Elliott, Thomas, Springfield, 1955, pp. 311–89). By 1938 Feldberg's colleague Trethewie (*Aust. J. exp. Bio. Med.* **16**, 225–32) was able to conclude that if acetylcholine, or histamine, is sequestered within a membrane-bound compartment, that compartment must be smaller than a whole cell. But real progress toward the identification of the storage organelles had to await the arrival of modern methods of cell fractionation, controlled by electron microscopy, in the early 1950s.

6　At this point a modern reader, having just admired Dale's shrewd suggestion that metabolically active compounds, like the adenosine esters, are normally held within the cell, though they can move inside more freely than some of the other vasodilators, may be startled by the contrast with his somewhat primitive ideas about the state and behaviour of intracellular potassium. Such a reader, however, should realize that the views Dale expresses in this passage are typical of those held in 1936 by many physiologists – especially, perhaps, by those who had not yet learned to think of bioelectric potentials as ultimately based on chemical diffusion gradients.

Here, for example, is the penultimate sentence of a classical paper (*J. Physiol.* **86**, 290–305, 1936) by Dale's colleagues Brown and Feldberg:

The liberation of ACh, by K ions, as established in these experiments, and the probable association of the K ion with the propagated disturbance in nerve, made the assumption not unreasonable that the wave of mobilized K ions accompanying the nerve impulse liberates ACh at the nerve terminals.

It is interesting that in the same issue of the *Journal of Physiology*, a paper by Feldberg and Guimarãis states similar views of the relationship between potassium and acetylcholine, while in another paper from Dale's laboratory, by M. Vogt, on the loss of potassium by ganglia subjected to prolonged stimulation, the discussion accords more closely with the advanced concepts put forward by W. O. Fenn in the important review of muscle metabolism he published in that year.

Modern views of the relationship between ions and excitability, based on the established concepts of physical chemistry, were only beginning to emerge in 1936; the insights of Fenn and Vogt were exceptional. (The argument over electrical versus chemical transmission in ganglia, which was just becoming lively in that year, might have been still-born if it had

been realized that transmission by acetylcholine was not refuted by
experiments in which eserine failed to cause obvious prolongation of the
post-synaptic response to a single pre-synaptic volley; the experimenters
did not appreciate the possibility that physical diffusion alone might
remove the acetylcholine from its scene of action almost as quickly as
cholinesterase could.) Electrophysiologists began to acquire more
physico-chemical sophistication a few years later, when Boyle and
Conway gave an account of the behaviour of potassium in resting muscle
that is still basically acceptable (*J. Physiol.* **100**, 1–63, 1941). But it was
not until 1948, following the re-discovery of the action-potential
overshoot, that Hodgkin and his colleagues elucidated the respective roles
of sodium and potassium in the conducted impulse, and thus initiated the
modern era of electrophysiology.

7 'Phosphagen' was the name given by P. and G. P. Eggleton to
phosphocreatine when they discovered it in 1927. (Fiske and Subbarow
independently made the same discovery.)

8 The source of Popielski's 'vasodilatin' was in some cases a tissue extract,
in other cases Witte's peptone, beloved by bacteriologists as an ingredient
of their media.

9 See Note 1, above. The present volume's Bibliography gives titles and
references for the following lectures: Croonian Lectures (3), London,
1929; Dohme Lectures (3), Baltimore 1933a; Linacre Lecture, Cambridge,
1934; Walter Ernest Dixon Memorial Lecture, Cambridge, 1935a;
Nothnagel Lecture (in German), Vienna, 1935b.

II. The pharmacological analysis of tissue fluids and extracts: general principles

1 During the period between the two world wars, pharmacological methods
of analysis became the basis of a sophisticated technology that led to
important advances in biomedical science. This list of endogenous
drug-like substances was growing steadily longer, and the importance of
many of them was obvious, or at least suspected. Chemical methods for
detecting them, let alone measuring them, were too insensitive to be
really useful, and the pharmacologists came to the rescue. With
experience, they were able to find biological test-objects – isolated tissue
preparations in most cases, whole-animal preparations occasionally – that
had the necessary sensitivity, selectivity and ruggedness. By the 1930s a
number of the more interesting substances could be measured with fair
accuracy, and new substances of comparable potency were brought to
light. Gaddum was the principal theorist of this approach and one of its
leading practitioners. This chapter of his monograph states the basic
principles.

When still newer technologies – biophysical, biochemical, radiochemical
and morphological – arrived, in the early 1950s and later, the

pharmacological methods receded from the foreground and were often the subject of patronizing comments; but they retained much of their value, as indeed they still do. This seems to me a suitable time and place for offering some comments on what might be called the golden age of bioassay, and on Gaddum's contributions to it. To start with, I will quote three characteristic passages from his writings: they will be familiar to some readers.

'New chemical methods have been described in recent years which are a great improvement on older methods, and there is little doubt that as time goes on biological methods will be less used, but they are still important and chemical methods will only inspire universal confidence if they are shown to give the same results as the biological methods.'

'If properly carried out, pharmacological methods of estimation are very sensitive and very specific. Their main disadvantages are that they are not very accurate and are comparatively slow; it is generally not possible to make more than about 10 estimates in a day. Another disadvantage is that there are fewer pharmacologists in the world than chemists, so it is more difficult to find someone else who will do the tests for you.'

(These two quotations are from the proceedings of a symposium (*Pharmac. Rev.* **11**, 241–9, 1959).)

'The pharmacologist has been a jack of all trades, borrowing from physiology, biochemistry, pathology, microbiology and statistics — but he has developed one method of his own and that is the technique of bioassay.'

(This quotation is from a book on *Drugs in our Society*, published by Johns Hopkins Press, Baltimore, 1959, pp. 17–26.)

I should add that there is a good section on pharmacological assay in Gaddum's textbook, which is called simply *Pharmacology*; the section is followed, appropriately, by one on clinical drug trials. And I should also mention that Feldberg's account of Gaddum's life and work (*Biog. Mem. Fell. R. Soc.* **13**, 57–77, 1967) gives further quotations, and pays tribute to Gaddum's 'endearing personality', as well as to his scientific career.

Gaddum's writing was generally brisk and terse, as the quoted examples illustrate. But in person, although he was a large man, he was gentle and sometimes seemed modest to the point of diffidence. In the early part of his career, even people who knew him well were apt to underestimate him. One of these, I think, was Dale, who never shared Gaddum's liking for statistics and quantitative comparisons. But Gaddum's calculations, like the ingenious gadgets he designed, always had a clear purpose. His reading ranged widely over the scientific and medical literature, and he found much to interest him. (It was Gaddum, according to Feldberg, who noticed in *Pflügers Archiv* the paper by Kibjakow on perfusion of the

superior cervical ganglion. Gaddum suggested to Feldberg that they
should use the Russian's ingenious technique to test the possibility of
acetylcholine release at ganglionic synapses. The idea was immediately
pursued, and the experiments gave the first clear evidence for chemical
transmission at any neuronal synapse.)

Gaddum was competent in mathematics. He attempted no flights into
higher theory, but he made one major contribution and several minor
ones to biostatistics. The major one was his Normal Equivalent Deviation,
a statistic which allowed him to linearize dose–response curves derived
from quantal data. The American statistician C. I. Bliss turned this into
the 'probit' by adding five to it. Gaddum himself used the same principle
as the basis of his '2×2 assay', which simplified the comparison of a
standard preparation with an unknown by using only two doses of each.
Later, following A. J. Clark, Gaddum devised a valuable formula for
quantitating drug antagonism, an approach that was further developed
by his one-time colleague Heinz Schild. Later, in his *Who's Who* entry,
Gaddum would list his recreations as 'reading, writing and arithmetic'.

Gaddum's interests were broad – his fine textbook of pharmacology was
a one-man effort — and he returned at times to his theoretical interests.
But his two continuing themes, often interwoven, were
neurotransmission and bioassay. His development of these themes was
characteristically his own, though his perception had been sharpened by
his association with Dale, Feldberg and von Euler. Early in his career he
had formed clear ideas about the principles that should guide research in
these new fields, and he must have welcomed the opportunity of giving
them formal expression, as he was able to do in this monograph, and
especially in the present chapter. He had already exemplified these
principles in his research, particularly in two classical papers: one with
Chang (1933) on the measurement and distribution of tissue
acetylcholine, and one with Barsoum (1935a) on the measurement and
distribution of histamine and adenosine.

The monograph and the papers just mentioned gained in authority
because they followed the chemical isolation, in Dale's laboratory, of two
of the three agents. But even apart from that, Gaddum could convince his
readers that all three agents could be identified, and measured, on the
basis of their pharmacological effects alone. Evidence that was
particularly decisive was obtained by 'parallel bioassay' on several
different tissues; once this had been achieved for a particular sort of
experiment on particular tissues, routine measurements could be made
with fewer precautions. This being recognized, able investigators were
encouraged to work on substances of this type. Some of Gaddum's
influence may have been exerted through his colleagues. Ulf von Euler,
who had shared the discovery of substance P with him, and who later

initiated the first effective attack on the prostaglandins and their relatives, made brilliant use of Gaddum's principle of parallel bioassay to prove that noradrenaline is the major sympathetic transmitter. Later on, the high-powered parallel bioassays devised by Vane and his colleagues · helped to unravel the complex biochemical interrelationships of the prostaglandin family. (The influence of scientists on their successors is hard to disentangle: still one must regret that Gaddum didn't live to hear about the mode of action of aspirin, a discovery in the Gaddum tradition.)

It must be admitted that for scientists of today, to whom the researches described in Gaddum's book are ancient history, Gaddum often seems to be stressing the obvious when he goes into detail about elementary precautions. In the 1930s, however, many of his points were by no means immediately apparent. It was not that the people whose errors he noted were stupid. More often, their imagination had over-fuelled their enthusiasm. But until Gaddum had expressed the basic principles of a good assay in simple and cogent language, it was all too easy to take treacherous short-cuts.

During World War II, Gaddum's research was side-tracked into military problems. After the war was over he went back to chemical transmission, working productively in the catecholamine field in collaboration with Goodwin, Peart, Vogt, Lembeck and others, before joining another group of colleagues to study 5-hydroxytryptamine (5-HT, serotonin). This substance was not Gaddum's discovery, but he and his colleagues, by demonstrating its uneven distribution in the brain, helped to provide evidence for its role as a central transmitter. One provocative finding, not yet fully analysed, was the antagonism (on smooth muscle) between 5-HT and the hallucinogen lysergic acid diethylamide (LSD). Gaddum's last researches were done while he was burdened with the directorship of an institute, but it included studies on kallikrein and related agents, and on the distribution of substance P, soon to be recognized as one of the group of neuropeptides, that were rapidly growing in number and respectability.

A point that has not yet been mentioned is Gaddum's talent as a designer of apparatus. One device that is now coming into its own is his 'push–pull cannula' of 1961, which allows information to be gathered about the metabolism of catecholamines and other agents in the unanaesthetized brain.

2 This is an impressive kind of test when it works. But rather few people seem to have used it, and in my limited experience it doesn't always work. I think that tests of this sort deserve further study, in order to learn whether they depend on true receptor desensitization or on failure of the response at a later stage, e.g. in the components of a second-messenger system.

3 In 1937 Burn published an enlarged and updated version of his book in

order to meet the growing demand, especially in the drug industry, for an authoritative manual of assay techniques. The new book, called *Biological Standardization* (Oxford University Press) had a theoretical introduction · that was strongly influenced by Gaddum's ideas. Written in Burn's lively style, the book became popular. It helped to introduce pharmacologists to Gaddum's ways of thought, and incidentally to elementary biostatistics, which had usually not formed part of their training.

4 Perhaps Gaddum's mood when he wrote these lines was an unusually sombre one! He had moved to Cairo not long before, and in a new environment it may take an experimenter some time to eliminate unsuspected disturbing factors from his procedures. Apart from this it is certainly common experience that the sensitivity of the assay preparation can change fairly rapidly over the first few doses, before it becomes constant or shows only a gradual slow drift. Fortunately, however, one rarely observes, when using a well-tried method, 'sudden changes in sensitivity due to unknown causes', unless one neglects such fairly obvious precautions as not smoking over the tissue bath that holds the strip of leech muscle or guinea pig gut whose contractions one means to record.

5 In the early days of chromatographic separation, some bizarre conclusions were drawn from experiments in which the mobility of an unidentified active component of a tissue extract failed to match the mobility of the pure substance with which it was being compared. Sometimes a single active substance turned up in two spots on a paper chromatogram, because the extract contained some material that complexed with the substance and altered its mobility. For example one laboratory reported that the acetylcholine-like activity of brain extracts was due to something other than acetylcholine itself. It cost time and money to eliminate such misconceptions.

6 Biologically active natural products that were believed to be single substances have often turned out to be mixtures of closely related compounds. The classical example is the supposedly pure 'Adrenalin' of the pharmacopoeias, which, after it had been on the market for nearly 50 years, was found to contain some 20 per cent of noradrenaline. It is of interest that the reports that supplied chemical evidence for this contamination were published in 1949 (Goldenberg *et al*; Tullar; Auerbach and Angell: *Science* 109, 534–5, 536–7, 537–8), about three years after von Euler had supplied cogent pharmacological evidence for the presence of noradrenaline in sympathetic neurones. Of course today's chemists have more powerful methods at their disposal.

III. Histamine

1 For the titles of more recent surveys, consult any large textbook of pharmacology. By 1980, in his excellent article in *The Pharmacological Basis of Therapeutics* (6th edn, ed. A. G. Gilman, L. S. Goodman and A. Gilman, Macmillan, New York, 1980, pp. 608–40), W. W. Douglas was able to list about sixty reviews that were mainly concerned with various aspects of the physiology and pharmacology of histamine.

2 While 'β-iminazolylethylamine' seems to be the chemical name that has found most favour, at least six or seven others appear in the literature. But in all the cases I have seen, the chemical name is displayed once in an opening paragraph and then is not used again. Dale recalled in his *Adventures in Physiology* (Pergamon Press, Oxford, 1953; The Wellcome Press, London, 1965) that he and Laidlaw felt obliged to write 'β-iminazolylethylamine' rather than 'histamine' in order to avoid a possible trade-mark infringement. 'Later somebody else called it histamine', Dale added, 'and then the way was clear.'

3 'Chloramine-T' was the name under which *p*-toluenesulphochloramide was listed in the British Pharmaceutical Codex and marketed as an antiseptic in the United Kingdom.

4 The classical papers of Dale, Lewis and their colleagues report a great variety of reproducible and important observations, which are well summarized in these pages by Gaddum. These early workers were aware that histamine has both vasoconstrictor and vasodilator activities, and that the balance between these depends on (a) which species, (b) which vascular bed and (c) what experimental conditions have been chosen for the observations. Some of their findings can now be seen to require reinterpretation at the cellular level. For example, it is now apparent that the 'capillaries' of the older literature must have included the venules, which can contract under the influence of the sympathetic, and which in fair people are mainly responsible for the colour of the skin; that 'increased capillary permeability' (for example whealing) involves the outward leakage of plasma protein through channels opened by histamine between the cells of the *venular* endothelium; that at least in some kinds of vessel the primary action of histamine is on the endothelial cells, which respond by releasing a factor that relaxes the vascular smooth muscle, and that this is probably an important mechanism of histamine-provoked vasodilatation; and finally that there are two sorts of histamine receptor, known as H_1 and H_2, on the surface of histamine-sensitive cells, with the former participating in vasoconstrictor responses and both types helping to produce vasodilator responses.

5 'Ergotoxine', often mentioned in this volume, was shown by Stoll of Basel to be a mixture of several ergot alkaloids, but as used by Dale and others it seems to have been a fairly selective α_1 partial agonist and blocker.

6 Keeton, Koch and Luckhardt (1920) seem to have made this observation independently.

7 Gaddum guessed wrong here. But the opinion he expressed was probably the majority one of its time. In retrospect it is apparent that at least part of the activity in the extracts obtained by Edkins must have been due to gastrin and not to histamine. Komarov (*Proc. Soc. exp. Biol. Med.* **38**, 514–16, 1938) made extracts of pyloric mucosa that contained no detectable histamine (I worked in the next lab, and did the tests for histamine on the guinea pig ileum myself, using the method of Barsoum and Gaddum), but scepticism persisted for many years and was not finally dissipated until gastrin was isolated in pure form (Gregory, *Harvey Lectures*, 1970). It was also in 1938 that a different role was suggested for the histamine of the gastric mucosa, based on its presence in gastric juice: the role of a local hormone, rather than a circulating one. This proposal too was greeted with scepticism, but C. F. Code's campaign on its behalf, combined with the discovery of the H_2 blockers by Black and his colleagues, eventually won acceptance for it (Code, *New Engl. J. Med.* **296**, 1459–62, 1977). The tangled relationship between gastrin, histamine, acetylcholine, and other denizens of the gastric mucosa such as noradrenaline, bombesin and somatostatin, has not yet been fully unravelled.

8 For reference to, and discussion of, Popielski's papers on 'vasodilatin', look up the articles by Keeton and Koch (*Am. J. Physiol.* **37**, 481–504, 1920) and by Lim (*Quart. Jl exp. Physiol.* **13**, 79–103, 1922).

9 The discovery that a large part of the body's histamine is located in the granules of the widely distributed mast cells was not made until 1953 (Riley and West, *J. Physiol.* **120**, 528–37). It initiated a new era in histamine research. Curiously, it had been known since 1937 that the special staining properties of the mast cell granules are due to their high content of heparin (Holmgren and Wilander, *Z. mikrosk.-anat. Forsch.* **42**, 242–78), and Wilander's thesis, published in the following year *Skand. Arch. Physiol.* suppl. 15), showed that the incoagulability of the blood in peptone or anaphylactic shock of the dog is due to the release of heparin. When Gaddum's monograph was published in 1936 there was already plenty of evidence for the release of histamine in anaphylaxis, and Dale recapitulates it in the final section (pp. 179–84) of the text. Why then was it left to Riley and West so many years later to postulate a common cellular origin for histamine and heparin? There is no simple answer, I think. The possibility may have occurred to a number of the people who were interested in histamine release, but because it was only in the dog that the picture of shock included frank loss of blood coagulability, the idea of a cellular association between the two agents seemed rather far-fetched. Riley's monograph on *The Mast Cells* (E. & S. Livingstone,

Edinburgh, 1959) presents the historical background of the discovery. In my opinion it would have required great prescience, or luck, to relate mast cells and tissue histamine in 1936.

10 For a note on the partition of blood histamine between cells and plasma, see Chapter VI, Note 2.

11 A modern reader may be surprised to learn that the physiologists of the 1920s devoted so much attention to the absorption of histamine from the gut. My guess is that part at least of the reason was the popularity in some clinical quarters of the notion, which we now find naive, that 'autointoxication' may easily arise from the absorption of amines and other products of the bacterial decomposition of food residues. The clinicians who held such views were likely to be vigorous users of purgatives and enemas, or even to recommend surgical extirpation of the colon in patients whose ill-health they couldn't otherwise explain.

12 As E. A. Zeller pointed out in 1938 (*Helv. chim. Acta* **21**, 1645–65), preparations of partly purified 'histaminase' will degrade a number of aliphatic and aromatic diamines in addition to histamine. He supplied evidence that in the case of histamine the reaction involved oxidative deamination of the side-chain, and he proposed that the enzyme be re-named 'diamine oxidase'. Later work with labelled histamine showed that diamine oxidase is probably not a single enzyme, and that several enzyme systems contribute to the degradation of free histamine.

IV. Acetylcholine

1 Some further important reviews appeared during the three-year period 1935–7. See Chapter VII, Note 1 for references.

2 Gaddum is not referring here to muscarinic actions on pre-synaptic terminals or on autonomic ganglion cells, for the good reason that these actions had not yet been discovered. He does mention (p. 58) the observation by Feldberg, Minz and Tsudzimura (1934) that though nicotine blocks most of the effect of acetylcholine on the suprarenal medulla, there is a residual effect that can only be blocked with atropine. Thus the cells of this tissue, which are modified sympathetic neurones, provided an early model for the dual sensitivity to acetylcholine that was later found in the sympathetic ganglion cells examined by R. M. Eccles and B. Libet (*J. Physiol.* **157**, 484–503, 1961).

For a comment by Gaddum on the earliest reported *central* action of acetylcholine, see Note 7.

3 Of course Gaddum knew that nicotine acts on parasympathetic, as well as on sympathetic, ganglia. But did he know that acetylcholine, too, acts at both these sites? I suppose he couldn't be sure of that. So far as I know the action on parasympathetic ganglia was first described in 1953, by Perry and Talesnik (*J. Physiol.* **119**, 455–69).

4 Hunt and Taveau seem to have worked mostly on rabbits. They often
   curarized them, and perhaps that is why they did not discover the
   nicotinic actions of acetylcholine, even though they injected quite large
   doses of it.

5 In some of the early attempts to use acetylcholine therapeutically, its
   rapid hydrolysis by blood caused its potency to be underestimated. Dale
   had an anecdote from this period about a clinical friend (I think it was
   F. R. Fraser) who obtained some acetylcholine from him, and complained
   that it had proved to be inactive when he gave it by vein to a patient.
   Dale was surprised, but his questions elicited a likely explanation: the
   physician had drawn some blood back into the syringe before he pushed
   the plunger home. When the trial was repeated, with care not to expose
   the drug prematurely to the blood, the effect of the injection on the heart
   was dramatic, though evanescent. The paper by Carmichael and Fraser
   (*Heart* **16**, 263–74, 1933) may be the one that refers to this early clinical
   trial.

6 The contractures evoked by acetylcholine were leisurely responses;
   typically they took minutes to wax and wane. Until 1936, with some
   irregular exceptions, they were the only responses that acetylcholine had
   been seen to produce in vertebrate skeletal muscle. As Gaddum notes,
   most normal mammalian muscles did not show them at all. A
   physiologist could give an anaesthetized cat or dog, after he had protected
   its circulation with atropine, many milligrams of acetylcholine by vein
   without eliciting the slightest movement of its major muscles. The
   exquisite sensitivity of denervated muscles was a piquant contrast. If that
   was a hint that acetylcholine might somehow be involved in normal
   neuromuscular transmission, it was a hint that nobody took seriously
   before 1934.

7 Dikshit (1933a, b), working in A. J. Clark's department, seems to have
   been the first, by several years, to report what looked like a direct effect of
   acetylcholine on the brain. He was also (1933b) the first to report that
   acetylcholine is unevenly distributed in that organ: the cerebral cortex, he
   said correctly, contains more than the cerebellum but less than the
   striatum. Further studies with acetylcholine, and parallel observations
   with other transmitters, came much later: see M. Vogt (*Brain* **102**,
   444–59, 1982) for references. Gaddum would have noted that Dikshit's
   extraction method was faulty, and gave acetylcholine values that were an
   order of magnitude too low. He may also have suspected that the
   inhibition of respiratory movements that Dikshit recorded when he
   injected acetylcholine into the third ventricle did not represent a direct
   action on neurones with a respiratory function. I think that Gaddum
   would have been aware of the recent discovery of a parasympathetic
   vasodilator innervation of the brain, through the seventh cranial nerve

(Chorobski and Penfield, *AMA Arch, Neurol. Psychiat.* 28, 1257–89, 1932); and if so, he would have wondered whether the changes in breathing were secondary to changes in blood flow. Dikshit's remarkable findings may thus have to be discounted to some degree. Nevertheless he was one of the pioneer observers in the acetylcholine field, and deserves more recognition than he received. There are several other early workers of whom the same might be said, for example Loewi's collaborators M. Witanowski, E. Navratil and E. Engelhardt.

8 For later experiments on acetylcholine and ciliary movement, see P. Cordik, E. Bülbring and J. H. Burn, *Br. J. Pharmacol.* 7, 67–79, 1952.

9 These recipes are largely based on those in the important paper by Chang and Gaddum (1933).

10 Minz (1932) developed the leech assay while he was Feldberg's student in Berlin, and Feldberg brought it to Dale's laboratory. The story of how it was used there has been delightfully told by Feldberg (in the *Pursuit of Nature*, Cambridge University Press, 65–83, 1977). The leech assay dominated cholinergic research throughout the 1930s. It was seldom required to give results of great precision, which was fortunate. The muscle, though it is splendidly sensitive, is not very strong, and it reacts unfavourably to sudden increments of tension. The experimenter who was recording its slow contractions on a smoked drum (this operation was seldom entrusted to a technician) had to be careful that the writing-point didn't stick in the soot. For many years the most characteristic sight and sound in a laboratory where acetylcholine was being assayed was the rhythmic tapping of the experimenter's fingers on the table where the leech bath sat as he generated the vibrations that assisted the muscle to complete its contraction or relaxation. The frog rectus preparation, introduced by Chang and Gaddum, is faster and doesn't require so much conscious artistry, but it is less sensitive.

11 When it became necessary to measure acetylcholine in tissue extracts more accurately, Feldberg and Hebb (*J. Physiol.* 106, 8–17, 1947) introduced the useful precaution of dissolving the acetylcholine standard in a similar extract whose acetylcholine had been destroyed by brief treatment with sodium hydroxide followed by neutralization. It was hoped, with some justification, that this procedure would not destroy any constituents of the extract that might sensitize or desensitize the leech muscle to acetylcholine.

12 As a long-time devotee of the blood-pressure assay I feel obliged to note its sensitivity (often responding to less than 1 ng of acetylcholine), its speed (fast enough to be used 'on line' in experiments on acetylcholine release), its stability (especially when arterial pressure is stabilized by feedback of an adrenaline or noradrenaline solution into an artery from a reservoir pre-set to the desired pressure) and its praiseworthy indifference

to many of the sensitizing and desensitizing materials that affect other assays. (The feedback stabilization of arterial pressure was introduced by D. M. J. Quastel and R. I. Birks.) The cat is anaesthetized with chloralose and subjected to abdominal evisceration in order to minimize endogenous acetylcholine release.

13 Chang and Gaddum did report one exception to this statement: their parallel bioassays could not distinguish acetylcholine from pyruvylcholine. But later it was found that this was only an apparent exception, because pyruvylcholine is unstable and breaks down to form acetylcholine.

14 Reinecke acid was the first of a number of complex organic acids to be used for separating acetylcholine from other water-soluble tissue constituents. The procedure devised by Kapfhammer and his colleagues was basically sound, but some of the results they reported were almost certainly fraudulent. No member of the group ever published a withdrawal of their extraordinary claim that blood contains large amounts of acetylcholine, but later it was rumoured that one of the members had taken his own life. It seems likely that what was originally a simple error grew into a tragedy.

15 The values quoted in Table 8 seem by current standards to be too low in the case of some tissues, especially brain, but this may reflect the way in which the animals were handled before the tissue was sampled. On the other hand, it is hard to believe that there are measurable amounts of acetylcholine in erythrocytes. Perhaps the values at the head of the table reflect the combined pharmacological activity of choline, potassium and other blood constituents. However, it is known that some kinds of leucocyte have acetylcholine receptors, and it seems just possible that they contain some acetylcholine as well. I am not aware that anyone has looked into this point with modern methods. (For a survey of older and newer data on tissue acetylcholine concentrations, see Appendix III, compiled by Saelens and Simke, of *Biology of Cholinergic Function*, ed. A. M. Goldberg and I. Hanin, Raven Press, New York, 1976.)

16 Some more incisive studies of acetylcholine synthesis by nervous tissue were beginning just as Gaddum's monograph was approaching completion. The key papers are those by Brown and Feldberg on ganglia (*J. Physiol.* **88**, 265–83, 1936) and by Quastel, Tennenbaum and Wheatley on brain tissue (*Biochem. J.* **36**, 1668–81, 1936).

17 Alles and Hawes (*J. biol. Chem.* **133**, 375–90, 1940) seem to have been the first to recognize a difference between the red-cell and the serum forms of cholinesterase. Augustinsson's article in *Cholinesterases and Anticholinesterase Agents* (ed. G. B. Koelle, Springer, Berlin, 1963, pp. 89–128) summarizes the early history of research in that field.

18 Holmstedt's contribution to the volume edited by Koelle (see Note 17, above) gives a fascinating account of the early research on the

organosphosphorus compounds. This had begun in Germany before
Gaddum's book was written, but for some time it was not realized that
the mechanism by which these compounds killed insects and other
animals was the preservation of endogenous acetylcholine. The mode of
action of these powerful toxins seems to have been discovered
independently by scientists in several countries doing secret research
during World War II. In Britain, Gaddum was involved in this kind of
research at an early stage, but it seems to have been the Cambridge group
of Adrian, Feldberg, Kilby and Kilby who, in 1942 first recognized that
the 'nerve gases' were anticholinesterases; their reports were published in
1947 (*Br. J. Pharmacol.* **2**, 56–8, 234–40). No doubt the same discovery
had already been made by German scientists. Probably there were some
information leaks: when the Belgian fortress of Liége fell after a brief
siege in May 1940, the London papers reported that the Germans had
used a new gas called acetylcholine!

Gaddum was apt to be rather casual in his use of poisons. After the
war he mentioned to me that on one occasion he had accidentally spilled
some DFP (diisopropylfluorophosphate) on the floor and had wiped it up
with a bit of cotton. Later he felt there was something odd about his
vision, and went to the mirror to see if his pupils were constricted. They
were so small that he could not see them at all, so he was satisfied then
that DFP must indeed be an anticholinesterase.

V. Adenosine compounds

1 R. A. Kekwick says in his account of Drury's career (*Biographical Memoirs
of Fellows of the Royal Society*, **27**, 173–98, 1981) that the start of this
research was 'fortuitous'. Szent-Györgyi, on a 1929 visit to Cambridge,
had brought with him a heart–muscle extract, which he had found would
inhibit the isolated heart of the frog. When he and Drury began trying to
identify the extract's active ingredient they thought for a while that it
was potassium, but their further tests, especially on intact animals,
revealed some features of its action that could not be explained on that
basis. In particular, the anaesthetized guinea pig responded with a
complete though transient heart-block, whose duration was
dose-dependent; and an assay based on that response enabled them to
isolate the active principle and to identify it as adenylic acid, which was
already a familiar substance to Szent-Györgyi as a biochemist.

As we now know, the muscle adenylic acid in Szent-Györgyi's extract
was an artefact, derived from the *post-mortem* hydrolysis of adenosine
triphosphate (ATP). ATP itself was also discovered in 1929. That was not
entirely a coincidence, for the physiologists and biochemists of that period
had a lively interest in the labile phosphate esters of muscle, and already
suspected that these compounds might be involved in the chemical

processes that underlie contraction and relaxation. The central role of
ATP in cellular energy transfers would not be clearly perceived for
another dozen years. But the discovery that it is a major constituent of
muscle must have encouraged Drury and Szent-Györgyi to continue their
exploration of the pharmacology of adenylic acid and its relatives. In fact,
it is apparent from their paper that they lost no time in getting a sample
of ATP from Fiske, one of its discoverers. They may have been
disappointed when they found that its pharmacological effects, probably
because of rapid *in vivo* hydrolysis, were generally like those of adenylic
acid, which in turn were like those of the parent base adenosine. The
Cambridge workers were nevertheless impressed by the intensity and
variety of these effects, and they were not surprised when they found
that adenosine-like activity was released into the blood from heart muscle
exposed to damage by heat. But they scrupulously refrained from
speculating that any physiological phenomena, such as functional or
reactive hyperaemia, depend on the release of adenosine or its esters.

   Their immediate successors were less cautious. Zipf (1931a), who had
already (1930) identified adenylic acid as the vasodilator material formed
in shed blood, suggested, though he did not prove, that it is responsible
for the hyperaemia that accompanies activity in skeletal and heart
muscle, and Rigler (1932) argues in the same direction. At that time little
was known about the intracellular roles of the adenosine compounds,
except that they contributed to zymase activity and ammonia generation
in muscle. Thus the history of these substances as suggested autacoids is
longer than their history as 'high-energy' compounds. But their supposed
significance in functional vasodilatation did not win widespread
acceptance. Even forty years after the pioneer observations just
mentioned, textbooks of physiology mentioned the vasodilator roles of
adenosine only as hypothetical, and they said nothing about any other
actions that the adenosine compounds might exert after they leave their
cells of origin. This situation has changed in the last ten or fifteen years,
and the textbooks are beginning to catch up (see Note 3).

2 Perhaps the first guess that adenosine and its relatives might have a role
   in blood-flow regulation was made by two men whose names aren't
   usually connected with these compounds. It was Joseph Barcroft and
   Walter Dixon who wrote in 1907 (*J. Physiol.* **35**, 182–204), in a paper
   on the gaseous metabolism of the heart: 'In our experiments the carbon
   dioxide levels may be only an index of other metabolic products such as
   lactic acid and purine bodies which have a similar vasodilatory action.'

   This quotation can be found in an article entitled 'Adenosine or
   adenosine triphosphate?', by T. Forrester, in the volume on *Vasodilatation*,
   edited by P. M. Vanhoutte and I. Leeusen (Raven Press, New York, 1981).
   The article gives a good terse summary of the present status of the

adenosine compounds as vasodilators, and together with the other contributions to the volume provides basic historical references to work in this field.

3 The chapter notes in this volume are not intended to update the information provided by Gaddum, but I will make a partial exception to that policy at this point, because I think that the history of a development is best appreciated by those who are familiar with its consequences. In the case of the autacoid functions of the adenosine compounds, much of the relevant information has been acquired so recently that it is only specialists who are really familiar with it. I therefore hazard some summary statements in the following paragraphs, realizing that experts in this field will recognize that my conclusions reflect the limitations of my own knowledge. I add a few further references to more authoritative sources of information.

As I understand it, we can now identify with confidence three extracellular roles, or autacoid functions, of these substances. They are: (1) metabolic vasodilator, (2) synaptic transmitter or co-transmitter, and (3) platelet activator. Role (1) belongs mainly to adenosine itself, role (2) mainly to ATP, and role (3) mainly to ADP, adenosine diphosphate. Not only the first role, but the last two as well, can contribute to vasodilator responses.

(1) Unesterified adenosine is presumably present only in low concentration in cells that are not starved of fuel or oxygen. But the parent base can escape through the plasma membrane more readily than can its esters, and it can also be formed extracellularly, by the hydrolysis of an adenosine ester that has been released by exocytosis or by cell damage. Its most striking effect is vasodilatation, and it is physiologically important in adjusting blood flow to tissue requirements; it is one of the major contributors to that adjustment in the coronary circulation, and probably in other vascular beds as well, though it is still true that vasodilator responses have always been easier to record than to interpret. It seems likely, for example, that at least part of adenosine's action on the microcirculation is an indirect one, depending either on pre-synaptic inhibition of noradrenergic vasoconstriction or on release of an unidentified mediator from the endothelium (R. F. Furchgott, *Ann. Rev. Pharmacol. Toxicol.* **24**, 175–97, 1984). The autacoid role of adenosine in relation to non-vascular smooth muscle is less well documented. Adenosine receptors on cell surfaces are widely distributed. The effects of adenosine mediated by such receptors are blocked by theophylline.

(2) Secretory granules and synaptic vesicles often contain ATP in high concentration. The ATP released by their exocytosis is quickly destroyed by ubiquitous ATPases, but not before it exerts some of its characteristic actions. That it functions sometimes as a principal transmitter and

sometimes as a co-transmitter is now widely accepted, in confirmation of the persistent researches of Burnstock, though of course ATP does not account for all the examples of non-cholinergic, non-adrenergic transmission at neuroeffector junctions. ATP has both pre-synaptic and post-synaptic actions; the former are generally inhibitory. Theophylline does not block the actions of unhydrolysed ATP.

(3) The role of ADP in platelet function is briefly considered in the next chapter.

A useful reference volume is *Physiology and Pharmacology of the Adenosine Derivatives* (ed. J. W. Daly, Y. Kuroda, J. W. Phillis, H. Shimizu, H. Shimizu and M. Ui, Raven Press, New York, 1983). Summaries that emphasize one or other of the above-mentioned roles of these substances include the following: *Regulatory Functions of Adenosine* (ed. R. M. Berne, T. W. Rall and R. Rubio, Martinus Nijhoff, Boston, 1983); *Purinergic Receptors*) ed. G. Burnstock, Chapman and Hall, London, 1981); and M. M. Frojmovic and J. G. Milton, *Physiol. Rev.* **62**, 186–261, 1982 (reviews relationship of adenosine compounds to platelet activation).

VI. Unidentified substances present in tissue extracts and body fluids

1 For a comment on the supposed presence of traces of acetylcholine in normal blood, see Chapter IV, Note 15.

Reports that free choline is present in blood date from the turn of the century (see Vincent and Cramer, 1904, who give still earlier references). The first dependable measurements of free choline in plasma were made by J. Bligh (*J. Physiol.* **117**, 234–40, 1952). Its level in most mammals is maintained at close to 10–15 μM, which is high enough to support acetylcholine and phospholipid synthesis, but not high enough to produce frank muscarinic or nicotinic effects.

2 There is no doubt that the method of Barsoum and Gaddum (1935a) allowed them to detect histamine in blood and measure it with reasonable accuracy. The method was therefore an important advance. But the value of applying it to samples of whole blood was soon questioned, because it was realized that when freshly collected blood is centrifuged, with care to avoid clotting, most of its histamine can be recovered from the white-cell/platelet layer (Code, *J. Physiol.* **90**, 485–500, 1937). (Eventually it was shown that the blood basophils, which can be regarded as circulating mast cells, are the richest source in most species.) The obvious implication of this finding was that any experimentally observed change in the blood histamine level might merely reflect a change in the leucocyte (or basophil) count, rather than in the level of free, physiologically active histamine. This sort of uncertainty may help to explain why some of the early reports that functional vasodilatation is associated with histamine release could not be confirmed. The negative findings by Code, Lovatt Evans and Gregory (*J. Physiol.* **92**, 344–54,

1938) encouraged a more critical attitude. Nevertheless, it is not in dispute that normal plasma does contain small amounts of free histamine, and under some conditions even smaller amounts of N-acetylhistamine (which would be converted to histamine in samples extracted by the Barsoum–Gaddum procedure). The continuing presence of such traces of histamine seems to have no significant effect on the microcirculation; the N-acetylhistamine is inactive.

A similar source of uncertainty may have influenced the data for blood adenosine obtained by the method described by Barsoum and Gaddum (1935a) in the same paper. For the authors soon became aware that the cells of the blood contain more adenosine than the plasma; and from the work of Zipf (1930, 1931a, b) they might have guessed that most of the extra adenosine is in the platelets.

Today's readers may find it surprising that Gaddum, usually so aware of all the variables that could influence the results of his assays, should have been content in this work with Barsoum to measure whole-blood rather than plasma levels of histamine and adenosine. He and Barsoum already knew when they wrote their paper that in rabbit blood most of the histamine is in the cells (though they did not know there is so much in the platelets). Why did they not separate the plasma and use that for their measurements?

My guess is that their principal reason was the unsatisfactory quality of the anticoagulants then available. Of these, citrate and oxalate would be likely to compromise their bioassays. There was some canine heparin on the market in 1935, but it was expensive, somewhat variable in potency, and likely to contain pharmacologically active impurities. Hirudin from the leech was open to similar objections, and the acidic azo dyes had only recently been introduced and had not been used in this sort of situation. By 1937, however, Code was able to obtain a satisfactory heparin preparation, and he made use of further precautions (e.g., the use of paraffined collecting tubes) to reduce the hazard of losing histamine from the cells to the plasma.

3 Another possible reason for these discrepancies, and one to which Gaddum pays little attention, was the use of different animal species to supply the blood or serum used as the source of the active materials. The distribution of both histamine and 5-hydroxytryptamine between the cellular elements of the blood varies from species to species.

Nevertheless, it has been clear for many years that the pharmacological activity of shed blood, in so far as it was revealed by the simple tests used by these early workers, is largely due to two of the substances released by activated platelets when their dense granules are extruded into the medium. These substances are 5-hydroxytryptamine (5-HT, serotonin) and adenosine diphosphate (ADP). To a first approximation, ADP, along with the adenosine monophosphate and adenosine derived from its partial hydrolysis, constituted *Frühgift*, while 5-HT was the principal constituent of *Spätgift*, though the dense granules can store and release other amines as well. The reason why *Spätgift* appears after *Frühgift* is mainly that the latter's effects are not obvious until the former's have disappeared.

The serum vasoconstrictor 5-HT was first isolated by Rapport and Page and their colleagues at the Cleveland Clinic, who were also the first to purify it; they called it serotonin. Soon afterward Erspamer and his co-workers in Italy recognized its identity with 'enteramine', the smooth-muscle-stimulating compound they had obtained from the intestinal mucosa. The presence of 5-HT in platelets was verified by the Cleveland group in 1954. The details of its pharmacology can be found in the textbooks; many of its actions are those that had been attributed to *Spätgift*. An additional one, discovered by Feldberg and Smith in 1953, is that it can liberate tissue histamine. As might have been expected from the observations of Verney and Starling (1922: see p. 100), it is inactivated in the pulmonary circulation (Vane, Second Gaddum Lecture, *Br. J. Pharmacol.* **35**, 209–42, 1969).

Gaddum was a major contributor to the research on 5-HT. He was the first to show that many of its effects are blocked by lysergic acid diethylamide (LSD) and other hallucinogens (*J. Physiol.* **121**, 15P, 1953). With Amin and Crawford (*J. Physiol.* **126**, 596–618, 1954) he provided early information about its uneven distribution in the central nervous system, and thus pointed the way to its recognition as a neurotransmitter; and with Picarelli (*Br. J. Pharmacol.* **12**, 323–8, 1957) he distinguished two principal types of the 'tryptamine receptor'.

4 Gaddum rightly suspected that the adenosine derivatives (*Frühgift*) and the compound or compounds with pressor activity (*Spätgift*) were not the only autacoids released when blood clots. Any attempt to consider all the pharmacologically active materials formed or released during the enzymatic cascades associated with haemostasis would exceed the limits of my space, and of my knowledge. Platelets, endothelium and tissue components contribute initiating factors and plasma proenzymes are required in the subsequent stages, in which eicosanoids like thromboxane A_2 and prostacyclin are generated along with active peptides. Experts in this field may amuse themselves by supplying their own interpretation of the phenomena described in the preceding sections. The same sort of exegesis can also be applied to the following sections, on the substances from tissues other than blood: I think that in most cases it will be even more difficult.

5 See Chapter VIII, Note 2. In retrospect it seems that the research by James *et al.* was valuable mainly because it led to the studies by Dale and his colleagues on the identification and distribution of tissue histamine.

6 Most of the material in 'vagotonin' must have been protein or peptide in

nature, but the activity described is not readily attributed to any of the known hormones or proenzymes present in the pancreas.

7 In contrast to 'vagotonin' and 'angioxyl', kallikrein eventually attained scientific respectability, though after a chequered career (see Note 12).

8 Substance P received its non-committal name because von Euler and Gaddum made a dry powder of their gut extract for use as a standard, and labelled the bottle 'P'. Later they were both somewhat apologetic about the name; but at least it didn't turn out to be a misnomer like 'kallikrein', which is not pancreas-specific, or 'prostaglandin', which (a) did not come originally from the prostate, and (b) might appropriately have been dubbed in the plural rather than in the singular.

The original preparation 'P' was of course very impure, and must have contained a number of biologically active substances, for example vasoactive intestinal polypeptide (VIP), and probably 5-HT as well. By 1936, however, von Euler (*Arch. exp. Path. Pharmak.* **181**, 181–97) had found that the material responsible for most of its activity behaved like a peptide; and this clue led to the development, especially by Pernow, of methods leading toward its purification, which was finally achieved in 1970 by Chang and Leeman. As noted in the text, von Euler and Gaddum in their 1931 paper had reported the presence of substance P activity in the brain as well as in the intestine, and that finding was amply verified after the pure undecapeptide became available. Substance P, therefore, has some claim to be the first neuropeptide to be recognized as such. (Although a number of gut and pituitary hormones had been characterized before 1931, they had not been looked for in brain.)

The best short history of substance P is in an article by von Euler (*Trends in Neurosci.* **4**, iv–ix, 1981). It contains some interesting references. One is to an almost forgotten paper by L. Jendrassik of Pécs (for an abstract see *Ber. ges. Physiol.* **63**, 654–5, 1931: the original publication was in 1929), who may have been the first researcher to observe the action of substance P obtained from brain. Von Euler also cites three symposium volumes published in the early 1960s, which vividly portray the status of research into substance P and other active peptides at a time when interest in these compounds had begun to accelerate. (The references are: *Polypeptides which affect Smooth Muscle and Blood Vessels*, ed. M. Schachter, Pergamon Press, Oxford, 1960; *Structure and Function of Biologically Active Peptides: bradykinin, kallidin and congeners*, ed. E. G. Erdös, *Ann. N.Y. Acad. Sci.* **104**, 1–464, 1963; and *Symposium on Substance P*, ed. P. Stern, Sarajevo, 1963 (the last of these three is hard to obtain).) More recent summaries that contain historically interesting material are those by V. Erspamer (*Trends in Neurosci.* **4**, 267–9, 1981) and S. Nakanishi (*Trends in Neurosci.* **9**, 41–4, 1986); these examine the relationship of substance P to the other members of the family of 'tachykinins'. In von Euler's own review, he briefly restates the principles that should be observed for identifying and characterizing the pharmacologically active substances in the tissues, as originally expressed by Gaddum in Chapter II of this monograph; and he points out that nearly all the conclusions about the properties and distribution of substance P, which had been drawn from studies of preparations that had been well characterized pharmacologically rather than chemically, had been validated by research based on the pure peptide.

As to the functional roles of substance P, much remains to be learned. But I think it is clear that most of the present workers in this field think that it is involved in the generation of the 'flare' of the cutaneous triple response (see Chapter VIII, Note 6), in the central mediation of nociceptive information, in transmission at some of the synapses in the gastro-intestinal tract and in autonomic ganglia, and probably also in central pathways other than those mentioned.

9 There have been no recent references that I know about to the work of Major and his colleagues, and I find it as hard as Gaddum did to identify their substance with any of the known autacoids. Perhaps it was a mixture, as Gaddum suspected.

10 Today's readers can readily recognize the historical significance of the five reports that Gaddum cites in this paragraph. They were harbingers of the remarkable developments that began – or more accurately, greatly accelerated – some 25 years later, and that eventually led to (a) the chemical and pharmacological characterization of the eicosanoids, and (b) the elucidation, which still continues, of their widespread influence on physiological and pathological processes. The current status of the members of this group has been summarized, with copious references, in the chapters of *Prostaglandins and Related Substances* (ed. C. Pace-Asciak and E. Granström, Elsevier, Amsterdam, 1983).

The history of these later researches lies outside the scope of these notes, but brief mention should be made here of a few other key papers that were published either before, or slightly after, the ones mentioned by Gaddum.

Some recent summaries of prostaglandin history credit the New York gynaecologist Kurzrok and his pharmacological colleague Lieb (*Proc. soc. exp. Biol. Med.* 28, 268–72, 1930) with the earliest observations of the effects of these substances. They were conducting an artificial insemination programme, and they noticed that in some of their patients the semen they had just instilled into the uterus was actively expelled. Pursuing these observations, they tested semen on human uterine strips suspended in a tissue bath, and they saw varied effects, contraction in some cases and relaxation in others. It is indeed likely that prostaglandins were responsible for most of these effects, but unfortunately a later paper by Kurzrok and his co-workers (*Am. J. Physiol.* **112**, 577–80, 1935), which has not been cited so often, ascribed them to acetylcholine.

Two other papers from 1930, which Gaddum would hardly have thought of mentioning even if he had read them, turned out long afterward to contain information of significance in the present context. These papers were by G. O and M. M. Burr, and reported their discovery that the polyunsaturated fatty acids are essential food factors in the rat. The Burrs suggested that one such acid, arachidonic, might be of special physiological importance.

Gaddum, when revising his book, probably reached this section before he saw the recent short report by von Euler (*Klin. Wschr.* **14**, 1182–3, 1935) on the active principles of prostatic tissue and fluid. (The activity really came from the seminal vesicles rather than the prostate, but that point was established later.) In this report von Euler introduced the term 'prostaglandin' (in the singular, although he already suspected that there was more than one active substance corresponding to his definition), and he distinguished it from other autacoids on the basis of its acidic nature and its solubility in ether as well as its pharmacological properties. Two years later he wrote a longer paper (*J. Physiol.* **88**, 213–34, 1937), in which he gave a more extended account of 'prostaglandin' and differentiated it from the similar 'vesiglandin' produced by the corresponding glands of the monkey. His collaboration with his biochemical colleague Sune Bergström began about 1945, and was steadily intensified as the number and versatility of the prostaglandins came to light.

Prostaglandin research had an impressively long gestation period. But it seems clear that von Euler knew from the outset that he was dealing with extremely powerful substances, which for reasons not yet obvious occur in uniquely high concentration in the male's accessory sex glands (something like a million times higher than in any other resting tissue), and that in spite of his other research interests he never lost his determination to solve some of the problems these agents presented. Eventually it was realized that the eicosanoids are unique in another respect: in contrast to the other agents that Gaddum surveyed in this monograph, they are formed and liberated on demand, not stored within granules or vesicles to await their release.

In the fullness of time, the separate research interests of von Euler and Feldberg recruited younger contributors, who were the first to draw attention to what turned out to be a new family of eicosanoids, the leukotrienes (see Chapter VIII, Note 13).

11 Later authors have generally agreed with Gaddum that the 'urohypertensin' of Abelous and Bardier (1908) was probably the same substance as the 'Kallikrein' of Kraut, Frey and Werle (1930a).

12 Kallikrein had to wait even longer than 'prostaglandin' did – 20 years at least – before it was widely recognized as existing *in vivo* and likely to be involved in normal physiology. The original idea of its discoverers (Frey and Kraut, 1928) was, as Gaddum explains, that it is a vasoactive hormone (*Kreislaufhormon*) made in the pancreas and excreted by the kidney. This hypothesis could not be sustained for long, nor could the companion one that circulating kallikrein is inactive because it is loosely combined with an inhibitor, also of pancreatic origin. It is now evident that kallikrein is a serine protease; that it is formed in several organs, including pancreas, kidney, intestine and salivary glands; and that it circulates as an inactive precursor, prekallikrein, which can be activated in a variety of ways. (Nevertheless it should be noted that the kallikrein inhibitor found in pancreas by Kraut, Frey and Werle (1930b) does exist, and in fact is identical with the trypsin inhibitor isolated from that organ a few years later by Northrop and Kunitz.) The key discovery was reported in 1937 by Werle and his associates (*Biochem. Z.* **289**, 217–33), who showed that kallikrein acts on a plasma-protein substrate to produce a powerful smooth-muscle-stimulating and hypotensive peptide, which they first called 'Substanz DK' (*Darmkontrahierende Substanz*) and later re-named 'kallidin'.

Kallidin can now be recognized as the first of the 'kinins' to be discovered, just as kallikrein was the first 'kininogenase'. The next kinin to make its début was kallidin's near relative 'bradykinin' (Rocha e Silva, Beraldo and Rosenfeld, *Am. J. Physiol.* **156**, 261–73, 1949); it is produced by the action of trypsin or a snake-venom enzyme on a plasma 'kininogen'. Between kallidin and bradykinin another vasoactive peptide was identified by researchers in the USA and Argentina and given the compromise name of 'angiotensin'. Its formation, by the action of the kidney enzyme renin on a plasma protein, parallels that of the kinins, but the rest of its physiology is curiously inverse to theirs. Thus it is activated, not inactivated, in the lungs, and its renal actions are generally opposed to those now attributed to kallidin and bradykinin.

There is no shortage of recent reviews in the kinin field, and several of them discuss the field's complicated history. As usual, Douglas in the Goodman and Gilman text (see Chapter III, Note 1) provides a competent survey with key references. Of special interest to the historically-minded reader is the report of a symposium held at a time when the pharmacologically active peptides had begun to attract widespread interest: it appeared under the title *Polypeptides which affect Smooth Muscles and Blood Vessels* (ed. M. Schachter, Pergamon Press, Oxford, 1960). Most of the later themes in the kinin field are introduced in two 1980 reviews by M. Schachter (*Pharmac. Rev.* **31**, 1–17) and by D. Regoli and J. Barabé, *ibid.*, **32**, 1–46); these deal with the kininogens and the kinins respectively, and refer to the complex interactions between the kinins and the eicosanoids.

Dr M. Schachter, who has contributed substantially to research in this field, has told me that he finds Gaddum's treatment of kallikrein accurate and complete for its time of writing, except for one omission. This was the second paper by A. H. Elliot and F. R. Nuzum (*Endocrinology* 18, 462–74, 1934), which reported on the excretion of kallikrein in hypertension. In view of the current interest in the renal actions of kallikrein (see H. S. Margolius, *Ann. Rev. Physiol.* 46, 309–26, 1984) this paper has some historical importance.

13 This suggestion by Komarow has been verified by later workers who used the assay methods developed by Barsoum and Gaddum (1935a). For references see the review by Code (*New Engl. J. Med.* 296, 1459–62, 1977), who was himself active in these researches, which helped to provide a basis for the discovery of the H_2-antihistamines by J. W. Black and his colleagues.

14 Dale and Feldberg (1934b) added water to the venous fluids they collected during their skin perfusion experiments, in order to make these fluids isotonic with the medium in which the leech muscle was suspended. The dilution was 1:1.4 by volume, and at first the mixing was done by simply inverting the graduated test-tube, with a finger-tip as a stopper. When it was found, unexpectedly, that the leech muscle sometimes contracted in response to a control sample, the procedure was checked and the experimenter's skin was seen to be the source of the activity. The active material was naturally named 'Fingerstoff' (though not for publication) and was eventually identified (A. Vartiainen, *Archs. int. Pharmacodyn.* 56, 349–62, 1937) as a mixture of potassium, choline and acetylcholine, which act synergistically on the leech.

An oft-repeated tale from Dale's laboratory was that when G. L. Brown visited there on Boxing Day of 1933, to confirm his appointment as Gaddum's successor, he found his prospective chief boiling his tennis socks. The extract, however, proved to be inactive on the eserinized leech muscle (W. Feldberg, personal communication, 1986).

15 Perhaps the findings of Grant and Jones have been confirmed and analysed with modern methods, but I have found no evidence of it. The activity of their frog skin extract was probably due at least in part to substance P, which does occur in the skin of English frogs (*Substance P*, Vol. 2, ed. P. Skrabanek and D. Powell, Eden Press, Montreal, 1980, p. 64), or to a related tachykinin. Substance P, like many other basic peptides, is a histamine releaser (Johnson and Erdös, *Proc. Soc. exp. Biol. Med.* 142, 1253–6, 1973) and is known to contribute to the triple response of human skin (Hagenmark, Hökfelt and Pernow, *J. invest. Dermatol.* 71, 233–5, 1978).

16 Collip's 1928 paper is not cited in any of the recent monographs or citation indexes that I have consulted. So I am inclined to think that neither he nor anyone else ever identified the material responsible for the pressor activity of his extracts. Perhaps it was a mixture of amines. Pharmacologically, as Gaddum's summary shows, his preparation behaved rather like tyramine, but there could hardly have been enough tyramine in the tissues to account for the activity observed.

In a later paper (*Trans. R. Soc. Can.*, Sect. 5, 3rd Ser. **23**, 165–8, 1929) Collip reported that the 'prostate' gland of the bull contained an adrenaline-like substance, and von Euler seems to have accepted that interpretation in his Sherrington Lectures (*Release and Uptake Functions in Adrenergic Nerve Granules*, Liverpool University Press, 1980).

VII. The release of specific active substances by nerve impulses

1 It may be useful to cite here some of the more important reviews published during the next few years. Cannon's Kober Lecture should certainly be added to Gaddum's list. Gaddum mentions it in a footnote, but his Bibliography omits the journal reference, which is *Am. J. Med. Sci.* **188**, 145–59, 1934. The 1934 reviews by G. A. Alles and by Z. M. Bacq, though not listed here, are referred to elsewhere in the text and in the Bibliography. Bacq published a second important review in 1935 (*Ergebn. Physiol.* **37**, 82–185); it was in French and was almost of book length.

By 1936 it was clear that chemical transmission was now a major theme in physiology and pharmacology. The massive review by J. C. Eccles (*Ergebn. Physiol.* **38**, 339–44) appeared in that year: it was the first to deal in detail with the electrophysiological aspects of chemical transmission; Gaddum brought out the present monograph at almost the same time; and to cap that remarkable year, Loewi and Dale received their expected Nobel prizes.

The pace of reviewing was to be further accelerated in 1937. There were five published lectures by Dale; a single issue of *Physiological Reviews* contained articles by G. L. Brown, Arturo Rosenblueth, and Eccles; and with the publication of *Autonomic Neuroeffector Systems*, by Cannon and Rosenblueth, the field of chemical transmission received its first monograph that was devoted primarily to that topic. The famous names just mentioned remind us that the time was one of keen controversy as well as dramatic progress.

2 I admire this sentence of Gaddum's. It is clear and succinct, and I wonder if he drafted it easily, or only after deep thought. I do have reservations about it myself, as will appear; but I realize that anything I could put in its place would be longer, less precise, and perhaps in the end no more accurate. Nevertheless, having given that warning, I propose to comment

at some length on the early history of the chemical transmission concept. I think that the origins of the concept are more complex than Gaddum implied them to be, and that some other early contributors deserve to be mentioned.*

(1) If Gaddum had written this section a year or two later, he would have been obliged to consider giving priority to a still earlier author, the German physiologist Emil Du Bois-Reymond, rather than to Bayliss or Elliot. For in 1937 Dale gave a short communication to the Physiological Society (*J. Physiol.* **91**, 4P), in which he gave Du Bois-Reymond credit for the first clear statement of the chemical transmission hypothesis. Many subsequent authors have followed Dale's lead. It should be noted, though, that Dale was not the first to cite Du Bois-Reymond in that context; for Langley, when he gave his Croonian Lecture of 1906 (*Proc. Roy. Soc.* B **87**, 170–94), had interpreted the German scientist in the same way that Dale did. Dale had apparently forgotten that, and Gaddum, writing in 1936, may have read neither Langley's lecture nor Du Bois-Reymond's article. (I think that Dale respected Langley but didn't like him much, which may help to explain his lapse of memory.)

Du Bois-Reymond's statement was first published in 1874 (*Monatsber. kgl. Preuss, Akad. Wiss.* 20 April), and again twice in 1877. Dale in his communication quoted, in the original German, what he thought to be the key passage. That passage, read by itself, certainly seems to be clear enough, and to specify electrical and chemical transmission as the two possible mechanisms by which the nerve impulse might bring about the response of the muscle fibre.

Dale's interpretation, however, has been challenged by Krnjević (*Physiol. Rev.* **54**, 418–540, 1974). As Krnjević points out, Du Bois-Reymond wrote at a time 'before general acceptance of the neuron theory securely established the existence of a protoplasmic discontinuity across which the neural signal must be carried'. And when the passage usually cited is considered together with its rather murky context, its clarity is not so impressive. It then seems to refer, not to neuromuscular transmission as we think of it today, but rather to the spread of excitation from the motor end-plate to the rest of the muscle fibre. (It is easy for us to forget that well into the present century there were respected histologists and physiologists who thought that when a postganglionic fibre supplies a smooth-muscle cell, it goes right through the latter's sarcolemma to end close to the nucleus. Cannon in his Kober Lecture of 1934 (reference in Note 1) was one who clearly favoured that view.) I therefore find Krnjević's argument persuasive, so I don't think that Du Bois-Reymond really deserves the priority given him by Langley and Dale.

*For an addendum see p. 235.

(Some of Gaddum's friends may know – I don't myself – what Gaddum's later opinion was on this historical point.)

(2) Another figure from the late nineteenth century worthy of mention is the Italian psychiatrist Ernesto Lugaro, many of whose views on brain function can still seem enlightened to a modern reader. Here is one of several passages that I might have chosen from his long article of 1899 (*Riv. Patol. nerv. ment.* 4, 481–547: my translation).

I believe rather that this fact shows how the nerve terminal exerts on the contiguous neurone an action that is not purely physical, comparable to the transmission of heat or electricity, but a chemical action that extends for a certain distance in every direction.

So I think that Lugaro, though he gave no hint of the sort of chemical agent that he supposed might be involved in this transmission, belongs on the list of Loewi's forerunners. (Lugaro has been quoted occasionally in this context; I heard about him from Krnjević.)

(3) I find no clear reference to the possibility of chemical transmission in Bayliss's (1901) paper on antidromic vasodilatation following stimulation of fibres whose cell bodies were in dorsal root ganglia, even though the long duration (more than a minute after the end of stimulation) of the response might have been thought to suggest a chemical mechanism.

(4) In chronological sequence, the story now presents a lively young German scientist, none other than Otto Loewi himself. The year is 1903. Loewi has recently visited England, where he left a vivid impression on his contemporaries. Now he is back in Marburg, and one of them is his guest: it is Walter Morley Fletcher, and the two young men have gone for a country walk. They are talking about antagonism between drugs, when suddenly Loewi makes a remark that Fletcher will never forget; isn't it possible, Loewi wonders, that instead of muscarine acting on vagal endings, these endings act by releasing a muscarine-like substance? Wouldn't that explain why atropine abolishes the effects of both nerve and the drug? The two men then discuss how the idea might be tested; but thereafter such ideas sink into Loewi's subconscious – not to re-emerge, perhaps, except in the disguise of his famous repeated dream of Easter 1920.

Some years after that dream and the epoch-making experiments it engendered, Loewi is in Boston for the 1929 Physiological Congress. He and others form a group round Walter Cannon, who asks him: 'How did you get the idea?' 'I don't know', Loewi replies, 'it just came.' But as it happens, Fletcher is also in the group, and he now recounts what Loewi had said in 1903.

Loewi published his only account of that Boston conversation in the 'Autobiographical Sketch' (*Persp. Biol. Med.* 4, 3–25, 1960) that he wrote in his late eighties. There he tells this story, and many others, with his usual dramatic force, but unfortunately with an occasional lapse of memory. For example, he has forgotten that in his classical experiments of the 1920s on the frog heart, he had to stimulate the combined vagosympathetic trunk, not the vagus and the sympathetic separately as he later thinks he did; in fact the two components are not easily separated by dissection. And by 1960, with Cannon and Fletcher dead, there was no one to confirm what was said in Boston.

Now, however, I think I can support Loewi's credibility on the Marburg incident. For I have found that he reported the Boston conversation in the same terms in 1934, in one of the many long letters he wrote to Dale (Dale Correspondence, Royal Society of London, letter dated 26 April 1934). Loewi wouldn't invent the story, I'm confident, and in 1934 his memory was fine. And it seems most unlikely that Fletcher would have imagined the conversation of 1903.

Shall we then conclude that Loewi was really the first to conceive a precise idea of chemical transmission at autonomic junctions, as well as the first to demonstrate the correctness of the idea? Perhaps. But here a further comment is called for. Since it was Loewi's delight throughout his life to exchange novel, even daring, ideas with this scientific friends, might he not have passed this idea on to others, or even heard it from others, before he went hiking with Fletcher? Might not such an exchange have occurred in 1902, during his visit to Langley's laboratory, when he became acquainted with Fletcher and Elliott, among others. The Cambridge conversations would surely have included the recent confirmation by Langley of Lewandowsky's discovery that the actions of adrenaline closely mimic those of the sympathetic.

Dale, in his memoir of Loewi (*Biog. Mem. Fell. Roy. Soc.* 8, 67–89, 1962), speculates about the Marburg incident. He does not doubt that it occurred, but he doubts its significance. 'It seems reasonable, in any case', he writes, 'to suppose that ideas of that kind were waiting, as it were, to emerge in the Cambridge community, during Loewi's visit there in 1902...But although Loewi, in later years, was not unnaturally disposed to attribute some retroactive importance to the incident, it seems clear to me that it amounted to nothing more than a casual suggestion.' Fair comment! Dale knew the Cambridge milieu of 1902 with the intimacy of a member, and he also knew Loewi. But when he wrote the passage I have just quoted, he had been firmly convinced for nearly sixty years that the first real approach to the modern concept of chemical transmission had been made by his admired friend Thomas Renton Elliott.

(5) Elliott's name certainly comes next in the chronological sequence. He owes his place, of course, to one sentence, much admired and often quoted, at the end of his first brief scientific publication (*J. Physiol.* 31, xx–xxi, 1904). The idea it expressed is one he never again stated with such clarity. Here is the sentence once more: 'Adrenaline might then be

the chemical stimulant liberated in each case when the sympathetic impulse reaches the periphery.'

I too admire the sentence. But is it heresy to suggest that it contains a concealed ambiguity, one that is nearly invisible to the modern reader? For Elliott does *not* say (though we may think he does) that the liberated adrenaline comes from a pre-synaptic site, from the nerve ending. It seems to me that a 1904 reader, even though he were to find Elliott's suggestion engaging, would be likely to think first of a post-synaptic site of origin: a micro-depot of chromaffin material, perhaps, located close to, or even inside, the effector cell. I do wish that Elliott had written 'liberated from the nerve'! Then his wording would have been unambiguous, and it would have clearly expressed the idea that, according to his friend Dale, had been in his mind all along.

Here I feel obliged to make a guess that has no documentary support. This is that Elliott *did* write 'liberated from the nerve', but that Langley, a compulsive editor of other people's papers, struck the last three words out. Langley would not have hesitated to make such small revisions as he thought would improve his student's report; and he was certainly subtle enough to hedge his bets, and Elliott's, in the way that I am now suggesting. Moreover he owned, as well as edited, the *Journal of Physiology*!

Whatever happened between the two men, Elliott never advanced his hypothesis again, not even in the full paper of 1905 that followed his preliminary note. That he did not do so was a severe disappointment to his slightly older friend, Dale, who, as I have said, never had any doubt about what Elliott was proposing in 1904: namely, adrenergic transmission in the modern sense of the term. Dale discusses the puzzle of Elliott's reticence in his memoir of the latter (*Biog. Mem. Fell. Roy. Soc.* 7, 53–74, 1961). That memoir, and the one Dale wrote about Loewi a year later, are important contributions to the early history of chemical transmission.

(6) In 1906 Langley himself appears as a proponent of Elliott's hypothesis, though without mention of Elliott. In his Croonian Lecture of that year (*Proc. R. Soc.* B 78, 170–94), which I have already referred to, he discusses (see especially page 183) the likelihood of chemical transmission at what we now call nicotinic synapses, in words that escape the sort of ambiguity that I think he may have imposed on Elliott. Had Langley's ideas changed since 1904, or had he all along, perhaps for years, been attracted by the idea of chemical transmission? The Russian authors Mikhelson and Zeitmal (*Acetylcholine – An Approach to the Molecular Mechanism of Action*, Pergamon Press, Oxford, 1983) and Kharkevich (*Handb. exp. Physiol.* 53, 1–9, 1980), perhaps influenced by A. G. Ginetzinsky, appear to favour the second possibility, and I don't

think we can exclude it. It would have been consistent with what we know about Langley to find him hesitant to express a concept that attracted him until he had buttressed it with a sufficient weight of fact. He may have felt that he had done so by 1906.

(7) Perhaps it is significant, however, that by 1906 Langley had accepted for the *Journal of Physiology* the second of two papers (*Brain*, **28**, 506–26, 1905; *J. Physiol.* **34**, 145–62, 1906) by F. H. Scott, in which this long-neglected author argued in favour of chemical transmission at *all* synapses, with axonal transport as a corollary hypothesis. Scott thought that the transmitters were probably proteolytic enzymes. In his *Journal of Physiology* paper he expresses his appreciation of Langley's editorial comments; Langley, however, does not refer to Scott in his Croonian Lecture, or, so far as I know, in any later paper. (For an account of Scott's career and scientific contributions, see H. Blaschko, *Notes and Records of the Royal Society* **37**, 235–47, 1983).

(8) A few years after accepting Scott's paper, Langley allowed his pharmacological colleague Walter Dixon to speculate in the *Journal of Physiology* (Dixon and Hamill, **38**, 314–36, 1909), as he had already done elsewhere in the two reports cited by Gaddum (Dixon 1906, 1907), about the possible involvement of a muscarine-like substance in parasympathetic neurotransmission. By that time, as Langley could see, the concept of chemical transmission was becoming familiar. Within a few years another of his ex-students, Dale, would make that concept the basis of his discreet speculations about the possible *physiological* importance of two substances, acetylcholine and noradrenaline, that had not yet been found in any animal tissue. And it is legitimate to guess that Dale, in the still earlier conversations that he surely had with most of the people named above, was not a mere listener to their novel ideas, but helped in their refinement.

Anyone who has read right through this lengthy note will, I hope, agree with me that the concept of chemical transmission was glimpsed by a number of people, including Loewi himself, before Loewi made it a reality. Perhaps such a reader will also agree that we have no firm basis now, and are not likely to have one in the future, for apportioning the credit for inventing the concept between the men whose names I have chosen: Lugaro, Loewi, Elliott, Langley, Scott, Dixon, Dale, and quite possibly others that I haven't heard of. (I don't think that Du Bois-Reymond, Bayliss or Howell, all deservedly famous for other researches, really belongs on my not-so-short list.) Indeed, anyone who tries to reconstruct a chapter in the history of science should remember that scientists who have bright ideas quite often talk about them before they write about them. Then the bright idea may seem to come out of the milieu, with no single inventor's name attached to it. That the chemical transmission concept may have had such a beginning need not cause us any distress.

3 In spite of all the attention I have just given to Otto Loewi's predecessors, I want to emphasize that there has never been any doubt in my mind that it was primarily Loewi who made chemical transmission a major theme in biology and medicine. This small, merry, voluble man with his genius for friendship is still remembered with affection as well as with admiration by his younger colleagues. The memorial volume edited by F. Lembeck and W. Giere, entitled *Otto Loewi – Ein Lebensbild in Dokumenten* (Springer-Verlag, Berlin, 1968) is a worthy tribute (it includes the autobiographical sketch already mentioned), and so is Dale's memoir of his friend (see Note 2 (4)).

4 In the corresponding paragraph of Gaddum's original typescript, and in its German translation, the words 'nerves' or 'Nerven' are used, where 'nerve fibres' or 'axons' would seem more appropriate today. I have used one of the latter terms where I thought it was called for.

5 Perhaps Loewi's ideas were rather better than his experimental realization of them. His basic experiment is simple enough in principle, but one can go wrong with it. Part of the problem is that the frog heart mounted on a Straub cannula is a delicate and somewhat temperamental preparation; I can testify myself that it is tricky to pipette Ringer solution from one cannula to the other without disturbing the beat of the recipient heart. Loewi might have saved himself a lot of controversy if he had arranged his set-up in a way that allowed the fluid from the donor heart to be delivered automatically to the recipient heart. R. H. Kahn's double cannula of 1926 was perhaps the most elegant of the devices used by Loewi's successors. But if Loewi had used such a cannula from the start, we should not have been able to enjoy reading his brisk polemical exchanges with Leon Asher of Bern, the most scathing and persistent of his critics, who was sure that the *Vagusstoff* effect was an artefact.

Here it is appropriate to mention the little book by Zénon Bacq of Liége, entitled *Chemical Transmission of Nerve Impulses: A Historical Sketch* (Pergamon Press, Oxford, 1975); it is part critical analysis and part personal memoir, and altogether delightful to read. Bacq points out another potential source of difficulty with Loewi's experiment, namely the seasonal variation in the balance between the cholinergic and adrenergic components of the frog's vagosympathetic nerve trunk. He thinks that Loewi was lucky to have started his experiments in the spring, because later in the year his *Acceleransstoff* would not have manifested itself well, in which case Loewi would have missed half of his discovery.

Loewi invited Asher to visit Graz and see the experiment being done properly, but apparently Asher refused. In 1926, however, Loewi had the opportunity of giving a demonstration of the *Vagusstoff* transfer at the

International Physiological Congress at Stockholm. There is a photograph
of the scene in Holmstedt and Liljestrand's *Reading in Pharmacology*
(Macmillan, New York, 1963). As Loewi reminisced in his
'Autobiographic Sketch': 'Like most experimenters, I had experienced
time and again that experiments before a large audience often failed
although they never did in the rehearsals. Fortunately, I was able to
demonstrate in Stockholm the experiment not less than eighteen times on
the same heart.'

6 I find it hard to judge retrospectively whether any of these enthusiastic
followers of Loewi really succeeded in demonstrating the release of a
Vagusstoff from mammalian hearts without using an anticholinesterase to
preserve it. Probably most of them saw *something* transferred from one
heart to another when they stimulated the vagus of the first. But was
that something acetylcholine? Or was it choline, or adenosine, or a
neuropeptide, or some other unknown material released by the heart in
response to a change in coronary flow?

Today's cholinergic physiologists should feel some compassion for their
predecessors who kept looking for *Vagusstoff* and had not yet learned how
vulnerable it is to cholinesterase. The experiment does work if you do it
on an amphibian heart, as Loewi fortunately did, or on an avian heart, as
Kilbinger and Löffelholz showed much later (*J. Neural Transm.* **38**, 9–14,
1976). But if you must choose a mammalian heart for your trial,
you should be grateful to Feldberg and Krayer (1933), who, by using
eserine, provided the indispensable paradigm for such experiments; and
going back a little further, to Engelhart and Loewi (1930) and Matthes
(1930), who demonstrated that eserine inhibits cholinesterase.
(Acetylcholine differs from most transmitters in that its precursor,
choline, is also its breakdown product, so without doing special controls
one can't assess its turnover by merely measuring the incorporation or
release of choline).

7 Of course, Howell and Duke were good experimenters. Their result was
the correct one: vagal stimulation does increase myocardial permeability
to potassium, hence the muscle is hyperpolarized, the action potential is
cut short, and the beat is weakened. With the colorimetric method they
used they did well to find the modest efflux of potassium and demonstrate
its atrial origin. They did not suggest that it came from the vagal endings,
and indeed they could hardly have considered that as a possibility: small
as the efflux was, it was too big for that. They could hardly have guessed,
in 1908, that the efflux was a sign, not the cause, of the inhibition.

8 Rudolf Magnus of Utrecht was the first experimenter to study the
responses of isolated smooth-muscle preparations, and most of the work
described in the next four paragraphs of the text came from his
laboratory. His Lane lectures (1930), which summarize the main findings

of twenty years of work, are hard to obtain but they are still worth reading. Magnus and his school regarded choline as the motor hormone of the gut and speculated about its probable conversion to acetylcholine years before Loewi reported on *Vagusstoff*. If they had used eserine in their experiments they would have discovered that the isolated intestine has a brisk acetylcholine metabolism.

9 The citations here should read 'Arai, 1922; Zunz and György, 1914'. Arai's findings helped to give Magnus the idea that post-surgical paralysis of the bowel (and bladder) might be due to a shortage of choline. The 1930 review by Magnus reports the results of clinical trials, stimulated by his laboratory findings, in which appropriately large doses of choline were administered to patients with ileus, in some cases with apparent relief. Plasma choline was not measured in these clinical trials, though it probably could have been with the techniques used by the Utrecht workers. I had an opportunity once to make such measurements, in about a dozen patients before and after major surgery, and was impressed to find that the level fell quickly after the operation and usually stayed down for a few days (*Canad. J. Biochem. Physiol.* **45**, 2555–71, 1963). My surgical friends and I wondered whether returning the plasma free choline to its normal level, which we knew was 10–15 μM, might help to restore the gut's activity, as Magnus had proposed. But we never tried it, and so far as I know no one else has done so yet. It would be a demanding trial, since it would probably require grams of choline, given by intravenous drip, with frequent measurements of plasma choline, to restore the level to normal; but I think that the problem posed by Magnus deserves investigation by modern methods. One should remember, of course, that acetylcholine is far from being the only transmitter in the enteric plexuses. Even if one supposes that paralytic ileus is the result of synaptic failure somewhere in those structures, the fault might still lie in under-supply of some other substance, e.g., a neuropeptide. (From our own tests we had some evidence that the post-surgical fall in plasma choline was due to glucocorticoid release.)

10 These old experiments of Le Heux's are interesting, and his interpretation of his findings has some plausibility. Perhaps these too are worth trying to confirm.

11 I have no information, perhaps because I haven't looked hard enough, about the identity of the salivary secretagogue. Its destruction by heat suggests that it might be one of a number of enzymes that are stored in the striated duct cells of the salivary glands of several species and secreted into the saliva, especially in response to adrenergic stimulation (see Schachter, *Pharmac. Rev.* **31**, 1–17, 1980 for a review). Kallikrein is one of these, but its identity with the secretagogue seems doubtful (for references see Burgen and Emmelin, *Physiology of the Salivary Glands,*

Edward Arnold, London, 1961, p. 87). The function of the secretagogue is also unknown to me; perhaps it is related to taste.

12 Throughout the 1930s, as well as later, many of Loewi's successors used perfusion techniques to collect the released transmitter. But there were always some investigators who eschewed that approach, because they thought that perfusion or irrigation of a tissue with a saline medium might cause an artificial increase in the permeability of the cells, and thus encourage outward leakage of materials whose normal metabolism was entirely intracellular. These investigators placed more trust in experiments in which the putative transmitter was released into the blood stream at one site, and carried like a hormone by the blood to another site where its action could be detected. Cannon and his colleagues were of that school of thought. In his Kober Lecture of 1934 (see Chapter VII, Note 1 for the reference) Cannon quotes with approval the 'prudent scepticism' of Demoor, who wrote (*Ann. Physiol. Physio-Chem. Bio.* 5, 42–129, 1929) that exposure to saline media 'may create new conditions of existence for the tissues, conditions accompanied by permeabilities which do not exist in physiological states...the escape of vagal and sympathetic substances, though proved experimentally, may not occur normally.' Similarly, my own supervisor Babkin felt that his experiments (Babkin *et al.* 1932a, b) on acetylcholine release from salivary glands *in situ* were superior to those of his friends Gibbs and Henderson (see pp. 133–5) because the only abnormalities he imposed on the tissues were the presence of the anaesthetic and of eserine.

An episode I have reason to recall myself came a few years later, when the admired neuroscientist R. Lorente de Nó attempted to repeat the ganglion-perfusion experiments of Feldberg and Gaddum (1934) and Feldberg and Vartiainen (1934). Lorente describes (*Am. J. Physiol.* 121, 331–49, 1948) the special precautions he took to keep his small piece of nervous tissue in good shape, but he could find no dependable relationship between the preganglionic stimulation and the appearance of acetylcholine in the perfusate. It even seemed that the release was least likely to appear in the experiments that he thought were technically the most successful. His experience helped to persuade other researchers, especially some senior electrophysiologists, that the acetylcholine release attributed by Feldberg and his colleagues to preganglionic terminals was really an artefact, which was exhibited only in the presence of an unphysiological increase in the permeability of some neural element.

13 Both Henderson and Roepke (1932, 1933b) and Gibbs and Szelöczey (1932) in their perfusion experiments seem to have demonstrated the release of a second active substance in addition to acetylcholine. The Henderson team found something that was set free in the absence of eserine, and the Gibbs team found something whose action was not

blocked by atropine. Could this have been a peptide, for instance VIP (vasomotor intestinal peptide), or an adenosine compound, or something different from either of these? Most readers of this book would now guess VIP as the best bet, I suppose. If so, and if Henderson and Gibbs had given higher priority to the characterization of the second substance, they might have been the first researchers to provide evidence of peptidergic transmission (or co-transmission) in an autonomic pathway. (See Notes 29 and 30, below.)

14 These experiments, of course, would not have been possible without the use of the leech muscle preparation, which is insensitive to adrenaline (and also to atropine). But the experiments of the following year, with Tsudzimura, were no mere confirmation, because in these the experimental cat itself was made to serve as the sole pharmacological test-object (cf. Note 12, above). For a modern reader, of course, the principal significance of all these experiments on the suprarenals is something that Feldberg and his co-workers were well aware of: namely, that they provided the first strong evidence for chemical transmission at any set of inter-neuronal synapses. (Admittedly the cells of the adrenal medulla could hardly be regarded as typical neurones.)

15 Kibjakow was not as fortunate as Loewi has been at Stockholm (see Note 5, above). I have been told that this experiment worked regularly for him until he undertook to demonstrate it at the Moscow Physiological Congress of 1935. It has never worked for anyone since then, except when an anticholinesterase is added to the perfusion fluid, and when that is done it always works. The most likely explanation, I think, is that by mischance some enzyme inhibitor had got into the ingredients of Kibjakow's medium. Another possibility is that the perfusion was slow enough to allow detectable amounts of a co-transmitter – ATP or a peptide – to accumulate in the venous fluid. Whatever the truth of the matter, Kibjakow's ingenious perfusion technique has been the delight of a succession of cholinergic physiologists for over fifty years.

16 H. C. Chang came from China in 1932 to work with Gaddum. The outcome was their paper on the measurement of choline esters in tissue extracts (Chang and Gaddum, 1933: surely one of the classics of chemical transmission literature). Professor Chang is still alive and active in Beijing. He has told his friends how he worked very hard during his months in London, but planned to take one day off for sightseeing before he returned to China. When the time came, however, he decided it would be better to measure the acetylcholine content of a sympathetic nerve from the horse than to view the Tower of London (H. C. Chang, personal communication, 1985). He found an impressively high content in both the cervical sympathetic trunk and the superior cervical ganglion, so he and Gaddum advanced 'the theory that acetylcholine might play a part in

the normal transmission of impulses through ganglia'. After Chang's return to China, he and Wong (1933) discovered that the human placenta is extraordinarily rich in acetylcholine, a phenomenon whose teleological significance is still obscure.

17 The *amounts* of acetylcholine discharged by the blood-perfused ganglia when they were stimulated were rather minute, but because the flow rate was much smaller when whole blood was used, the *concentration* of the transmitter was similar to that observed in experiments with a saline medium. The investigators did not comment on the discrepancy, perhaps because Gaddum was no longer present. It took some time to realize that when whole blood was used as the perfusion medium, the concentration of eserine had to be much higher than in the earlier tests with Locke's solution.

18 There are several engaging personal accounts of this strenuous but good-tempered debate. Especially recommended are those by Sir John Eccles (*Notes and Records of the Royal Society of London*, 30, 219–31, 1975–6: it is based on his talk at the Dale Centennial meeting in Cambridge) and Professor W. S. Feldberg (in *The Pursuit of Nature*, Cambridge University Press, 65–83, 1977: from his contribution to the centennial meeting of the British Physiological Society). Bacq's little history (see Note 5, above) gives a particularly colourful account.

19 From the brief passages (pp. 144 and 154) in which Gaddum refers to unpublished experiments by Dale, Feldberg and Vogt (1935) it is clear that Dale and his colleagues were nearly convinced that acetylcholine is truly the transmitter in muscles, just as it is in sympathetic ganglia. Their first short note (Dale and Feldberg, *J. Physiol.* 81, 39–40P, 1934) had in fact already appeared; and by the time that Gaddum's monograph was ready for the press, sometime in 1935, the evidence was nearly complete that impulses arriving at motor terminals do release acetylcholine. Gaddum, working in Cairo to put the finishing touches on his book, felt that he could at least hint at that conclusion. The full paper by Dale, Feldberg and Vogt (*J. Physiol.* 86, 353–80, 1936) appeared before the monograph did. Still to be presented, however, was the other half of the story. Could acetylcholine, having been released by an impulse, act fast enough (and transiently enough) to elicit a single twitch of the muscle fibre?

Within a few months that question could be answered in the affirmative (Brown, Dale and Feldberg, *J. Physiol.* 87, 394–424, 1936). It is clear that G. L. Brown played an essential part in that phase of the research. What the Hampstead workers now showed was that acetylcholine and other nicotine-like drugs, if properly applied, could evoke a contraction in a normal innervated muscle that was more powerful than its nerve-induced twitch, and almost as brief – in striking contrast to the protracted response of a chronically denervated muscle (Chapter 4, Note 6). The way to elicit this new type of response was to inject the acetylcholine as quickly as possible into the muscle's own empty artery, so that the drug's concentration at the junction rose steeply. Electrical recording then showed that the fast response was 'a brief asynchronous tetanus'. Somewhat later Brown and his co-workers showed that the typical 'contractures' of denervated muscles are electrically silent, which means that they are not conducted responses. One is entitled to wonder why the investigators did not at the time explicitly state what now seems to be the obvious corollary, namely that chronic denervation causes the receptor zone to spread. This last important generalization was first stated by Ginetzinsky and Sharmarina (*Usp. sovrem. Biol.* 15, 283–94, 1942). But 'receptor', or 'receptive substance', was not part of the vocabulary in Dale's laboratory.

20 Cannon and Rosenblueth, in their 1933 paper, presented strong evidence for a sympathin that was different from adrenaline. But their two-sympathin model went beyond the observed facts. They had no evidence for the release of a common intermediate which was transformed locally into either an excitatory or an inhibitory agent that could be carried by the blood to the organ used for its detection. Later, some of Cannon's Harvard colleagues suggested that the model was invented by Arturo Rosenblueth and reflected his liking for elegant theoretical structures. Whether or not that was so, Cannon adopted the model with enthusiasm. He was a very shrewd observer, and I think his critical judgement must have been satisfied by the experimental records that seemed to demonstrate the existence of two kinds of sympathin, one more inhibitory than the other. If that were so, of course, the sympathin effects he and his colleagues observed could not all have been due to noradrenaline (see Note 23, below). As to Arturo Rosenblueth, whose positive contributions to this field of research are not often mentioned nowadays, he was unquestionably a resourceful experimenter, and also a man of many-sided talent, great personal charm and a degree of intellectual arrogance. For sketches of Rosenblueth against his Harvard background, read Zénon Bacq (*Chemical Transmission of Nerve Impulses: A Historical Sketch*, Pergamon Press, Oxford, 1975) and Horace Davenport (*Physiologist*, 24, 1–5, 1981).

It may be appropriate to recall here that some, and probably nearly all, of the commercial preparations of 'adrenaline' used by experimenters before 1949 contained some 10 to 20 per cent of noradrenaline, even though the hormone had been recrystallized (Goldenberg *et al.*, *Science* 109, 537–8). The presence of the contaminant may have made it slightly harder for the physiologists of the 1930s to distinguish the released transmitter from adrenaline. It seems unlikely, though, that this was a factor that influenced the debates on the transmitter's chemical identity.

21 In his autobiography (*The Way of an Investigator*, W. W. Norton, New York, 1945) Cannon explains why he set aside this highly suggestive observation for eight years, in order to pursue his work on the physiology of the totally sympathectomized animal. It was not until Bacq came to his department that he felt ready to give priority to his study of the agent released by sympathetic stimulation. By that time, of course, Loewi's researches had added greatly to the potential interest of such a study.

22 Finkleman seems to have been the first student of chemical transmission to make regular use of a 'superfusion' technique. His successors have included H. Kwiatkowski (*J. Physiol.* **100**, 147–58, 1941); Gaddum, who coined the term (*Br. J. Pharmac.* **8**, 321–6, 1953); and J. R. Vane, in the *Second Gaddum Lecture* (*Br. J. Pharmac.* **35**, 209–42, 1969).

23 It is of interest that Gaddum makes no reference to Bacq's suggestion of 1934 that 'sympathin' might be noradrenaline. Indeed, there is no mention of noradrenaline anywhere in Gaddum's monograph. When Gaddum was writing it, of course, noradrenaline (Arterenol) was merely an interesting compound that had been made by pharmaceutical chemists in Germany; though some people would have remembered that Barger and Dale (*J. Physiol.* **41**, 19–59, 1910) had hinted long before at its apparent suitability for the transmitter role that Elliott envisaged for adrenaline. (Dale comments engagingly on this bit of history in his *Adventures in Physiology*, Pergamon, Oxford, 1953; Wellcome Trust, London, 1965, Comment on Paper 4.) Gaddum was, however, clearly impressed by Bacq's chemical experiments, whose findings he reported in detail. My guess is that Gaddum's opinion at that time was that 'sympathin', from whatever source, was mostly adrenaline mixed with the products of its partial inactivation (cf. Gaddum and Kwiatkowski, *J. Physiol.* **96**, 385–411, 1939).

   Perhaps Bacq in his later years may have unconsciously exaggerated the strength of his early attachment to the noradrenaline hypothesis. Originally his statement of it was somewhat tentative. In the 1934 paper, which Gaddum cites, and also in his impressive 1935 review (*Ergebn. Physiol.* **37**, 82–195), his preferred candidate for the role of sympathin was a partially oxidized derivative of adrenaline.

   (Bacq believed, correctly, that the susceptibility of adrenaline to oxidation must be of physiological significance, and he kept trying experiments based on that idea. I remember that about 1938 he thought he had found a mushroom enzyme that would convert adrenaline into a strong vasodilator, and he came to Dale's laboratory with the intention of demonstrating his finding to the Physiological Society. But unfortunately the mushroom extract, when it was tested in London, did not produce the vasodilator, nor did it ever do so again, even after Bacq's return to Liége.)

The subsequent work on the identity of 'sympathin' lies outside the scope of these notes. Others followed Bacq in proposing that noradrenaline is the transmitter, at least where the sympathetic is excitatory; but this identification was not widely accepted until 1946, when von Euler showed that the sympathin-like material in adrenergic nerves is mostly noradrenaline. It is of interest that Gaddum and his associates, who had rejected the noradrenaline hypothesis in their research reports of 1939, were among its early defenders later on (Gaddum and Goodwin, *J. Physiol.* **105**, 357–69, 1947; Peart, *J. Physiol.* **108**, 491–501, 1949), and that the 1947 paper just cited was based on experiments that had been completed in 1939 (L. G. Goodwin, personal communication).

Dr Goodwin has remarked in a recent letter to me:

I guess the Gaddum and Kwiatkowski papers published in 1939 reported work done in 1938; the noradrenaline idea superseded others during the 1939 work...All of the tracings in the 1947 paper were made at the 'Square' in 1938–9 and JHG drafted a paper more or less on the lines that eventually appeared. But he wasn't satisfied with it – and then all sorts of things happened. The Pharmacy School, because of the imminence of war, was planned to be evacuated to Cardiff, Gaddum was swept into the Ministry of Defence to work on gas warfare and I got a job with the Wellcome Bureau of Scientific Research and moved there during the late summer, 1939. Gaddum left all the tracings and results (in my notebook) with me and retrieved them when he went to Edinburgh after the war was over. He did a few more experiments there – with Grundy, I think – to make sure there had been no mistakes, and then wrote the 1947 paper. Certainly Euler's work had an influence – he also came to the Square occasionally – it was quite an international club.

Of the many reviews of the early work on 'sympathin' and noradrenaline, I shall cite only the brief one by L. L. Iversen in his book on *The Uptake and Storage of Noradrenaline in Sympathetic Nerves* (Cambridge University Press, 1967); it gives references to some of the other reviews, including those by von Euler himself.

There are still, no doubt, some exceptions to the rule that noradrenaline is the sole catecholamine to serve a transmitter function in peripheral adrenergic neurones. Thus Loewi seems to have been right when he identified the *Acceleransstoff* of the frog heart as adrenaline. And though mammalian sympathetic neurones do not usually contain adrenaline, their endings can be made to take it up and release it as a false transmitter (for references see Iversen's 1967 monograph cited above). I find it intriguing to suppose that this happened in some of the experiments of Cannon and Rosenblueth, whose animals underwent extensive operations, and may have had a high blood adrenaline level at some stage of the experiment. So perhaps in the hands of physiologists adrenaline can be made to act like the Harvard 'sympathin I', though we can never be sure that it behaved in that way in the experiments that led Cannon and Rosenblueth to propose that there are two sympathins.

A final reminiscence here is that Gaddum once told me that the prefix in 'noradrenaline' comes from 'N-Ohne-Radikal'. I was sceptical of his etymology at the time, but I suppose it must have been correct, since it is repeated in the preface to *Chemical Transmission 75 Years* (ed. L. Stjärne and others, Academic Press, London, 1981), a tribute to Ulf von Euler by his Stockholm colleagues.

24 This was an early, perhaps the earliest, attempt at a general classification of synaptically active drugs in terms of transmitter-related mechanisms. We note the presence of one dubious category (b, ii) and the absence of several of the categories and sub-categories that would be used by today's pharmacologists. It is significant that Gaddum can name only two drugs that have a pre-synaptic action: potassium (in addition to its other actions) and, perhaps too hesitantly, tyramine. (Five years later the relationship between potassium gradients, membrane potentials, and excitation was beginning to be better understood, and even pharmacologists were using phrases like the 'depolarizing action' of a drug with some confidence; by 1940 the release of acetylcholine induced by nerve stimulation or by potassium was known to be calcium-dependent.)

25 For the electrophysiologists of the late 1930s and early 1940s, especially J. C. Eccles, the inability of eserine to prolong the postganglionic response to a single preganglionic volley was hard to reconcile with the idea that acetylcholine is the principal ganglionic transmitter. The pharmacologists may have been satisfied on that point when Feldberg and Vartiainen (1934) showed that eserine could increase the effectiveness of repetitive *submaximal* stimulation, but the electrophysiologists preferred, not altogether unreasonably, to interpret that finding as due to non-specific sensitization of the ganglion cells. It took many years for physiologists generally to realize that at ganglionic junctions in particular, the transmitter can be removed from the immediate vicinity of the synapses by physical diffusion almost as rapidly as by enzymatic hydrolysis (Fatt, *Physiol. Rev.* **34**, 674–710, 1954; Ogston, *J. Physiol.* **128**, 222–3, 1955). Once that was understood, most of the 'soup' versus 'spark' controversy lapsed into silence. (That controversy was, however, put on a different basis by David Nachmansohn, who had made classical contributions to acetylcholine enzymology, but then argued tenaciously for years that acetylcholine is required for transmission along axons but not across synapses, where he believed that transmission is essentially electrical.)

26 It was still an almost universal belief among physiologists, dating back to the turn-of-the-century experiments of Ringer and Locke on the heart, that potassium ions and calcium ions are natural antagonists. Most elementary laboratory courses in physiology included an exercise that purported to prove this principle. So the results reported in this paragraph seemed to fit together comfortably. (Looking back, I think that Brown and Feldberg used calcium at too high a concentration; I cannot explain the curious findings of Aladar von Beznák.)

27 By the time that Brown and Feldberg (*J. Physiol.* **86**, 290–305, 1936) reported their studies in full, they had demonstrated that chronically decentralized ganglia have lost most of their original store of acetylcholine.

28 See Chapter I, Note 6. Neither Dale nor Gaddum at this time had been able to achieve what they thought was a clear understanding of the physiological significance of potassium and its relationship to acetylcholine.

29 In contrast to their admitted uncertainties about potassium ions (see preceding Note), Dale's group were satisfied that they had a tolerably good understanding of why atropine does not always block the effect of stimulating a parasympathetic nerve although it always blocks the effect of exogenous acetylcholine on the same tissue. Their favoured explanation of these 'atropine paradoxes', as I shall call them, was first offered in the paper by Dale and Feldberg on the gastric vagus (*J. Physiol.* **81**, 320–34, 1934). Gaddum reiterates it here, and Dale was still pleased with the explanation in 1953 when he wrote his commentary on the 1934 paper (*Adventures in Physiology*, Comment on Papers 26 and 27). So far as I know there was no serious objection to the explanation for something like forty years. Any case of atropine resistance in a parasympathetic pathway was interpreted in the same way, almost automatically, though sometimes the interpretation was made more complex by postulating that part of the effect of the released acetylcholine was to release a second agent that had actions of its own. During the last decade, however, there has been a great change in attitudes to the atropine paradoxes, with the result that they are now ascribed, again almost automatically, to the release of a transmitter, or co-transmitter, other than acetylcholine. One's first guess is likely to be that this substance is a neuropeptide, one's second guess that it is ATP.

30 I'm sure that I was one of those who found this argument convincing when I read it in 1936 or 1937. Now, with the benefit of hindsight, it is easy to find the flaws in it. The dilemma that Gaddum posed was real. One can account for the atropine paradoxes only in two ways: either one must suppose that in these tissues atropine can block the actions of exogenous, but not of endogenous, acetylcholine, or else one must admit the existence of a parasympathetic transmitter that is not acetylcholine.

But Gaddum went rather beyond his facts, I feel, when he thought that in the latter case 'a whole series of chemical transmitters' would be needed. A single additional transmitter for each paradox would have sufficed, and it could even have come from another set of fibres in the same nerve, different from the ones that released acetylcholine. (Even at that time there was never any objection to the idea that a nerve trunk could contain several kinds of fibre, with a different transmitter for each kind.)

It is perhaps a little disappointing to find Gaddum ready to postulate a variety of synaptic barriers, with different permeabilities to atropine, between the cholinergic ending and the effector cell, while at the same time he is unwilling to allow more than one kind of transmitter to operate in any parasympathetic pathway. We might expect to see Gaddum, especially in view of his later achievements, worrying about A. J. Clark's evidence (see Clark, *Mode of Action of Drugs on Cells*, Edward Arnold, London, 1933) that atropine is what we now call a competitive antagonist at muscarinic receptors, and that consequently the degree to which it blocks the action of a cholinergic nerve should depend primarily on the relative concentration of atropine and acetylcholine in the vicinity of the receptors. The idea of a barrier, as posed by Dale and Gaddum, really did not have much to recommend it, even at that time.

Obviously Gaddum was not the only physiologist of his period who resisted the idea that a neurone can make, and store and release, more than just one transmitter. In the 1930s, with evidence that the two putative transmitters on the scene each had an enormous repertoire of effects, the need for a large number of transmitters, or for two transmitters in a single neurone, must have seemed fanciful: too heretical to be entertained, let alone discussed. The converse principle, of 'one neurone, one transmitter', was never stated explicitly, so far as I'm aware. But it was widely, perhaps universally, accepted, as a corollary of Dale's classification of neurones as cholinergic and adrenergic (and no doubt, even then, as 'other-ergic', to accommodate clear misfits like the primary afferent neurones). It was a classification that had proved itself to be economical and fruitful; with Occam's advice in mind, nobody wanted to replace it with anything more complicated. All that was needed to make it universally applicable, it seemed, was one supplementary postulate to make the atropine anomalies fit into the scheme. (Some parallel anomalies of adrenergic transmission (see Section (*j*) on dioxane derivatives) could be handled in the same way). As has happened elsewhere in scientific history, the immense authority of Dale and his colleagues, and the continuing spectacular successes of their models, formed the basis of a new paradigm for explanations in this field. It was a paradigm that may have delayed the search for co-transmitter mechanisms by a decade or two.

(I think that the first proposal that one neurone may liberate two transmitters was made by J. H. Burn and M. J. Rand in 1959 (*Nature* **184**, 163–5). Their hypothesis, which never won wide acceptance, was that adrenergic neurones contain acetylcholine as well as noradrenaline; an impulse arriving at the periphery releases the former, which then acts on the terminal to release the latter. The Burn–Rand hypothesis was vigorously defended for a long time, and it helped to generate much valuable research, but it seems now to be only of historical interest.)

Here I think I should add a comment about 'Dale's Principle', about which there have been some misconceptions. By now, however, everyone should know that Dale's Principle, so named by J. C. Eccles, 'is simply that the same transmitter is liberated at all the terminals of a neuron' (Eccles, *Physiology of Synapses*, Springer-Verlag, Berlin, 1964). The statement is based on a passage in Dale's Nothnagel Lecture (Dale, 1935b), and expressed thus, it seems almost immune to challenge. For an axonal bifurcation is a very simple structure, which one can hardly suppose to possess machinery for sorting out the materials transported along the axon and directing them down different axonal branches. Dale's first research project, given him by Langley, may have biassed him towards that point of view. His task, in 1900, was to compare the fibre counts in a dorsal spinal root on either side of the ganglion, so probably he looked at many bifurcations and thought about their significance. (That first research experience did not give him a liking for detailed neuroanatomy or for laborious quantitative comparisons.)

In recent years some authors have confused 'Dale's Principle' with the related, but quite different, generalization that is expressed in the phrase 'one neurone, one transmitter'. This latter generalization, we now know, is false; we have abundant evidence that many neurones, perhaps most, make and release more than one transmitter. For many years after Gaddum wrote his monograph physiologists kept looking for transmitters other than the two they were familiar with. But few, if any, of them suspected that there might be novel transmitters in the cholinergic and adrenergic neurones with which they were so familiar. (Recall that Eccles used 'transmitter' in the singular when he re-stated Dale's Principle, and that the Burn–Rand hypothesis received a generally unsympathetic reception, although it apparently accounted for a number of otherwise unexplained phenomena.)

31 As we now know, the question posed in this paragraph had no simple answer. By the mid-1950s, however, it was possible to state that most of the store of preformed transmitter at both cholinergic and adrenergic endings was available for release by nerve impulses, and also that local synthesis of the transmitter was greatly accelerated by stimulation. The work of J. H. Quastel and his colleagues (Mann *et al.*, *Biochem. J.* **32**. 243–61, 1938), which was the first to show that transmitter metabolism

can be studied in brain slices, pointed clearly in that direction. Further evidence came from the use of drugs (e.g. reserpine, hemicholinium-3) that interfere with transmitter synthesis or storage. With the availability of isotopic labelling techniques this kind of research could be greatly accelerated.

VIII. The release of active substances in other tissues

1 With the discovery of new vasodilator substances and new mechanisms of action, the field of Gaddum's monograph long ago became too complex for a single author to survey. Unfortunately there has not been a matching yield of simplifying generalizations. I think that no one would claim today that we are close to a full understanding of how blood flow is regulated in any tissue of any species, including our own.

Good short accounts of most of the known tissue vasoactive agents, with useful historical summaries, will be found in *The Pharmacological Basis of Therapeutics* (6th edn, ed. A. G. Gilman, L. S. Goodman and A. Gilman, Macmillan, New York, 1980). The chapters by W. W. Douglas, in the section on autacoids, deal with a number of the substances discussed in this monograph, including histamine, serotonin or 5-hydroxytryptamine (*Spätgift*), substance P, and kallikrein. Douglas also lists many of the most useful reviews that deal with research on these agents and on the mediation of local responses to trauma and allergy. Another chapter in the Goodman and Gilman text, by S. Moncada, R. J. Flower and J. R. Vane, surveys the prostaglandins and their relatives – compounds which, when Gaddum was writing, had just given the first hints of their existence (Goldblatt, 1933, 1935; von Euler, 1935).

For a more detailed treatment of the physiological significance of these and other vasodilator agents, there are several excellent multi-author volumes that deal with vascular regulation. Examples of these are: *Vasodilatation* (ed. P. M. Vanhoutte and I. Leusen, Raven Press, New York, 1981), and the series of volumes on *The Cardiovascular System* published by the American Physiological Society (II, *Vascular Smooth Muscle*, ed. D. F. Bohr, A. P. Somlyo and H. V. Sparks, Jr, 1980; III, *Peripheral Circulation and Organ Blood Flow*, ed. J. T. Shepherd and F. M. Abboud, 1983; IV, *Microcirculation*, ed. E. M. Renkin and C. C. Michel, 1984: American Physiological Society, Bethesda, MD). These volumes deal in detail with the agents considered by Gaddum, as well as with the blood and tissue components whose significance for blood-flow regulation came to light later, e.g., kinins and other peptides, endothelial mediators, metabolites of arachidonic acid, and presynaptic receptors.

The careers of J. H. Gaddum and U. S. von Euler intersected at many points, and each must have strongly influenced the other. Von Euler, one of the outstanding pioneers of autopharmacology, wrote engaging

historical accounts of substance P (*Trends in Neurosci.* **4** (10), iv–ix, 1981)
and of the prostaglandins. Gaddum's sense of scientific history was keen
but he did not live long enough to reminisce at length.

2 Gaddum's comments here are still appropriate. But it was also possible to
err in the opposite direction. Some of Gaddum's contemporaries had little
use for research on the pharmacological activity of tissue extracts,
because they thought it likely that most of the active materials were
either artefacts of the extraction procedures, or else substances whose
functions are confined in life to the interior of the cells. These people are
likely to have approved of the ferocious comments by a senior
physiologist, Swale Vincent, in a review he wrote a few years earlier
(*Physiol. Rev.* **7**, 288–319, 1927); I can't resist the temptation to quote
some passages from it.

There is, however, one method which has been so extensively used and
has been so fertile both in important facts and in groundless
hypotheses, that we must make reference to it in this place. The
method in question is the injection of extracts made in various ways
from the different organs and tissues of the body. The method was not
much employed till Schäfer, working in conjunction with Oliver,
discovered that extracts made from the adrenal bodies give rise (when
injected into the veins of an animal) to a very great rise in the arterial
blood pressure. It is worthy of note that all the fundamental facts in
the whole subject were noted by Schäfer in these early papers... It
gradually became clear that the rise of blood pressure given by adrenal
and pituitary extracts was a very special result peculiar to these two
organs, but that the fall of pressure first observed in the case of brain
extract is a phenomenon obtained with extracts made from all kinds of
organs and tissues. Although this was fully realized by workers
conversant with the subject, yet from time to time papers have
appeared from authors who carefully described the depressor action of
this or that tissue. Among recent examples are those of McDonald, and
James, Laughton and Macallum, who have recently made the
observation that extracts of liver have the power of lowering the blood
pressure... In the case of the majority of tissues at any rate there is no
reason to regard the presence of such active substances as evidence
that the tissues in question furnish an internal secretion. It would for
example be preposterous to assert that the presence of depressor
substances in brain extracts has any direct bearing upon the problems
of brain physiology... The whole line of investigation derived from the
action of extracts has been carried out with a sad lack of critical
judgment. (References cited by number have been omitted.)

There is much more of this sort of thing in Vincent's article, and some
of his strictures were certainly justified. One supposes that the recent
discovery of insulin had encouraged an excess of optimism about the

value of the materials that could be got out of tissues by simple means. What more logical, for example, than to find a tissue vasodilator that could be used to treat high blood pressure?

Dale's remarks on this point are illuminating: see especially his second Croonian Lecture (*Lancet, i*, 1233–37, 1929) and his 'Comment on Paper 20' in *Adventures in Physiology* (Pergamon, Oxford, 1953; Wellcome Trust, London, 1965). There he recalls how he was urged to examine the possible scientific basis for the claims of therapeutic value made on behalf of the liver extracts referred to by Vincent in the passage cited above. Dale's intervention had two important results. Firstly, he and his able colleagues established for the first time the presence of histamine in many fresh tissues (Best, Dale, Dudley and Thorpe, 1927). Secondly, in what I think may have been the earliest double-blind trial in this field of medicine, the efficacy of the liver extract was compared with that of a placebo and was found to be indistinguishable from zero. It is appropriate to add that this clinical assessment was carried out in Sir Thomas Lewis's wards by Dr Iris Harmer, later Lady Gaddum.

3 As already mentioned (Chapter VI, Note 2), the findings of these two reports from Cairo were soon to be regarded as suspect on technological grounds. Most later investigators have concluded that histamine release makes little, if any, contribution to the continuing adjustment of blood flow to metabolic needs in healthy mammals, i.e., to either active or reactive hyperaemia.

Gaddum spent only a year-and-a-half in Cairo as Professor of Pharmacology, and during much of his time there he must have been busy with administrative matters and with the writing of this monograph. I doubt therefore that he was deeply involved, except at the outset, in the work on blood histamine in reactive hyperaemia or after cutaneous burns (see pp. 170, 176). (The former project was completed by Barsoum and Smirk, who recognized that the histamine of blood is mostly in the cells but did not examine the white-cell layer; their papers and the one by Barsoum and Gaddum on burned patients are all in Volume 2 of *Clinical Science*.)

After Gaddum's departure, however, his former colleagues had the merit of being the first to document an example of drug-induced histamine release, a phenomenon of practical importance (Alam, Anrep, Barsoum, Talaat and Wieninger, *J. Physiol.* **95**, 148–58, 1939). The drug they investigated was tubocurarine. It was realized only much later that the histamine released by curare alkaloids, and by many other basic drugs, comes from the mast cells, whose granules contain most of the bound histamine of the tissues (Chapter III, Note 9). J. F. Riley, who with G. B. West discovered the relationship between these cells and histamine, tells in his book on *The Mast Cells* (E. and S. Livingstone, Edinburgh, 1959) how the use of a fluorescent histamine liberator made the discovery possible.

The mast cells liberate their histamine, and other active substances, in response to a variety of physical and chemical insults, one of the commonest of the latter being the formation of antigen–antibody complexes on the surfaces of these cells. The study of these responses perhaps belongs more to pathology than to physiology. But there is no doubt that some physiological phenomena also depend on the local release of histamine. One of these is the formation of gastric juice, which usually involves the cooperative action of histamine and other endogenous secretagogues (cf. Chapter III, Note 7). A second example, not yet fully documented, is synaptic transmission at a small minority of the sites in the central nervous system.

4 See Chapter VI, Note 12.

5 I think that Gaddum uses 'diffusion' loosely here. The spread of the flush must have been partly due to convection generated by pulsation and temperature gradients.

6 Lewis's identification of the 'H-substance' of human skin as almost certainly histamine has never been seriously challenged, and it received further powerful support from the demonstration, just as his own book was being prepared for the press, that mammalian skin contains chemically identifiable histamine (Best, Dale, Dudley and Thorpe, 1927). But Lewis's 'flare', which is due to a substance released through an axon reflex in afferent fibre collaterals, has been blamed at one time or another on almost every vasodilator substance mentioned in this book. That the flare-producing substance is not histamine has long been known, and Lewis showed that it is not rapidly destroyed following its release. At the time of writing, substance P seems to have been cast for this role.

7 Urticaria factitia is now recognizable as a form of mastocytosis. If Lewis and Harmer's histological analysis had been as penetrating as their clinical analysis they might perhaps have discovered how histamine and heparin are associated in mast cells. But that would have been a giant step to take in 1927. Ehrlich had identified the mast cells on the basis of the distinctive staining reaction given by their granules on the application of certain basic dyes such as toluidine blue. But the chemical basis of this 'metachromasia' was unknown until Lison (*Arch. Biol., Paris,* **46,** 599–668, 1935) showed that the colour change of the dye was a test for strongly acidic mucopolysaccharides of high molecular weight. The tissue constituents that most strongly display the metachromatic tint are the granules of basophil leucocytes and mast cells, because of their heparin content. (Heparin contains up to 40 per cent of esterified sulphuric acid.)

Lewis's colleague in this research was Dr Iris Harmer (see Note 2, above). Lady Gaddum and Professor Wilhelm Feldberg have much to recall about the excitements of histamine research sixty years ago.

8 See Chapter VI, Note 2.

9 Fashions in shock research have had their vagaries during the last fifty years. The blood-loss theme has rightly continued to be dominant, but the neurogenic and toxaemic themes have won the attention of some investigators, and there is no doubt that they are important in some circumstances. Among the chemical factors that can contribute to shock, various investigators have stressed the importance of bacterial toxins, vasodilators released by hepatic anoxia, and proteins whose release from damaged tissue can plug renal tubules or trigger enzymatic cascades leading to the 'sludging' of blood or the release of kinins and other vasodilators.

10 Gaddum's Cairo typescript, which still survives, ends with a section on anaphylaxis. The present section was written by Dale to replace it. It is about twice as long as Gaddum's and has three times as many references, with the additional ones coming mainly from the older literature. Dale retained Gaddum's short account (p. 182) of the perfused-lung experiments by Feldberg and his colleagues.

Reading their respective sections, I could see no great difference between their views. Both authors stress the resemblance between anaphylactic shock and histamine shock in a number of mammalian species; and both are confident that histamine release accounts for many, though not for all, the manifestations of anaphylaxis. Both authors, too, seem to ascribe the tissue injury resulting from the antigen–antibody combination to the presence of fixed antibody on, or even inside, the cells generally; they imply that this injury is widespread, and do not suppose that the damage takes place, or that the preformed histamine is stored, in a particular kind of cell. While they believe that histamine is unlikely to be the only active substance released, they do not implicate heparin or SRS-A (the slow-reacting smooth-muscle stimulating substance of anaphylaxis), since neither of these materials had as yet been associated with the 'anaphylactic symptom-complex'. They were, however, aware that an injection of Witte's peptone could reproduce more of that complex than an injection of histamine could. Finally, both authors were also aware that the anatomical distribution of the features of the complex, in any species, was related at least in part to the distribution of histamine sensitivity; but I think that Dale suggests more clearly than Gaddum does that, in addition, the cellular damage that leads to histamine release, either in anaphylaxis or peptone shock, may have a similar distribution.

11 Feldberg's Sherrington Lectures have been published in book form under the title *Fifty Years On: Looking back on some developments in neurohumoral physiology* (Liverpool University Press, 1982). The lectures contain much of his mature wisdom and some of his best stories. His reminiscence of this investigation, and the respective roles of the three researchers, is a small classic in the psychology of collaboration.

12 Many peptides, especially basic ones, can release histamine from mast cells. But it has not been shown, so far as I am aware, that Witte's peptone, as used traditionally in bacteriological culture media, contains any peptides that are particularly potent in this respect (Paton, *Pharmac. Rev.* **9**, 259–328, 1957).

13 Here Dale's first guess seems to have been the better one. As he emphasizes, the release of histamine could not account for the whole of the 'anaphylactic symptom-complex' as seen in the whole animal. But he had no basis for incriminating any other identifiable tissue substance, even heparin, for Holmgren and Wilander had not yet published their mast-cell studies; and in any case, it was only in dogs that the blood would not clot in anaphylaxis.

Kellaway's anaphylactic response of the rat uterus, which could not have been due to histamine, can now be seen as a forerunner of some interesting later developments. These seem to have accelerated after Feldberg joined Kellaway in Melbourne. Following their discovery that a 'slow-reacting substance' is released from tissues treated with snake venom, Kellaway and Trethewie (*Quart. J. exp. Physiol.* **30**, 121–45, 1940) found that a similar agent is liberated along with histamine in anaphylaxis; and even earlier, Schild (*J. Physiol.* **86**, 51–52P, 1936) had shown that sensitized smooth muscle can contract strongly in response to an antigen after it has been made insensitive to histamine. These and other observations pointed to the existence of at least one more anaphylactic mediator. When potent antihistaminic drugs became available, the pharmacology of the anaphylactic response could be examined more efficiently, and it then became possible for Brocklehurst, Uvnäs and their respective collaborators to characterize SRS-A, the slow-reacting substance of anaphylaxis, as a 'polar lipid'. Some challenging chemistry followed, but ultimately this remarkable substance was characterized as a member of the new family of 'leukotrienes', which are formed, like the prostaglandins, from arachidonic acid, although through the action of a different oxygenase. (For an account of these developments, whose pace has not yet slackened, consult *The Leukotrienes: Chemistry and Biology* (ed. L. W. Chakrin and D. M. Baily, eds., Academic Press, Orlando, 1984).)

Addendum to Chapter VII, note 2 (p. 211)

It can be argued that the notion of chemical transmission was glimpsed by some of the early physiologists, even before the Cell Theory had been enunciated. Sir William Paton, who read the foregoing Notes at my request, sent me the following comment and has allowed me to include it here.

'Compared with these claimants, a much earlier passage by G. A. Borelli (1608–1679), in his *De Motu Animalium* (1680–1), may be worth mentioning. Proposition 24 in Part 2 includes the following (as cited by A. Mosso in *Fatigue*, English translation, 1906):

"In the production of muscular contraction two causes concur, of which one resides in the muscles themselves, and the other comes from without. The excitation of movement can be transmitted from the brain by no other route than by the nerves...One rejects, however, the hypothesis that there is any question here of the action of any immaterial power or spirit; one must admit that some material substance is transmitted from the nerves to the muscles, or that a shock is communicated which is able in the twinkling of an eye to produce the swelling of the muscles...The substance or property which is transmitted by the nerves is incapable of producing a contraction by itself; it is necessary for it to unite with something which is found in the muscles and is there distributed abundantly, and from these substances there results something which may be compared to fermentation or ebullition and which produces the sudden swelling of the muscle."

As an illustrative possibility Borelli cites the 'fermentation' produced by adding acid to a fixed salt.'

Bibliography

(The titles of some articles have been shortened. Numbers in italics are those of text pages where the article is cited.)

Abderhalden, E. and Paffrath, H. (1925). Uber die Synthese von Cholinestern aus Cholin und Fettsäuren mittels Fermenten des Dünndarms. *Fermentforschung* 8, 299. *80, 82, 130*

Abderhalden, E., Paffrath, H. and Sickel, H. (1925). Wirkung des Cholins auf die motorischen Funktionen des Verdauungskanals. *Pflügers Arch. ges. Physiol.* 207, 241. *80, 82, 130*

Abe, K. (1920). The cause of the fall of arterial blood pressure due to the so-called paradoxical vasodilatory substances. *Tohoku J. exp. Med.* 1, 398. *26*

Abel, A. L. (1933). Acetylcholine in paralytic ileus. *Lancet* 2, 1247. *64*

Abel, J. J. and Geiling, E. M. K., (1924). Some hitherto undescribed properties of Witte's peptone. *J. Pharmac. exp. Ther.* 23, 1. *122*

Abel, J. J. and Kubota, S. (1919). On the presence of histamine (β-iminazolylethylamine) in tissues etc. *J. Pharmac. exp. Ther.* 13, 243. *46, 121*

Abel, J. J. and Macht, D. I. (1919). Histamine and pituitary extract. *J. Pharmac. exp. Ther.* 14, 279. *30*

Abelous, J. E. and Bardier, E. (1908). Urohypertensine und Urohypotensine. *J. Physiol. Path. gen.* 10, 627; (1909) 11, 777. *113, 114*

Ackermann, D. (1910). Uber den bakteriellen Abbau des Histidins. *Hoppe-Seyler's Z. physiol. Chem.* 65, 504. *19, 46, 52*

Ackermann, D. and Kutscher, Fr. (1910). Die physiologische Wirkung einer Secalebase und des Imidazolyläthylamins. *Z. Biol.* 54, 387. *5, 18, 39*

Adler, L. (1918). Beiträge zur Pharmakologie der Beckenorgane. *Arch. exp. Path. Pharmak.* 83, 248. *29*

Aeschlimann, J. A. and Reinert, M. (1931). The pharmacological action of some analogues of physostigmine. *J. Pharmac. exp. Ther.* 43, 413. *84*

Aldrich, R. H. (1933). Role of infection in burns. The theory and treatment with special reference to gentian violet. *New Eng. J. Med.* 208, 299. *176*

Alles, G. (1934). Physiological significance of choline derivatives. *Physiol. Rev.* 14, 276. *54*

Ammon, R. (1933). Die fermentative Spaltung des Acetylcholins. *Pflügers Arch. ges. Physiol.* 233, 486. *82, 85*

Ammon, R. and Kwiatkowski, H. (1934). Die Bildung von Acetylcholin in Serum und Embryonalextrakt. *Pflügers Arch. ges. Physiol.* 234, 269. *80, 81, 131*

Amsler, C. and Pick, E. P. (1920). Pharmakologische Studien am isolierten Splanchnikusgefäßgebiet des Frosches. *Arch. exp. Path. Pharmak.* 85, 61. *59*

Anderson, H. K. (1905). The paralysis of involuntary muscle. III. On the action of pilocarpine, physostigmine, and atropine upon the paralysed iris. *J. Physiol.* 33, 414. *156, 157, 166*

236

Andrus, E. C. (1924). The effect of certain changes in the perfusate upon the isolated auricles of the rabbit. *J. Physiol.* **59**, 361. *60*

Anochin, P. and Anochina-Ivanova, A. (1929). Über die vasomotorische und sekretorische Reaktion der Speicheldrüse auf Einführung von Azetylcholin. *Pflügers Arch. ges. Physiol.* **222**, 478. *68*

Anrep, G.V. (1926). The regulation of the coronary circulation. *Physiol. Rev.* **6**, 596. *27, 62*

Anrep, G. V. and Barsoum, G. S. (1935). Appearance of histamine in the venous blood during muscular contraction. *J. Physiol.* **85**, 409. *170*

Anrep, G. V., Blalock, A. and Samaan, A. (1934). Blood-flow in skeletal muscle. *Proc. R. Soc.* B. **114**, 223. *169*

Anrep, G. V., Cerqua, S. and Samaan, A. (1934). Blood-flow in muscle. *Proc. R. Soc.* B. **114**, 245. *169*

Anrep, G. V. and de Burgh Daly, I. (1924). The output of adrenaline in cerebral anaemia as studied by means of crossed circulation. *Proc. R. Soc.* B. **97**, 450. *35*

Antoniazzi, E. (1931). Cited in *Ber. ges. Physiol. exp. Pharm.* **64**, 820. *62*

Arai, K. (1922). Cholin als Hormon der Darmbewegung VII. Mitt. *Pflügers Arch. ges. Physiol.* **195**, 390. *130*

Armstrong, P. B. (1935). The action of acetylcholine on embryonic heart. *J. Physiol.* **84**, 20. *61*

Asher, L. (1921). Über die chemischen Vorgänge bei den antagonistischen Nervenwirkungen. *Pflügers Arch. ges. Physiol.* **193**, 84. *126*

Asher, L. (1923). Studien über antagonistische Nerven. Prüfung der angeblichen humoralen Übertragbarkeit der Herznervenwirkung. *Z. Biol.* **78**, 297. *126, 128*

Asher, L. (1925). Über die chemischen Wirkungen der Herznervenreizung. *Pflügers Arch. ges. Physiol.* **210**, 689. *126*

Asher, L. (1931). Die methodische Unsicherheit der experimentellen Begründung eines Vagushormons am Herzen. *Ber. ges. Physiol. exp. Pharm.* **61**, 337. *126*

Atzler, E. and Lehmann, G. (1927). Reaktionen der Gefäße auf direkte Reize. In Bethe, A. (ed.), *Handb. norm. path. Physiol.* **7**, Vol. 2, 963. *22*

Atzler, E. and Müller, E. (1925). Über humorale Übertragbarkeit der Herznervenwirkung. *Pflügers Arch. ges. Physiol.* **207**, 1. *128*

Auer, J. and Lewis, P. A. (1910). The physiology of the immediate reaction of anaphylaxis in the guinea-pig. *J. exp. Med.* **12**, 151. *180*

Babkin, B. P., Alley, A. and Stavraky, G. W. (1932). Humoral transmission of Chorda tympani effect. *Trans. R. Soc. Can.* **26**, 89. *79*

Babkin, B. P., Gibbs, O. S. and Wolff, H. G. (1932). Die humorale Übertragung der Chordatympanireizung. *Arch. exp. Path. Pharmak.* **168**, 32. *133*

Babkin, B. P., Stavraky, G. and Alley, A. (1932). Humoral transmission of chorda tympani hormone. *Am. J. Physiol.* **101**, 2. *133*

Babkin, B. P. and McLarren, P. D. (1927). The augmented salivary secretion. *Am. J. Physiol.* **81**, 143. *33*

Bacq, Z. M. (1933a). Recherches sur la physiologie du système nerveux autonome. III. Les proprietés biologiques et physico-chemiques de la sympathine comparées à celles de l'adrenaline. *Archs. int. Physiol.* **36**, 167. *103, 151, 152, 164*

Bacq, Z. M. (1933b). Inhibition des mouvements de l'intestin par la sympathine. *C. r. Séanc. Soc. Biol.* **112**, 211. *151*

Bacq, Z. M. (1934). La pharmacologie du système nerveux autonome et particulièrement du sympathique d'après la théorie neurohumorale. *Annls. Physiol. Physicochim. biol.* **10**, 467. *145*

Bacq, Z. M. (1935). Nature cholinergique et adrenergique des diverses innervations vagomotrices du pénis chez le chien. *Archs. int. Physiol.* **40**, 311. *143*

Bacq, Z. M. and Bovet, D. (1935). Action des dérivés de l'aminoethylbenzodioxane. *Archs. int. Pharmacodyn. Ther.* **50**, 315. *162*

Bacq, Z. M. and Brouha, L. (1932). Recherches sur la physiologie du système nerveux autonome. 1. La transmission humorale des excitations nerveuses sympathiques. *Archs. int. Physiol.* **35**, 163. *150*

Baehr, G. and Pick, E. P. (1913). Pharmakologische Studien an der Bronchialmuskulatur der überlebenden Meerschweinchenlunge. *Arch. exp. Path. Pharmak.* **74**, 41. *29, 32*

Baeyer, A. (1867). Über das Nervin. *Justus Liebigs Annaln Chem.* **142**, 325. *54*

Baer, R. and Rössler, R. (1926). Beiträge zur Pharmakologie der Lebergefäße. I. Mitteilung. Über die Abhängigkeit der Histaminwirkung von der Durchströmungsrichtung. *Arch. exp. Path. Pharmak.* **119**, 204. *28*

Bain, W. (1914). The pressor bases of normal urine. *Q. Jl. exp. Physiol.* **8**, 229. *113*

Bain, W. A. (1932a). Method of demonstrating humoral transmission of effects of cardiac vagus stimulation in frog. *Q. Jl. exp. Physiol.* **22**, 269. *126, 128*

Bain, W. A. (1932b). On the mode of action of vasomotor nerves. *J. Physiol.* **77**, 3P. *141*

Bain, W. A. (1933). The mode of action of vasodilator and vasoconstrictor nerves. *Q. Jl. exp. Physiol.* **23**, 381. *141*

Barbour, H. G. (1913). Note on the action of histamin upon surviving arteries. *J. Pharmac. exp. Ther.* **4**, 245. *27*

Barger, G. and Dale, H. H. (1910). β-Iminazolylethylamine and the other active principles of ergot. *Proc. chem. Soc.* **26**, 128; (1910) *Trans. chem. Soc.* **97**, 2592; (1910) *Zentbl. Physiol.* **24**, 885. *5, 18, 39, 45, 46, 111*

Barger, G. and Dale, H. H. (1911). β-Iminazolylethylamine, a depressor constituent of intestinal mucosa. *J. Physiol.* **41**, 499. *45, 46, 108*

Barger, G. and Walpole, G. S. (1909). Isolation of the pressor principles of putrid meat. *J. Physiol.* **38**, 343. *15, 114*

Bareenscheen, H. K. and Filz, W. (1932). Zur Chemie der Adenosintriphosphosäuren. *Biochem. Z.* **250**, 281. *82*

Barsoum, G. S. (1935). The acetylcholine equivalent of nervous tissues. *J. Physiol.* **84**, 259. *79, 110*

Barsoum, G. S. and Gaddum, J. H. (1935a). The pharmacological estimation of histamine and adenosine in blood. *J. Physiol.* **85**, 1. *31, 41, 43, 44, 47, 64, 90, 91, 95, 102*

Barsoum, G. S. and Gaddum, J. H. (1935b). The liberation of histamine during reactive hyperaemia. *J. Phsyiol.* **85**, 13P. *170, 171*

Barsoum, G. S. and Gaddum, J. H. (1935c). The effect of cutaneous burns on the blood histamine. Unpublished experiments. *176*

Barsoum, G. E., Gaddum, J. H. and Khayyal, M. A. (1934). The liberation of a choline ester in the inferior mesenteric ganglion. *J. Physiol.* **82**, 9P. *137*

Bartosch, R. (1935). Über die Herkunft des Histamins. *Klin. Wschr.* **14**, 307. *182*

Bartosch, R., Feldberg, W. and Nagel, E. (1932a). Das Freiwerden eines histaminähnliches Stoffes bei der Anaphylaxie des Meerschweinchens. *Pflügers Arch. ges. Physiol.* **230**, 129. *32, 43, 44, 182*

Bartosch, R., Fedlberg, W. and Nagel, E. (1932b). Die Übertragung der anaphylaktischen Lungenstarre auf die Lunge normaler Meerschweinchen. *Pflügers Arch. ges. Physiol.* **230**, 674. *32, 43, 45, 182*

Bartosch, R., Feldberg, W. and Nagel, E. (1933). Weitere Versuche über das Freiwerden eines histaminähnlichen Stoffes aus der durchströmten Lunge sensililisierter Meerschweinchen beim Auslösen einer anaphylaktischen Lungestarre. *Pflügers Arch. ges. Physiol.* **231**, 616. *43, 182*

Bartosch, R. and Nagel, E. (1932). Unpublished experiments, cited in Feldberg and Minz (1932). *58*

Bauer, W., Dale, H. H., Poulsson, L. T. and Richards, D. W. (1932). The control of circulation through the liver. *J. Physiol.* **74**, 343. *24, 28, 62*

Baur, M. (1928). Versuche am Amnion von Huhn und Gans. *Arch. exp. Path. Pharmak.* **134**, 49. *65*

Bayliss, L. E. and Ogden, E. (1933). 'Vaso-tonins' and the pump–oxygenator–kidney preparation. *J. Physiol.* **77**, 34P. *100*

Bayliss, W. M. (1901a). On the origin from the spinal cord of the vasodilator fibres of the hind-limb, and the nature of these fibres. *J. Physiol.* **26**, 173. *124, 139*

Bayliss, W. M. (1901b). The action of carbon dioxide on blood vessels. *J. Physiol.* **26**, xxxii. *169*

Bayliss, W. M. (1908). On reciprocal innervation in vaso-motor reflexes and the action of strychnine and chloroform thereon. *Proc. R. Soc.* B **80**, 339. *169*

Bayliss, W. M. and Starling, E. H. (1902). The mechanism of pancreatic secretion. *J. Physiol.* **28**, 325. *108*

Bennet, D. W. and Drury, A. N. (1931). Further observations relating to the physiological activity of adenine compounds. *J. Physiol.* **72**, 288. *88, 89, 90, 92, 93, 94, 104, 105, 106, 108, 175*

Berthelot, A. and Bertrand, D. M. (1912a). Isolement d'un microbe capable de produire de la β-imidazoléthylamine aux dépens de l'histidine. *C. r. hebd. Séanc. Acad. Sci., Paris.* **154**, 1643. *52*

Berthelot, A. and Bertrand, D. M. (1912b). Sur quelques propriétés biochimiques du Bacillus aminophilus intestinalis. *C. r. hebd. Séanc. Acad. Sci., Paris.* **154**, 1826. *52*

Best, C. H. (1929). The disappearance of histamine from autolysing lung tissue. *J. Physiol.* **67**, 256. *50*

Best, C. H., Dale, H. H., Dudley, H. W. and Thorpe, W. V. (1927). The nature of the vaso-dilator constituents of certain tissue extracts. *J. Physiol.* **62**, 397. *13, 15, 39, 42, 45, 104, 181*

Best, C. H. and Huntsman, M. E. (1932). The effects of the components of lecithine upon deposition of fat in the liver. *J. Physiol.* **75**, 405. *69*

Best, C. H. and McHenry, E. W. (1930). The inactivation of histamine. *J. Physiol.* **70**, 349. *46, 49, 50, 51*

Best, C. H. and McHenry, E. W. (1931). Histamine. *Physiol. Rev.* **11**, 371. *18, 19, 34, 37, 39, 40, 46, 48*

Beznák, A., von (1932). Die autakoide Aktivität des venösen Blutes von sezernierenden Submaxillardrüsen. *Pflügers Arch. ges. Physiol.* **229**, 719; (1932). **231**, 400. *79, 133*

Beznák, A. B. L., von (1934). On the mechanism of the autacoid function of parasympathetic nerves. *J. Physiol.* **82**, 129. *60, 74, 79, 80, 154, 158, 159, 163, 164*

Biedl, A. and Kraus, R. (1910). Experimentelle Studien über Anaphylaxie. *Wien. klin. Wschr.* **23**, 385. *179*

Billingsley, P. R. and Ranson, S. W. (1918). Branches of the ganglion cervicale superius. *J. comp. Neurol. and Psychol.* **29**, 367. *137*

Binet, L. and Minz, B. (1934). Sur une substance sensibilisant à l'acétylcholine formée par le tronc du nerf vague. *C. r. Séanc. Soc. Biol.* **115**, 1669; (1934). **116**, 107; (1934). **117**, 1029. *165*

Bischoff, C., Grab, W. and Kapfhammer, J. (1931a). Azetylcholin im Rinderblut. 2. Mitteilung (see also Kapfhammer and Bischoff, 1930) *Hoppe-Seyler's Z. physiol. Chem.* **199**, 135. *78*

Bischoff, C., Grab, W. and Kapfhammer, J. (1931b). Azetylcholin im Rinderblut. 3. Mitteilung. *Hoppe-Seyler's Z. physiol. Chem.* **200**, 153. *78*

Bischoff, C., Grab, W. and Kapfhammer, J. (1932). Azetylcholin im Warmblüter. 4. Mitteilung. *Hoppe-Seyler's Z. physiol. Chem.* **207**, 57. *78*

Bishop, G. H. and Kendall, A. I. (1928). The effects of histamine, formaldehyde and anaphylaxis upon the responses to electrical stimulation of guinea-pig intestinal muscle. *Am. J. Physiol.* **85**, 546 and 561. *31*

Blalock, A. (1930). Experimental shock: cause of low blood pressure produced by muscle injury. *Arch. Surg.* **20**, 959. *178*

Boehm, R. (1908). Über Wirkungen von Ammoniumbasen und Alkaloiden auf den Skelettmuskel. *Arch. exp. Path. Pharmak.* **58**, 265. *66*

Boeke, J. (1927). Die morphologische Grundlage der sympathischen Innervation der quergestreiften Muskelfasern. *Z. mikr.-anat. Forsch.* **8**, 561. *141*

Bohn, H. (1931). Untersuchungen zum Mechanismus des blassen Hochdrucks. *Z. klin. Med.* **119**, 100. *100*

Bohnenkamp, H. (1924). Über die Zusammenwirkung von Organen durch humorale Übertragung. *Klin. Wschr.* **3**, 61. *126*

Borgert, H. and Keitel, K. (1926). Über die vasokonstriktorischen Substanzen im Blutserum. *Biochem. Z.* **175**, 1. *98*

Boruttau, H. and Cappenberg, H. (1921). Die wirksamen Bestandteile des Hirtentäschelkrautes. *Arch. Pharm., Berl.* **259**, 33. *77*

Bouckaert, P. G. and Heymans, C., personal communication.

Bourdillon, R. B., Gaddum, J. H. and Jenkins, R. G. C. (1930). The production of histamine from histidine by ultra-violet light and the absorption spectra of these substances. *Proc. R. Soc. B.* **106**, 388. *20*

Bourne, A. and Burn, J. H. (1927). Dosage and action of pituitary extract and of ergot alkaloids on uterus in labour, with note on the action of adrenalin. *J. Obstet. Gynaec. Br. Commonw.* **34**, 249. *30*

Boyd, T. E., Tweedy, W. L. and Austin, W. C. (1928). Some effects of histamine on the acid–base balance. *Proc. Soc. exp. Biol. Med.* **25**, 451. *38*

Bremer, F. and Rylant, P. (1924). Phénomènes pseudo-moteurs et vasodilatation antidrome. Chronaxie des fibres pseudo-motrices de la corde du tympan. *C. r. Séanc. Soc. Biol.* **90**, 982. *141*

Brinkman, R. and van Dam, E. (1922). Die chemische Übertragbarkeit der Nervenreizwirkung. *Pflügers Arch. ges. Physiol.* **196**, 66. *128, 147*

Brinkman, R. and Ruiter, M. (1924). Die humorale Übertragung der neurogenen Skelettmuskelerregung auf dem Darm. *Pflügers Arch. ges. Physiol.* **204**, 766. *144*

Brinkman, R. and Ruiter, M. (1925). Die humorale Übertragung der Skelettmuskelreizung eines ersten auf den Darm eines zweiten Frosches. *Pflügers Arch. ges. Physiol.* **208**, 58. *144*

Brinkman, R. and v. d. Velde, J. (1925a). Humorale Übertragung der Vaguswirkung beim Kaninchen. *Pflügers Arch. ges. Physiol.* **207**, 488. *127*

Brinkman, R. and v. d. Velde, J. (1925b). Nachweis einer momentanen Zunahme der kapillaraktiven Substanzen des Kaninchenblutes unmittelbar nach direkter oder reflektorischer Vagusreizung. *Pflügers Arch. ges. Physiol.* **207**, 492. *127, 128, 129*

Brinkman, R. and v. d. Velde, J. (1925c). Die humorale Übertragbarkeit der Magen-Vagusreizung beim Kaninchen. *Pflügers Arch. ges. Physiol.* **209**, 383. *127*

Brodie, T. G. (1900). The immediate action of an intravenous injection of blood-serum. *J. Physiol.* **26**, 48. *99*

Brown, G. L. (1934). Conduction in the cervical sympathetic. *J. Physiol.* **81**, 228. *138*

Brown, G. L. and Feldberg, W. (1935a). Effect of potassium chloride on a sympathetic ganglion. *J. Physiol.* **84**, 12P. *99*

Brown, G. L. and Feldberg, W. (1935b). Unpublished experiments. *75, 79, 159*

Bruce, A. N. (1910). Über die Beziehung der sensiblen Nervenendigungen zum Entzündungsvorgang. *Arch. exp. Path. Pharmak.* **63**, 424. *169*

Brücke, E. T. (1929). Allgemeines über Tatsachen und Probleme der Physiologie nervöser Systeme. In Bethe, A. (ed.), *Handb. norm. path. Physiol.* **9**, 25. *153*

Brugsch, T. and Horsters, H. (1926). Cholagoga und Cholagogie. *Arch. exp. Path. Pharmak.* **118**, 267. *29, 32*

Brummelkamp, R. (1933). Das Magengeschwür bei der Ratte. *Nederl. Tijdschr. Geneesk.*, 5261. Cited in *Ber. ges. Physiol. exp. Pharm.* (1934) **78**, 263. *34*

Bücher, F., Siebert, P. and Molloy, P. J. (1929). Über experimentell erzeugte akute peptische Geschwüre des Rattenvormagens. *Beit. path. Anat.* **81**, 391. *34*

Bülbring, E. and Burn, J. H. (1934). Cholinergic nature of sympathetic vaso-dilator fibres. *J. Physiol.* **81**, 42P. *142, 156*

Bülbring, E. and Burn, J. H. (1935). Sympathetic vasodilators. *J. Physiol.* **83**, 483. *142, 156*

Buell, M. V. and Perkins, M. E. (1928). Adenine nucleotide content of blood with a micro-analytical method for its determination. *J. biol. Chem.* **76**, 95. *90, 92*

Buell, M. V., Strauss, M. B. and Andrus, E. C. (1932). Metabolic changes involving phosphorus and carbohydrate. *J. biol. Chem.* **98**, 645. *92*

Burn, J. H. (1928). *Methods of Biological Assay.* Oxford University Press. *12*

Burn, J. H. (1930). The cardio-vascular action of tyramine. *Quart. J. Pharm.* **3**, 187. *158*

Burn, J. H. (1932). On vaso-dilator fibres in the sympathetic, and on the effect of circulating adrenaline in augmenting the vascular response to sympathetic stimulation. *J. Physiol.* **75**, 144. *157*

Burn, J. H. (1934). Verbal communication. *156*

Burn, J. H. and Dale, H. H. (1922). Report on biological standards. Pituitary extracts. *Spec. Rep. Ser. med. Res. Counc.* no. **69**. *14*

Burn, J. H. and Dale, H. H. (1926). The vaso-dilator action of histamine, and its physiological significance. *J. Physiol.* **61**, 185. *22, 26, 35*

Burn, J. H. and Tainter, M. L. (1931). An analysis of the effect of cocaine on the actions of adrenaline and tyramine. *J. Physiol.* **71**, 169. *158*

Calvery, H. O. (1930). The isolation of adenosine from human urine. *J. biol. Chem.* **86**, 263. *91, 93, 114*

Cannon, W. B. (1919). Some characteristics of shock induced by tissue injury. Traumatic toxaemia as a factor in shock. *Spec. Rep. Ser. med. Res. Comm.*, no. **26**, 27. *177*

Cannon, W. B. (1923). *Traumatic shock.* New York: Appleton. *177*

Cannon, W. B. (1931). Recent studies on chemical mediation of nerve impulses. *Endochrinology* **15**, 473. *124*

Cannon, W. B. (1933). Chemical mediators of autonomic nerve impulses. *Science* **78**, 43.

Cannon, W. B. (1934). (Kober lecture.) The story of the development of our ideas of chemical mediation of nerve impulses. *Am. J. med. Sci.* **188**, 145. *124*

Cannon, W. B. and Bacq, Z. M. (1931). Studies on the conditions of activity in endocrine organs; a hormone produced by sympathetic action on smooth muscle. *Am. J. Physiol.* **96**, 392. *146, 149, 151, 161*

Cannon, W. B. and Bayliss, W. M. (1919). Traumatic toxaemia as a factor in shock. *Spec. Rep. Ser. med. Res. Comm.*, no. **26**, *177*

Cannon, W. B. and De La Paz. (1911). Emotional stimulation of adrenal secretion. *Am. J. Physiol.* **28**, 64. *101*

Cannon, W. B. and Griffith, F. R. (1922). Studies on the conditions of activity of the endocrine glands. X. The cardio-accelerator substance produced by hepatic stimulation. *Am. J. Physiol.* **60**, 544. *148*

Cannon, W. B. and Rosenblueth, A. (1933). Studies on conditions of activity in endocrine organs; sympathin E and sympathin I. *Am. J. Physiol.* **104**, 557. *145, 146, 148, 149, 150, 151, 158*

Cannon, W. B. and Rosenblueth, A. (1935). A comparative study of sympathin and adrenine. *Am. J. Physiol.* **112**, 268. *145*

Cannon, W. B. and Uridil, J. E. (1921). Studies on the conditions of activity in endocrine glands. VIII. Some effects on the denervated heart of stimulating the nerves of the liver. *Am. J. Physiol.* **58**, 353. *146, 148*

Carmichael, E. A. and Fraser, F. R. (1933). The effects of acetylcholine in man. *Heart* **16**, 263. *69*

Carnot, P., Koskowski, W. and Libert, E. (1922). L'influence de l'histamine sur la sécrétion des sucs digestifs chez l'homme. *C. r. Séanc. Soc. Biol.* **86**, 575. *33*

Carrier, E. B. (1922). The reaction of the human skin capillaries to drugs and other stimuli. *Am. J. Physiol.* **61**, 528. *24, 56*

ten Cate, J. (1922). L'action des ions K, Ca et Mg sur le nerf sympathetique du cœur. *Archs. néerl. Physiol.* **6**, 269. *147*

ten Cate, J. (1924). Sur la question de l'action humorale du nerf vague. *Archs. néerl. Physiol.* **9**, 588. *126, 128*

Cattell, M., Wolff, H. G. and Clark, D. A. (1934). The liberation of adrenergic and cholinergic substances in the submaxillary gland. *Am. J. Physiol.* **109**, 375. *151*

Chambers, E. K. and Thompson, K. W. (1925). Quantitative changes in canine histamine shock. *J. infect. Dis.* **37**, 229. *38*

Chang, H. C. (1935). The formation of placental acetylcholine. *Proc. Soc. exp. Biol. Med.* **32**, 1001. *81*

Chang, H. C. and Gaddum, J. H. (1933). Choline esters in tissue extracts. *J. Physiol.* **79**, 255. *14, 71, 72, 73, 74, 75, 76, 78, 79, 83, 95, 103, 104, 106, 108, 110, 131, 136, 164*

Chang, H. C. and Wong, A. (1933). Studies on tissue acetylcholine: origin, significance and fate of acetylcholine in human placenta. *Chin. J. Physiol.* **7**, 151. *79*

Chittenden, R. H., Mendel, L. B. and Henderson, Y. (1899). Certain derivatives of the proteids. *Am. J. Physiol.* **2**, 142. *121*

Clark, A. J. (1913). The action of ions and lipoids upon the frog's heart. *J. Physiol.* **47**, 66. *101*

Clark, A. J. (1924). Some active principles of peptone. *J. Pharmac. exp. Ther.* **23**, 45. *26, 122*

Clark, A. J. (1926a). The reaction between acetylcholine and muscle cells. *J. Physiol.* **61**, 530. *59*

Clark, A. J. (1926b). The antagonism of acetylcholine by atropine. *J. Physiol.* **61**, 547. *67*

Clark, A. J. (1927). The reaction between acetylcholine and muscle cells. Part II, *J. Physiol.* **64**, 123. *60, 82*

Clark, A. J. (1933). *Mode of action of drugs on cells.* London: Edward Arnold. *59*

Coca, A. F. (1914). The site of reaction in anaphylactic shock. *Z. ImmunForsch. exp. Ther.* **20**, 622. *180*

Colle, J., Duke-Elder, P. M. and Duke-Elder, W. S. (1931). Studies on the intra-ocular pressure. *J. Physiol.* **71**, 1. *63*

Collip, J. B. (1928). A non-specific pressor principle derived from a variety of tissues. *J. Physiol.* **66**, 416. *112, 122*

Corkill, A. B. and Tiegs, O. W. (1933). The effect of sympathetic nerve stimulation on the power of contraction of skeletal muscle. *J. Physiol.* **78**, 161. *150*

Cornell, B. S. (1928). An immediate fall of blood cholesterol after eating or after histamine injections. *J. Lab. clin. Med.* **14**, 209. *38*

Cowan, S. L. (1934). The action of potassium on Maia nerve. *Proc. R. Soc. B.* **115**, 216. *165*

Crile, G. W. (1921). *A Physical Interpretation of Shock, Exhaustion and Restoration.* London: Hodder & Stoughton. *177*

Cruickshank, E. W. H. and Subba Rau, A. (1927). Reactions of isolated systemic and coronary arteries. *J. Physiol.* **64**, 65. *27*

Cullis, W. and Tribe, E. M. (1913). Distribution of nerves in the heart. *J. Physiol.* **46**, 141. *60*

Cushny, A. R. and Gunn, J. A. (1913). The action of serum on the perfused heart of the rabbit. *J. Pharmac. exp. Ther.* **5**, 1. *100*

Dakin, H. D. (1916). The oxidation of amino acids to cyanides. *Biochem. J.* **10**, 319. *19*

Dale, A. S. (1930). The relation between amplitude of contraction and rate of rhythm in the mammalian ventricle. *J. Physiol.* **70**, 455. *60*

Dale, H. H. (1913a). The anaphylactic action of plain muscle in the guinea-pig. *J. Pharmac. exp. Ther.* **4**, 167. *179, 182*

Dale, H. H. (1913b). The effect of varying tonicity on the anaphylactic and other reactions of plain muscle. *J. Pharmac. exp. Ther.* **4**, 517. *30*

Dale, H. H. (1914a). The action of certain esters and ethers of choline and their relation to muscarine. *J. Pharmac. exp. Ther.* **6**, 147. *55, 57, 59, 64, 65, 68, 75*

Dale, H. H. (1914b). The occurrence in ergot and action of acetylcholine. *J. Physiol.* **48**, 3P. *70*

Dale, H. H. (1919a). Supplementary note on histamine shock. Traumatic toxaemia as a factor in shock. *Spec. Rep. Ser. med. Res. Comm.*, no. **26**, 15. *177*

Dale, H. H. (1919b). The biological significance of anaphylaxis. (Croonian lecture.) *Proc. R. Soc. B.* **91**, 126. *179*

Dale, H. H. (1920a). Conditions which are conducive to the production of shock by histamine. *Br. J. exp. Path.* **1**, 103. *35*

Dale, H. H. (1920b). Anaphylaxis. *Johns Hopkins Hosp. Bull.* **31**, 310. *179*

Dale, H. H. (1929). Some chemical factors in the control of the circulation. (Croonian lectures.) *Lancet* **1**, 1179, 1233, 1285. *29, 121, 124, 141, 181*

Dale, H. H. (1933a). Progress in autopharmacology. A survey of present knowledge of the chemical regulation of certain functions by natural constituents of the tissues. *Johns Hopkins Hosp. Bull.* **53**, 297. *124*

Dale, H. H. (1933b). Nomenclature of fibres in the autonomic system and their effects. *J. Physiol.* **80**, 10P. *125, 138*

Dale, H. H. (1934). Chemical transmission of the effects of nerve impulses. *Br. med. J.* **1**, 835. *124*

Dale, H. H. (1935a). Pharmacology and nerve endings. *Proc. R. Soc. Med.* **28**, 15. *166*

Dale, H. H. (1935b). *Reizübertragung durch chemische Mittel im peripheren Nervensystem.* 4th Nothnagel Lecture. Vienna: Urban and Schwarzenberg. *166*

Dale, H. H. and Dudley, H. W. (1929). The presence of histamine and acetylcholine in the spleen of the ox and the horse. *J. Physiol.* **68**, 97. *7, 46, 55, 70, 77, 78, 79, 105*

Dale, H. H. and Dudley, H. W. (1931). Enthält das normale Blut Acetylcholin? (Mit Berücksichtigung der Arbeit von Kapfhammer und Bischoff.) *Hoppe-Seyler's Z. physiol. Chem.* **198**, 85. *78*

Dale, H. H. and Feldberg, W. (1934a). The chemical transmitter of vagus effects to the stomach. *J. Physiol.* **81**, 320. *79, 131*

Dale, H. H. and Feldberg, W. (1934b). The chemical transmitter of nervous stimuli to the sweat glands of the cat. *J. Physiol.* **81**, 40P. *120, 138*

Dale, H. H., Feldberg, W. and Vogt, M. (1936). Release of acetylcholine at voluntary nerve endings. *J. Physiol.* **86**, 353. *144, 154*

Dale, H. H. and Gaddum, J. H. (1930). Reactions of denervated voluntary muscle, and their bearing on the mode of action of parasympathetic and related nerves. *J. Physiol.* **70**, 109. *68, 83, 140, 141, 154, 156, 157*

Dale, H. H. and Gasser, H. S. (1926). The pharmacology of denervated mammalian muscle. Part I. The nature of the substances producing contracture. *J. Pharmac. exp. Ther.* **29**, 53. *67*

Dale, H. H. and Laidlaw, P. P. (1910). The physiological action of β-iminazolylethylamine. *J. Physiol.* **41**, 318. *20, 24, 25, 26, 27, 29, 30, 31, 32, 37, 50, 51, 121, 180*

Dale, H. H. and Laidlaw, P. P. (1911). Further observations on the action of β-iminazolylethylamine. *J. Physiol.* **43**, 182. *26, 38*

Dale, H. H. and Laidlaw, P. P. (1912). A method of standardising pituitary (infundibular) extracts. *J. Pharmac. exp. Ther.* **4**, 75. *44*

Dale, H. H. and Laidlaw, P. P. (1919). Histamine shock. *J. Physiol.* **52**, 355. *37, 38, 177*

Dale, H. H. and Richards, A. N. (1918). The vaso-dilator action of histamine and of some other substances. *J. Physiol.* **52**, 110. *20, 21, 23, 24, 35, 56, 57, 169, 170*

Dale, H. H. and Richards, A. N. (1927). The depressor (vaso-dilator) action of adrenaline. *J. Physiol.* **63**, 201. *22*

Daly, I. de B. (1927). Negative pressure pulmonary ventilation in the heart–lung preparation. *J. Physiol.* **63**, 81. *101*

Daly, I. de B. and Schild, H. (1934). Inactivation by histaminase preparations of the histamine-like substance recovered from lungs during anaphylactic shock. *J. Physiol.* **83**, 3P. *182*

Daly, I. de B. and Schild, H. (1935). The release of a histamine-like substance from the lungs of guinea-pigs during anaphylactic shock. *Q. Jl. exp. Physiol.* **25**, 33. *32, 44, 182*

Daly, I. de B. and Thorpe, W. V. (1932). An isolated mammalian heart preparation capable of performing work for prolonged periods. *J. Physiol.* **77**, 10P. *101*

Davidson, E. C. (1925). Tannic acid in the treatment of burns. *Surgery, Gynec. Obstet.* **41**, 202. *176*

Davis, E. (1931). Relations between the actions of adrenaline, acetylcholine, and ions, on the perfused heart. *J. Physiol.* **79**, 431. *60*

De Vleeschhouwer, G. (1935). Un sujet de l'action du diéthylaminométhyl-3-benzodioxane (F 833) et du pipéridométhyl-3-benzodioxane (F 933) sur le système circulatoire. *Archs. int. Pharmacodyn. Ther.* **50**, 251. *162*

Demoor, J. (1911). Action du sérum sanguin au point de vue de la sécrétion salivaire. *Archs. int. Physiol.* **10**, 377. *132*

Demoor, J. (1912). A propos du mécanisme de la sécrétion salivaire. (3e note.) (Action de la pilocarpine.) *Archs. int. Physiol.* **12**, 52. *132*

Demoor, J. (1913). Le mécanisme intime de la sécrétion salivaire. *Archs. int. Physiol.* **13**, 187. *119, 132*

Demoor, J. (1922). Contribution à la physiologie générale du cœur. II. Influences des substances extraites de l'oreillette et du ventricule du chien sur le cœur isolé du lapin. *Archs. int. Physiol.* **20**, 29. *103*

Demoor, J. (1923). Contribution à la physiologie générale du cœur. III. L'action des substances actives du cœur du chien sur l'oreillette droite isolée du lapin. *Archs. int. Physiol.* **20**, 446. *103*

Derer, L. and Steffanutti, P. (1930). Über das Verhalten des Serumeiweißes im Histaminschock. *Biochem. Z.* **223**, 408. *38*

Deuticke, H. J. (1932a). Uber den Einfluß von Adenosin und Adenosinphosphorsäuren auf dem isolierten Meerschweinchenuterus. *Pflügers Arch. ges. Physiol.* **230**, 537. *90*

Deuticke, H. J. (1932b). Über die Einwirkung von Adenosinphosphorsäuren auf Dehydrierungsvorgänge durch pflanzliche und tierische Fermente. *Pflügers Arch. ges. Physiol.* **230**, 556. *91*

Dikshit, B. B. (1933a). The presence in the brain of a substance resembling acetylcholine. *J. Physiol.* **79**, 1P. *79, 110*

Dikshit, B. B. (1933b). Action of acetylcholine on the brain and its occurrence therein. *J. Physiol.* **80**, 409. *69, 110*

Dikshit, B. B. (1934). Cardiac irregularities from the hypothalamus. *J. Physiol.* **81**, 382. *153*

Dingemanse, E. and Freud, T. (1933). Identifikation des Katatonins. *Acta brev. neerl. Physiol.* **3**, 49. *114*

Dirner, Z. (1929). Pharmakologische Untersuchungen an überlebenden Froschlungen. *Arch. exp. Path. Pharmak.* **146**, 232. *65*

Dittler, R. (1914). Über die Wirkung des Blutes auf den isolierten Dünndarm. *Pflügers Arch. ges. Physiol.* **157**, 453. *101*

Dittler, R. (1918). Über die Wirkung des Blutes auf den isolierten Dünndarm. II. Mitteilung. *Z. Biol.* **68**, 223. *101*

Dixon, W. W. (1906). Vagus inhibition. *Br. med. J.* **2**, 1807. *124, 162*

Dixon, W. E. (1907). On the mode of action of drugs. *Med. Mag.* **16**, 454. *124, 162*

Dixon, W. E. and Hoyle, J. C. (1930). Studies in the pulmonary circulation. III. The action of histamine. *J. Physiol.* **70**, 1. *27*

Dixon, W. E. and Ransom, F. (1912). Broncho-dilator nerves. *J. Physiol.* **45**, 413. *132, 156*

Doan, C. A., Zerfas, L. G., Warren, S. and Ames, O. (1928). A study of the mechanism of nucleinate-induced leucopenic and leucocytic states. *J. exp. Med.* **47**, 403. *90*

Doerr, R. (1913). Allergie und Anaphylaxie. In W. Kolle and A. von Wassermann, *Handb. d. pathogen. Mikroorg.* **2**, 947. *179*

Doi, Y. (1920). Studies on muscular contraction. *J. Physiol.* **54**, 218. *59*

Donnomae, J. (1934). Das Auftreten eines azetylcholinartigen Stoffes im Pfortaderblut der Katze. *Pflügers Arch. ges. Physiol.* **234**, 318. *131*

Donnomae, J. and Feldberg, W. (1934). Die Beeinflussung des arteriellen Blutdruckes der Katze durch vorübergehendes Umleiten des Pfortaderblutes in die Vena iliaca. *Pflügers Arch. ges. Physiol.* **234**, 325. *131*

Drake, T. G. H. and Tisdall, F. F. (1926). The effect of histamine on the blood chlorides. *J. biol. Chem.* **67**, 91. *38*

Drinker, C. K., Drinker, K. D. and Lund, C. C. (1922). The circulation in the mammalian bonemarrow. *Am. J. Physiol.* **62**, 1. *22*

Drury, A. N. and Szent-Györgyi, A. (1929). The physiological activity of adenine compounds. *J. Physiol.* **68**, 213. *87, 88, 91, 103*

Dudley, H. W. (1920). Some observations on the active principles of the pituitary gland. *J. Pharmac. exp. Ther.* **14**, 295. *43*

Dudley, H. W. (1929). Observations on acetylcholine. *Biochem. J.* **23**, 1064. *54, 77*

Dudley, H. W. (1931). Coordination compounds of the chloroplatinates of choline and its esters. *J. chem. Soc.* p. 763. *54, 77*

Dudley, H. W. (1933). The alleged occurrence of acetylcholine in ox blood. *J. Physiol.* **79**, 249. *78*

Duke-Elder, W. S. and Duke-Elder, P. M. (1930). The contraction of the extrinsic muscles of the eye by choline and nicotine. *Proc. R. Soc. B.* **107**, 332. *67*

Duschl, L. (1923). Über die humorale Beeinflussung der Herzaktion im Warmblüterorganismus nach Versuchen an Einzeltieren. *Z. ges. exp. Med.* **38**, 268. *126*

Duschl, L. and Windholz, F. (1923). Über die humorale Beeinflussung der Herzaktion im Warmblüterorganismus nach Versuchen an parabiosierten Ratten. *Z. ges. exp. Med.* **38**, 261. *126*

Ebbecke, U. (1922). Über elektrische Hautreizung. *Pflügers Arch. ges. Physiol.* **195**, 300. *47*

Eccles, J. C. (1933a). Action potentials from the superior cervical ganglion. *J. Physiol.* **80**, 23P. *138*

Eccles, J. C. (1933b). The effects of eserine and atropine on the vagal slowing of the heart. *J. Physiol.* **80**, 25P. *138, 156*

Eccles, J. C. (1934a). Synaptic transmission through a sympathetic ganglion. *J. Physiol.* **81**, 8P. *138*

Eccles, J. C. (1934b). Inhibition in the superior cervical ganglion. *J. Physiol.* **82**, 25P. *138*

Eggleton, P. (1926). The action of pure phosphatides on the perfused heart of the frog. *Biochem. J.* **20**, 395. *101*

Eichholtz, F. and Verney, E. B. (1924). On some conditions affecting the perfusion of isolated mammalian organs. *J. Physiol.* **59**, 340, *100*

Einis, W. (1913). Über die Wirkung des Pituitrins und Histamins auf die Herzaktion. *Biochem. Z.* **52**, 96. *26*

Ellinger, F. (1928). Über die Entstehung eines den Blutdruck senkenden und den Darm erregenden Stoffes aus Histidin durch Ultraviolettstrahlung. *Arch. exp. Path. Pharmak.* **136**, 129. *19, 175*

Ellinger, F. (1929). Das Absorptionsspektrum von Histidin und Histamin im Ultraviolett. *Biochem. Z.* **215**, 279. *175*

Ellinger, F. (1930). Weitere Untersuchungen über die Entstehung des Lichterythems. *Arch. exp. Path. Pharmak.* **149**, 343. *175*

Ellinger, F. (1932). Über die Absorption des Histidins im Ultraviolett. *Biochem. Z.* **248**, 437. *20*

Elliot, A. H. and Nuzum, F. R. (1931). The pharmacologic properties of an insulin-free extract of pancreas, and the circulatory hormone of Frey. *J. Pharmac. exp. Ther.* **43**, 463. *107, 116*

Elliott, T. R. (1904). On the action of adrenalin. *J. Physiol.* **31**, 20P; (1905) **32**, 401. *124, 139*

Ellis, L. B. and Weiss, S. (1932). A study on the cardiovascular responses in man to the injection of acetylcholine. *J. Pharmac. exp. Ther.* **44**, 235. *69*

Embden, G. and Deuticke, H. J. (1930). Über die Isolierung von Muskeladenylsäure aus der Niere. *Hoppe-Seyler's Z. physiol. Chem.* **190**, 62. *91*

Embden, G. and Schmidt, G. (1930). Herkunft des Muskelammoniaks. *Hoppe-Seyler's Z. physiol. Chem.* **186**, 205. *93*

Embden, G. and Zimmermann, M. (1927). Das Vorkommen von Adenylsäure in der Skelettmuskulatur. *Hoppe-Seyler's Z. physiol. Chem.* **167**, 137. *91*

Enderlen, Prof. Dr and Bohnenkamp, H. (1924). Über das Fehlen der Übertragbarkeit der Herznervenwirkung bei Gefäßparabiose an Hunden. *Z. ges. exp. Med.* **41**, 723. *126*

Engelhart, E. (1930). Die Vagusstoffverteilung auf Vorhof und Kammer bei Frosch und Säuger. *Pflügers Arch. ges. Physiol.* **225**, 721. *60, 74, 79, 80, 103, 162*

Engelhart, E. (1931). Der humorale Wirkungsmechanismus der Okulomotoriusreizung. *Pflügers Arch. ges. Physiol.* **227**, 220. *71, 79, 113, 132, 163, 164*

Engelhart, E. and Loewi, O. (1930). Fermentative Azetylcholinspaltung im Blut und ihre Hemmung durch Physostigmin. *Arch. exp. Path. Pharmak.* **150**, 1. *55, 81, 82, 84*

Eppinger, H. (1913). Über eine eigentümliche Hautreaktion, hervorgerufen durch Ergamin. *Wien med. Wschr.* **63**, 1413. *22, 37*

Eppinger, H. and Gutmann, J. (1913). Zur Frage der vom Darm ausgehenden Intoxikationen. (I. Mitteilung.) *Z. klin. Med.* **78**, 399. *46*

Eppinger, H. and Hess, L. (1909). Versuche über die Einwirkung von Arzneimitteln auf überlebende Coronargefäße. *Z. exp. Path. Ther.* **5**, 622, *62*

Eppinger, H., Laszlo, D. and Schürmeyer, A. (1928). Über die mutmaßlichen Ursachen der Unökonomie im Herzfehlerorganismus. *Klin. Wschr.* **7**, 2231. *24*

Epstein, D. (1932). The action of histamine on the respiratory tract. *J. Physiol.* **76**, 347. *32*

Erbsen, H. and Damm, E. (1927). Untersuchungen zur Funktion der extrahepatischen Gallenwege. *Z. ges. exp. Med.* **55**, 748. *32*

Ernould, H. (1931). Influence de l'acétylcholine sur le métabolisme des hydrates de carbone. *C. r. Séanc. Soc. Biol.* **108**, 434, 436. *69*

Esveld, L. W. van, (1928). Verhalten von plexushaltigen und plexusfreien Darmmuskelpräparaten. *Arch. exp. Path. Pharmak.* **134**, 347. *31, 65*

Ettinger, G. H. and Hall, G. E. (1934). Acetylcholine in ox and dog blood. *J. Physiol.* **82**, 38. *78*

Euler, U. S. von (1932). A vasoconstrictor action of acetylcholine on the rabbit's pulmonary circulation. *J. Physiol.* **74**, 271. *58, 62, 65, 132, 156*

Euler, U. S. von (1934a). An adrenaline-like action in extracts from the prostatic and related glands. *J. Physiol.* **81**, 102. *112*

Euler, U. S. von (1934b). Natursekreten und Extrakte männlicher akzessorischer Geschlecthsdrüsen. *Arch. exp. Path. Pharmak.* **175**, 78. *112*

Euler, U. S. von (1935). A depressor substance in the vesicular gland. *J. Physiol.* **84**, 21P. *113*

Euler, U. S. von and Gaddum, J. H. (1931a). An unidentified depressor substance in certain tissue extracts. *J. Physiol.* **72**, 74. *108, 110*

Euler, U. S. von and Gaddum, J. H. (1931b). Pseudomotor contractures after degeneration of the facial nerve. *J. Physiol.* **73**, 54. *141*

Ewins, A. J. (1914). Acetylcholine, a new active principle of ergot. *Biochem. J.* **8**, 44. *70, 76*

Ewins, A. J. and Pyman, F. L. (1911). Experiments on the formation of 4- (or 5-) β-aminoethylglyoxaline from histidine. *J. chem. Soc.* **99**, 339, *19, 43, 109*

Feldberg, W. (1927). The action of histamine on the blood vessels of the rabbit. *J. Physiol.* **63**, 211. *23, 25*

Feldberg, W. (1928). Das Verhalten des Pfortaderdruckes nach Injektion von Histamin und Pepton in den Kreislauf der Katze. *Arch. exp. Path. Pharmak.* **140**, 156. *29*

Feldberg, W. (1931). Die Wirkung von Histamin und Azetylcholin auf die glatte Muskulatur und ihre Beeinflussung durch Antropin. *Rev. de pharmacol. et de thérapie exp. Paris* **2**, 311; [(1934) *Rona-Berichte* **76**, 562.] *29, 64*

Feldberg, W. (1932). Die Empfindlichkeit der Lungengefäße des Hundes auf Lingualisreizung und auf Azetylcholin. *Pflugers Arch. ges. Physiol.* **232**, 75. *66, 142*

Feldberg, W. (1933a). Der Nachweis eines azelylcholinähnlichen Stoffes im Lungenvenenblut des Hundes bei Reizung des Nervus lingualis. *Pflügers Arch. ges. Physiol.* **232**, 88. *141*

Feldberg, W. (1933b). Die blutdrucksenkende Wirkung der Chorda-Lingualisreizung und ihre Beeinflussung durch Antropin. *Arch. exp. Path. Pharmak.* **170**, 560. *133*

Feldberg, W., Flatow, E. and Schilf, E. (1929). Die Wirkung von Blut und Serum auf Warmblütergefäße. *Arch. exp. Path. Pharamak.* **140**, 129. *22, 99*

Feldberg, W. and Gaddum, J. H. (1933). The chemical transmitter at synapses in a sympathetic ganglion. *J. Physiol.* **80**, 12P. *136*

Feldberg, W. and Gaddum, J. H. (1934). The chemical transmitter at synapses in a sympathetic ganglion. *J. Physiol.* **81**, 305. *136*

Feldberg, W. and Guimarãis, J. A. (1935a). Some observations on salivary secretion. *J. Physiol.* **85**, 15. *119, 133*

Feldberg, W. and Guimarãis, J. A. (1935b) Unpublished experiments. *154, 159*

Feldberg, W. and Krayer, O. (1933). Das Auftreten eines azetylcholinartigen Stoffes im Herzvenenblut von Warmblütern bei Reizung der Nervi vagi. *Arch. exp. Path. Pharmak.* **172**, 170. *72, 127, 129*

Feldberg, W. and Kwiatkowski, H. (1934). Das Auftreten eines azetylcholinartigen Stoffes in der Durchströmungsflüssigkeit beim Durchströmen des isolierten Katzendünndarmes. *Pflügers Arch. ges. Physiol.* **234**, 333. *131*

Feldberg, W. and Minz, B. (1931). Die Wirkung von Acetylcholin auf die Nebennieren. *Arch. exp. Path. Pharmak.* **163**, 66. *57*

Feldberg, W. and Minz, B. (1932). Die blutdrucksteigernde Wirkung des Azetylcholins an Katzen nach Entfernen der Nebennieren. *Arch. exp. Path. Pharmak.* **165**, 261. *57, 58, 61, 68*

Feldberg, W. and Minz, B. (1933). Das Auftreten eines azetylcholinartigen Stoffes im Nebennierenvenenblut bei Reizung der Nervi splanchnici. *Pflügers Arch. ges. Physiol.* **233**, 657. *135*

Feldberg, W., Minz, B. and Tsudzimura, H. (1934). The mechanism of the nervous discharge of adrenaline. *J. Physiol.* **81**, 286. *58, 135, 136, 156*

Feldberg, W. and Rempel, F. (1933). Cited in Feldberg and Krayer (1933). *82*

Feldberg, W. and Rosenfeld, P. (1933). Der Nachweis eines azelylcholinartigen Stoffes im Pfortaderblut. *Pflügers Arch. ges. Physiol.* **232**, 212. *79, 131*

Feldberg, W. and Schild, H. (1934). Distribution of choline and acetylcholine in suprarenal glands. *J. Physiol.* **81**, 37P. *79, 80, 136*

Feldberg, W. and Schilf, E. (1930). *Histamin.* Berlin: Springer. *18, 24, 26, 27, 28, 31, 32, 34, 37, 104, 168*

Feldberg, W. and Schriever, K. Unpublished experiments. *144*

Feldberg, W and Vartiainen, A. (1934). The physiology and pharmacology of a sympathetic ganglion. *J. Physiol.* **81**, 39P; (1931) **83**, 103. *83, 137, 154, 155, 156*

Felix, K. and Putzer-Reybegg, A. von (1932). Physiologisch–chemische Analyse der blutdrucksendenden Wirkung von Organextrakten. *Arch. exp. Path. Pharmak.* **164**, 402; (1932) **169**, 214. *19, 109*

Fenyes, (1930). *Wien. Arch. f. i. Med.* **20**, 287. (Cited in Best and McHenry (1931).) *37, 50*

Ferdmann, D. (1933). Eine Methode für Adenosintriphosphorsäurebestimmung. *Hoppe Seyler's Z. physiol. Chem.* **216**, 205. *90*

Finkelman, B. (1930). On the nature of inhibition in the intestine. *J. Physiol.* **70**, 145. *146, 149*

Fiske, C. H. and Subbarow, Y. (1929). Phosphorus compounds of muscle and liver. *Science* **70**, 381. *86, 91*

Flatow, E. (1929). Über die verschiedene Wirkung von Histamin und Adrenalin auf die Ohrgefäße des Kaninchens. *Arch. exp. Path. Pharmak.* **141**, 161. *23*

Fleisch, A. (1931). Die Wirkung von Histamin, Azetylcholin und Adrenalin auf die Venen. *Pflügers Arch. ges. Physiol.* **228**, 351. *62*

Fleisch, A. and Sibul, I. (1933). Über nutritive Kreislaufregulierung. II. Die Wirkung von pH, intermediären Stoffwechselprodukten und anderen biochemischen Verbindungen. *Pflügers Arch. ges. Physiol.* **231**, 787. *15*

Florey, H. (1930). The secretion of mucus by the colon. *Br. J. exp. Path.* **2**, 348. *32, 48, 50*

Florey, H. W. and Carleton, H. M. (1926). Rouget cells and their function. *Proc. R. Soc. B.* **100**, 23. *24*

Forbes, H. S., Wolff, H. G. and Cobb, S. (1929). The cerebral circulation. X. The action of histamine. *Am J. Physiol.* **89**, 266. *25*

Forst, A. W. and Weese, H. (1926). Über die uteruswirksamen Substanzen im Mutterkorn. II Teil. Histamin. *Arch. exp. Path. Pharmak.* **117**, 232. *44*

Fox, H. and Lynch, F. B. (1917). Effect of nuclein injection upon the leukocytes of dogs. *Am. J. Med.* **153**, 571. *90*

Frank, E., Nothmann, M. and Guttmann, E. (1923). Über die tonische Kontraktion des quergestreiften Säugetiermuskels nach Ausschaltung der motorischen Nerven. *Pflügers Arch. ges. Physiol.* **199**, 567. *61*

Frank, E., Nothmann, M. and Hirsch-Kaufmann, H. (1922). Über die 'tonische' Kontraktion des quergestreiften Säugetiermuskels nach Ausschaltung der motorischen Nerven. *Pflügers Arch. ges. Physiol.* **197**, 270. *66*

Frank, E., Nothmann, M. and Hirsch-Kaufmann, H. (1923). Über die tonische Kontraktion des quergestreiften Säugetiermuskels nach Ausschaultung der motorischen Nerven II. *Pflügers Arch. ges. Physiol.* **198**, 391. *66*

Franklin, K. J. (1926). The pharmacology of the isolated vein ring. *J. Pharmac. exp. Ther.* **26**, 215. *62*

Franklin, K. J. (1932). The actions of adrenaline and of acetylcholine on the isolated pulmonary vessels. *J. Physiol.* **75**, 471. *62*

Fredericq, H. (1927). La transmission humorale des excitations nerveuses. *C. r. Séanc. Soc. Biol.* **97**, 3. *124*

Freedlander, S. O. and Lenhart, C. H. (1932). Traumatic shock. *Archs. Surg., Chicago* **25**, 693. *177*

Freeman, N. E., Phillips, R. A. and Cannon, W. B. (1931). Unsuccessful attempt to demonstrate humoral action of 'vagus substance' in circulating blood. *Am. J. Physiol.* **98**, 435. *131*

Freund, H. (1920). Über die pharmakologischen Wirkungen des defibrinierten Blutes. *Arch. exp. Path. Pharmak.* **86**, 266; **88**, 39. *97, 99, 101, 102*

Freund, H. (1921). Studien zur unspezifischen Reiztherapie. *Arch. exp. Path. Pharmak.* **91**, 272. *97*

Frey, E. (1928). Giftwirkungen an dem quergestreiften Schleiendarm. *Arch. exp. Path. Pharmak.* **138**, 228. *64*

Frey, E. K. and Kraut, H. (1928). Ein neues Kreislaufhormon und seine Wirkung. *Arch. exp. Path. Pharmak.* **133**, 1. *115*

Frey, E. K., Kraut, H. and Schultz, F. (1930). Über eine neue innersekretorische Funktion des Pankreas. V. *Arch. exp. Path. Pharmak.* **158**, 334. *119*

Fröhlich, A. and Loewi, O. (1910). Über eine Steigerung der Adrenalinempfindlichkeit durch Kokain. *Arch. exp. Path. Pharmak.* **62**, 159. *158*

Fröhlich, A. and Paschkis, K. (1926). Verstärkung pharmakologischer Reaktionen durch gereinigtes Eiweiß. *Arch. exp. Path. Pharmak.* **117**, 169. *65*

Fröhlich, A. and Pick, E. P. (1912). Die Folgen der Vergiftung durch Adrenalin, Histamin, Pituitrin, Pepton, sowie der anaphylaktischen Vergiftung. *Arch. exp. Path. Pharmak.* **71**, 23. *33*

Fryer, A. L. and Gellhorn, E. (1933). On the principle of autonomic nervous action; observations on resistance to temperature of endings of vagus and sympathetic in heart. *Am. J. Physiol.* **103**, 392. *154*

Fühner, H. (1916). Pharmakologische Untersuchungen über die Wirkung des Hypophysins. *Biochem. Z.* **76**, 232. *64, 65, 68*

Fühner, H. (1918a). Die chemische Erregbarkeitssteigerung glatter Muskulatur. *Arch. exp. Path. Pharmak.* **82**, 51. *64*

Fühner, H. (1918b). Ein Vorlesungsversuch zur Demonstration der erregbarkeitssteigernden Wirkung des Physostigmins. *Arch. exp. Path. Pharmak.* **82**, 81. *72, 83, 155*

Fühner, H. (1918c). Der toxikologische Nachweis des Physostigmins. *Biochem. Z.* **92**, 347. *72, 83, 155*

Fühner, H. (1932). Die giftigen und tödlichen Gaben einiger Substanzen für Frösche und Mäuse. *Arch. exp. Path. Pharmak.* **166**, 437. *70*

Fühner, H. and Starling, E. H. (1913). Experiments on the pulmonary circulation. *J. Physiol.* **47**, 286. *26*

Fürth, O. von and Schwarz, C. (1908). Zur Kenntnis der 'Sekretine'. *Pflügers Arch. ges. Physiol.* **124**, 427. *108*

Fulton, J. F. (1926). *Muscular Contraction and the Reflex Control of Movement.* Baltimore, MD: Williams. *153*

Gaddum, J. H. (1933). Unpublished experiments. *22*

Gaddum, J. H. and Holtz, P. (1933). The localisation of the action of drugs on the pulmonary vessels of dogs and cats. *J. Physiol.* **77**, 139. *28, 58, 62, 89*

Gaddum, J. H. and Khayyal, M. A. Unpublished experiments. *165*

Gaddum, J. H. and Schild, H. (1934). Depressor substances in tissue extracts. *J. Physiol.* **83**, 1. *19, 41, 43, 108, 109*

Galehr, O. and Plattner, F. (1928). Über das Schicksal des Azetylcholins im Blute. *Pflügers Arch. ges. Physiol.* **218**, 488 and 506. *55, 71, 81, 82*

Garrelon, L. and Santenoise, D. (1924). Action de l'insuline sur l'excitabilité du pneumogastrique. *C. r. Séanc. Soc. Biol.* **90**, 470. *106*

Garrelon, L., Santenoise, D., Verdier, H. and Vidacovitch, M. (1930). Pancréas et l'excitabilité pneumogastrique. *C. r. Séanc. Acad. Sci., Paris* **190**, 213. *107*

Gaskell, W. H. (1880–82). On the tonicity of the heart and blood vessels. *J. Physiol.* **3**, 48. *169*

Gaskell, W. H. (1884). On the augmentor (accelerator) nerves of the heart of cold-blooded animals. *J. Physiol.* **5**, 46. *124, 146*

Gaskell, W. H. (1920). (new edn). *The Involuntary Nervous System.* London: Longmans. *139*

Gasser, H. S. (1926). Plexus-free preparations of the small intestine. *J. Pharmac. exp. Ther.* **27**, 395. *31, 65*

Gasser, H. S. (1930). Contractures of skeletal muscle. *Physiol. Rev.* **10**, 35. *66*

Gavin, G., McHenry, E. W. and Wilson, M. J. (1933). Histamine in canine gastric tissues. *J. Physiol.* **79**, 234. *34*

Gebauer-Fuelnegg, E. (1930). Zur Kenntnis der Paulyschen Diazoreaktion. *Hoppe-Seyler's Z. physiol. Chem.* **191**, 222. *42*

Gebauer-Fuelnegg, E., Dragstedt, C. A. and Mullenix, R. B. (1932). A physiologically active substance appearing during anaphylactic shock. *Proc. Soc. exp. Biol. Med.* **29**, 1084. *182*

Gebhardt, F. and Klein, J. (1933). Über Azetylcholin und Magensaft. *Klin. Wschr.* **12**, 535. *68*

Gerrard, R. W. (1922). I. The presence and significance of histamine in an obstructed bowel. *J. biol. Chem.* **52**, 111. *42*

Gibbs, O. S. (1926). The effects of atropine, physostigmine, and pilocarpine on the cardiac vagus of the fowl. *J. Pharmac. exp. Ther.* **27**, 319. *156*

Gibbs, O. S. (1935). On the alleged occurrence of acetylcholine in saliva. *J. Physiol.* **84**, 33. *119, 133*

Gibbs, O. S. and Szelöczey, J. (1932a). Die humorale Übertragung der Chorda tympani-Reizung. *Arch. exp. Path. Pharmak.* **168**, 64. *133, 134*

Gibbs, O. S. and Szelöczey, J. (1932b). Humoral transmission and the chorda tympani, *J. Physiol.* **76**, 15P. *133, 134*

Girndt, O. (1925). Cholin als Hormon der Darmbewegung; Die Unfähigkeit der isolierten Darmwand, Cholin neu zu bilden. *Pflügers Arch. ges. Physiol.* **207**, 469. *130*

Gley, P. and Kistinios, N. (1928). Sur l'action hypotensive des extraits pancréatiques. *C. r. Séanc. Soc. Biol.* **99**, 1840. *107*

Gley, P. and Kistinios, N. (1929). Recherches sur la substance hypotensive du pancréas. *Presse méd.* **37**, 1279. *107*

Gley, P. and Quinquaud, A. (1923). Effet vasoconstricteur de l'excitation du nerf splanchnique. *C. r. Séanc. Soc. Biol.* **88**, 1174. *149*

Goldblatt, M. W. (1933). *Chemistry and industry* **52**, 1056. *112*

Goldblatt, M. W. (1935). Properties of human seminal plasma. *J. Physiol.* **84**, 208. *112*

Gollwitzer-Meier, K. (1934). Zur Frage des Azetylcholins im Rinderblut. *Arch. exp. Path. Pharmak.* **174**, 456. *78, 95*

Gollwitzer-Meier, K. and Bingel, A. (1933). Der Nachweis eines azetylcholinartigen Stoffes in der Haut. *Arch. exp. Path. Pharmak.* **173**, 173. *79*

Gollwitzer-Meier, K. and Otte, M. L. (1933). Über den Nachweis einer azetylcholinartigen Substanz bei der reflektorischen Gefäßerweiterung. *Arch. exp. Path. Pharmak.* **171**, 1. *143, 164*

Govaerts, P., Cambier, P. and van Dooren (1931). Vitesse de destruction de l'acétylcholine par le sang des individus sensibles óu resistants à cette substance. *C. r. Séanc. Soc. Biol.* 108, 1178. *83*

Govaerts, P. and van Dooren (1931). Effets de fortes doses d'acétylcholine en injections intraveineuses chez l'homme. *C. r. Séanc. Soc. Biol.* 106, 934. *69*

Grabe, F., Krayer, O. and Seelkopf, K. (1934). Beitrag zur Erklärung der kreislaufwirksamen (adrenalinähnlichen) Stoffe in Leberextrakten. *Klin. Wschr.* 13, 1381. *105*

Granberg, K. (1925). Action de la physostigmine sur la partie motrice de l'innervation sympathetique. *C. r. Séanc. Soc. Biol.* 93, 1167. *155*

Grant, R. T. (1930). Observations on direct communications between arteries and veins in the rabbit's ear. *Heart* 15, 281. *24, 56*

Grant, R. T. and Bland, E. F. (1932). In co-operation with Camp, P. D. Observations on the vessels and nerves of the rabbit's ear with special reference to the reaction to cold. *Heart* 16, 69. *24*

Grant, R. T. and Duckett Jones, T. (1929). The effect of histamine and of local injury on the blood vessels of the frog; a vasodilator substance in extract of frog's skin. *Heart* 14, 339. *26, 120*

Gruber, C. M. (1928). A note on the rhythmic contractions of the ureter. *J. Pharmac. exp. Ther.* 34, 203. *29*

Gruber, C. M. (1929). The blood pressure in unanaesthetized animals. *J. Pharmac. exp. Ther.* 36, 155. *57, 68*

Guggenheim, M. (1913). Proteinogene Amine. *Biochem. Z.* 51, 369. *30*

Guggenheim, M. (1914). Wirkung des β-Imidazolyläthylamine (Imido 'Roche') am menschlichen Uterus. *Therap. Monatschr.* 28, 174. *30*

Guggenheim, M. (1924). *Die Biogenen Amine.* Berlin: J. Springer. *18*

Guggenheim, M. and Löffler, W. (1916a). Biologischer Nachweis proteinogener Amine in Organextrakten und Körperflüssigkeiten. *Biochem. Z.* 72, 303. *31, 44, 50, 51*

Guggenheim, M. and Löffler, W. (1916b). Über das Vorkommen und Schicksal des Cholins im Tierkörper. *Biochem. Z.* 74, 208. *64, 114, 119*

Guimaräis, J. A. (1930). Influence du sang et de la salive sur la sécrétion salivaire. *C. r. Séanc. Soc. Biol.* 104, 804. *119, 132*

Guimaräis, J. A. (1936). The secretagogue and depressor substances in saliva and pancreatic juice. *J. Physiol.* 86, 95. *119*

Gulewitsch, W. (1899). Über die Leukomatine des Ochsengehirns. *Hoppe Seyler's Z. physiol. Chem.* 27, 50. *110*

Gunn, J. A. (1926). The action of histamine on the heart and coronary vessels. *J. Pharmac. exp. Ther.* 29, 325. *26*

Gutowski, B. (1924). Sécrétion du suc gastrique sous l'influence de l'histamine. *C. r. Séanc. Soc. Biol.* 91, 1346. *34*

Gutentag, O. E. (1931). Histamin und histaminartige Substanzen im Blut. *Arch. exp. Path. Pharmak.* 162, 727. *95, 101, 102*

Haake, E. (1930). Die Wirkung der Frühgiftlösungen an überlebenden Organen. *Arch. exp. Path. Pharmak.* 150, 119. *98, 101, 102*

Haberlandt, L. (1924). Über ein Sinus-Hormon des Froschherzens. *Klin. Wschr.* 3, 1631. *103*

Haberlandt, L. (1927). *Das Hormon der Herzbewegung.* Berlin: Urban and Schwarzenberg. *103*

Haberlandt, L. (1928). Über ein Hormon der Herzbewegung. *Pflügers Arch. ges. Physiol.* 221, 576. *104*

Hadjimichalis, S. (1931). Die Wirkung von Azetylcholin und Histamin auf die Irismuskulatur des enukleierten Froschauges. *Arch. exp. Path. Pharmak.* 160, 49. *29, 63*

Halpert, B. and Lewis, J. H. (1930). Experiments on the isolated whole gall-bladder of the dog. *Am. J. Physiol.* **93**, 506. *32*

Hamet, R. (1926). Sur l'action vaso-dilatatrice renale de l'acétylcholine. *C. r. Séanc. Soc. Biol.* **94**, 727. *57*

Handowsky, H. and Pick, E. P. (1912a). Untersuchungen über die pharmakologische Beeinflußbarkeit des peripheren Gefäßtonus des Frosches. *Arch. exp. Path. Pharmak.* **71**, 89. *98*

Handowsky, H. and Pick, E. P. (1912b). Über die Entstehung vasokonstriktoricher Substanzen durch Veränderung der Serumkolloide. *Arch. exp. Path. Pharmak.* **71**, 62. *98*

Hanke, M. T. and Koessler, K. K. (1920). (See also papers by Koessler and Hanke.) The quantitative colorimetric estimation of histamine in protein and protein-containing matter. *J. biol. Chem.* **43**, 527, 543. *41, 95, 121*

Hanke, M. T. and Koessler, K. K. (1922). XII. The production of histamine and other imidazoles from histidine by the action of micro-organisms. *J. biol. Chem.* **50**, 131. *52*

Hanke, M. T. and Koessler, K. K. (1924a). XVII. On the faculty of normal intestinal bacteria to form toxic amines. *J. biol. Chem.* **59**, 835. *52*

Hanke, M. T. and Koessler, K. K. (1924b). XVIII. On the production of histamine, tyramine, and phenol in common laboratory media by certain intestinal micro-organisms. *J. biol. Chem.* **59**, 855. *52*

Hanke, M. T. and Koessler, K. K. (1924c). XX. On the presence of histamine in the mammalian organism. *J. biol. Chem.* **59**, 879. *40, 41*

Hansen, K. and Rech, W. (1931). Über humorale Herznervenwirkung. *Z. Biol.* **92**, 191. *126*

Hanzlik, P. T. and Karsner, H. T. (1920). Effects of various colloids and other agents which produce anaphylactoid phenomena on bronchi of perfused lungs. *J. Pharmac. exp. Ther.* **14**, 449. *32*

Harde, E. (1932). Lésions gastriques des souris après injections répétées d'histamine. *C. r. Séanc. Soc. Biol.* **109**, 1326. *34*

Harmer, I. M. and Harris, K. E. (1926). Observations on the vascular reactions in man in response to histamine. *Heart* **13**, 381. *37*

Harris, K. E. (1927). Observations upon a histamine-like substance in skin extracts. *Heart* **14**, 161. *175, 181*

Hartman, F. A., Evans, J. I. and Walker, H. G. (1929). Control of capillaries of skeletal muscle. *Am. J. Physiol.* **90**, 668. *24, 57*

Hartman, F. A., Rose, W. J. and Smith, E. P. (1926). The influence of burns on epinephrin secretion. *Am. J. Physiol.* **78**, 47. *176*

Hashimoto, H. (1925). Blood chemistry in acute histamine intoxication. *J. Pharmac. exp. Ther.* **25**, 381. *38*

Haynes, F. W. (1932). Factors which influence the flow and protein content of subcutaneous lymph in the dog. *Am. J. Physiol.* **101**, 612. *26, 59*

Heffter, A. (1923–27). *Handbuch der Experimentellen Pharmakologie.* Berlin: Julius Springer. **55**. *160*

Heidenhain, R. (1883). Über pseudomotorische Nervenwirkung. *Arch. Anat. Physiol.,* Suppl. Band. 133. *140*

Heinekamp, W. J. R. (1925). The mechanism of vagus inhibition as produced by adrenaline. *J. Pharmac. exp. Ther.* **26**, 385. *155, 156*

Heller, H. and Kusonoki, G. (1933). Dis zentrale Blutdruckwirkung des neurohypophysäuren Kreislaufhormons (Vasopressin). *Arch. exp. Path. Pharmak.* **173**, 301. *47*

Hemingway, A. (1931). A comparison of methods used for oxygenating blood in perfusion experiments. *J. Physiol.* **72**, 344. *100*

Hemingway, A. and McDowall, R. J. S. (1926a). On the survival of striped mammalian muscles. *J. Physiol.* **61**, vi. *22*

Hemingway, A. and McDowall, R. J. S. (1926b). The chemical regulation of capillary tone. *J. Physiol.* **62**, 166. *22*

Henderson, V. E. and Roepke, M. H. (1932). On the mechanism of salivary secretion. *J. Pharmac. exp. Ther.* **47**, 193. *134, 164*

Henderson, V. E. and Roepke, M. H. (1933a). On the mechanism of erection. *Am. J. Physiol.* **106**, 441. *142, 156*

Henderson, V. E. and Roepke, M. H. (1933b). Über den lokalen hormonalen Mechanismus der parasympathikusreizung. *Arch. exp. Path. Pharmak.* **172**, 314. *134, 135, 160, 164*

Henderson, V. E. and Roepke, M. H. (1934). The role of acetylcholine in bladder contractile mechanisms and in parasympathetic ganglia. *J. Pharmac. exp. Ther.* **51**, 97. *135*

Henking, G. von and Szent-Györgyi, A. von (1923). Über die Wirkung des defibrinierten Blutes auf das isolierte Säugetierherz. *Pflügers Arch. ges. Physiol.* **197**, 516. *101*

Herrick, J. and Markowitz, J. (1929). The toxic effects of defibrinated blood when perfused through the isolated mammalian heart. *Am. J. Physiol.* **88**, 698. *101*

Hess, W. R. (1923). Über die Wirkung von Acetylcholin auf den Skelettmuskel. *Q. J. exp. Physiol.*, Suppl. Band, 144. *67, 144*

Heymann, W. (1927). Untersuchungen über die pharmakologische Wirksamkeit des Blutserums. *Arch. exp. Path. Pharmak.* **125**, 77. *98, 101*

Heymans, C., Bouckaert, J. J. and Moraes, A. (1932a). Inversion par l'ergotamine de l'action constrictice des 'vasotonines' du sang défibriné. Au sujet de l'action vasculaire de l'ergotamine. *Archs int. Pharmacodyn. Ther.* **43**, 468. *97, 99*

Heymans, C., Bouckaert, J. H. and Moraes, A. (1932b). Au sujet de l'action vasculaire de l'ergotamine. Inversion de l'action vaso-constrictice des 'vasotonines' du sang défibriné par l'ergotamine. *C. r. Séanc. Soc. Biol.* **110**, 993. *97, 99*

Hiller, A. (1926a). The effect of histamine on the acid–base balance. *J. biol. Chem.* **68**, 833. *38*

Hiller, A. (1926b). The effect of histamine on protein catabolism. *J. biol. Chem.* **68**, 847. *38*

Hinsey, J. C. and Cutting, C. C. (1933). The Sherrington phenomenon. V. Nervous pathways. *Am. J. Physiol.* **105**, 535. *140, 141, 156*

Hinsey, J. C. and Gasser, H. S. (1930). The component of the dorsal root-mediating vasodilatation and the Sherrington contracture. *Am. J. Physiol.* **92**, 679. *140, 141*

Hirose, Y. (1932). Über die Wirkung großer Azetylcholindosen auf die Darm-, Nieren-, Lungen- und Extremitätengefäße. *Arch. exp. Path. Pharmak.* **165**, 401. *58, 62*

Hirschberg, E. (1931). Über nervöse Hemmungen. *Z. Biol.* **91**, 117. *54, 124, 126, 128*

Hochrein, M. and Keller, C. J. (1931). Die Beeinflussung des Kreislaufes durch Organextrakte. *Arch. exp. Path. Pharmak.* **159**, 438. *116*

Hoet, J. C. (1929). On traumatic shock and death by burns. *Am. J. Physiol.* **90**, 392. *177*

Hoffman, W. S. (1925). The isolation of crystalline adenine nucleotide from blood. *J. biol. Chem.* **63**, 675. *91*

Hofmann, E. (1930). Über die Hydrolyse von Azetylcholinchloridlösungen. *Helv. chim. Acta* **13**, 138. *55*

Holt, R. L. and Macdonald, A. D. (1934). Observations on experimental shock. *Br. med. J.* **1**, 1070. *177, 178*

Holtz, P. (1934). Die Entstehung von Histamin aus Histidin durch Bestrahlung. *Arch. exp. Path. Pharmak.* **175**, 97. *19*

Honey, R. M., Ritchie, W. T. and Thomson, W. A. R. (1930). The action of adenosine upon the human heart. *Q. Jl. Med.* **23**, 485. *89*

Hooker, D. R. (1912). The effect of carbon dioxide and of oxygen upon muscular tone in the blood vessels and alimentary canal. *Am. J. Physiol.* **31**, 47. *169*

Hooker, D. R. (1920). The functional activity of the capillaries and venules. *Am. J. Physiol.* **54**, 30. *24*

Hooker, D. R. (1921). The capillary circulation in the cat's ear. *Am. J. Physiol.* **55**, 315. *24*

Hosoya, K. (1931). Der Einfluß der Narkose auf die Gefäßwirkung des Histamins beim Kaninchen. *Arch. exp. Path. Pharmak.* **159**, 41. *43*

Houssay, B. A. and Cruciani, T. (1929). Etude des actions centrales ou périphériques sur les bronches chez le chien. *C. r. Séanc. Soc. Biol.* **101**, 246. *32*

Howell, W. H. (1906). Vagus inhibition of the heart in its relation to the inorganic salts of the blood. *Am. J. Physiol.* **15**, 280. *61, 127*

Howell, W. H. and Duke, W. W. (1908). The effect of vagus inhibition on the output of potassium from the heart. *Am. J. Physiol.* **21**, 51. *124, 128*

Hunt, R. (1899). Note on a blood-pressure lowering body in the suprarenal gland. *Am. J. Physiol.* **3**, xviii. *70, 79*

Hunt, R. (1901). Further observations on the blood-pressure lowering bodies in extracts of the suprarenal gland. *Am. J. Physiol.* **5**, vi. *70, 79*

Hunt, R. (1915). Some physiological actions of the homocholins and of some of their derivatives. *J. Pharmac. exp. Ther.* **6**, 477. *83*

Hunt, R. (1915b). A physiological test for cholin and some of its applications. *J. Pharmac. exp. Ther.* **7**, 301. *114, 119*

Hunt, R. (1918). Vasodilator Reactions I and II. *Am. J. Physiol.* **45**, 197; (1918) **45**, 231. *58, 84*

Hunt, R. (1934). Note on acetyl-β-methylcholine. *J. Pharmac. exp. Ther.* **52**, 61. *83*

Hunt, R. and Renshaw, R. R. (1934). Further studies of the methylcholines and analogous compounds. *J. Pharmac. exp. Ther.* **51**, 237. *70*

Hunt, R. and Taveau, R. de M. (1906). On the physiological action of certain choline derivatives and new methods for detecting choline. *Br. med. J.* **2**, 1788. *54, 56, 68*

Hunt, R. and Taveau, R. de M. (1910). On the relation between the toxicity and chemical constitution of a number of derivatives of choline and analogous compounds. *J. Pharmac. exp. Ther.* **1**, 303. *70*

Hunt, R. and Taveau, R. de M. (1911). The effects of a number of derivatives of choline and analogous compounds on the blood pressure. *Hyg. Bull., Washington,* no. 73. *70*

Hunter, G. (1922a). Note on Knoop's test for histidine. *Biochem. J.* **16**, 637. *42*

Hunter, G. (1922b). The estimation of carnosine in muscle extract. *Biochem. J.* **16**, 640. *42*

Inchley, O. (1923). The action of histamine on the veins. *Br. med. J.* **1**, 679. *27*

Inchley, O. (1926). Histamine shock. *J. Physiol.* **61**, 282. *27*

Ivy, A. C. and Javois, A. J. (1924). The stimultion of gastric secretion by amines and other substances. *Am. J. Physiol.* **71**, 604. *34, 49*

Ivy, A. C. and McIlvain, G. B. (1923). The excitation of gastric secretion by application of substances to the duodenal and jejunal mucosa. *Am. J. Physiol.* **67**, 124. *50*

Ivy, A. C., McIlvain, G. B. and Javois, A. J. (1923). The stimulation of gastric secretion by histamine. *Science* **58**, 286. *50*

Ivy, A. C. and Oldberg, E. (1928). Hormone mechanism for gall-bladder contraction. *Am. J. Physiol.* **86**, 599. *32*

Jackson, H., Jr. (1923). Adenine nucleotide in human blood. *J. biol. Chem.* **57**, 121. *91*

Jackson, H., Jr. (1924). The isolation of a nucleotide from human blood. *J. biol. Chem.* **59**, 529. *91*

Jacobsen, E. (1931). Über eine spezifische Adenylpyrophosphatase. *Biochem. Z.* **242**, 292. *93*

Jäger, F. (1913a). Versuche zur Verwendung des β-Imidazolyläthylamins in der Geburtshilfe. *Zentbl. Gynäk.* **37**, 265. *30, 37*

Jäger, F. (1913b). Ein neuer für die Praxis brauchbarer Sekaleersatz (Tenosin). *Münch. med. Wschr.* **60**, 1714. *30, 37*

James, A. A., Laughton, N. B. and Macallum, A. B. (1925). Studies on the control of blood pressure with hepatic extract. *Am. J. Physiol.* **75**, 392. *104*

Janeway, T. C., Richardson, H. B. and Park, E. A. (1918). Experiments on the vaso-constrictor action of blood serum. *Archs. intern. Med.* **21**, 565. *98, 101*

Jendrassik, L. (1924). Humorale Übertragbarkeit von Nervenreizen beim Warmblüter. *Biochem. Z.* **144**, 520. *127, 129, 148*

Jones, W. and Kennedy, R. P. (1919). Adenine mononucleotide. *J. Pharmac. exp. Ther.* **12**, 253. *91*

Jullien, A. and Morin, G. (1931). Action comparée de l'atropine et de l'acétylcholine sur le ventricule isolé de l'escargot et du Murex. *C. r. Séanc. Soc. Biol.* **106**, 187. *61*

Jürgensohn, E. (1931). Die Wirkung von frishcem und defibriniertem Blut auf den Meerschweinschendünndarm. *Arch. exp. Path. Pharmak.* **162**, 739. *101*

Justin-Besançon, L. (1929). *Recherches physiologiques sur acétylcholine.* Paris: Masson and Cie. *70*

Kahlson, G. (1932). Zum biologischen Nachweis von Cholin und Azetylcholin. *Arch. exp. Path. Pharmak.* **169**, 34. *60*

Kahlson, G. (1934). Nachweis und Vorkommen präformierten Azetylcholins in Blut und Geweben. *Arch. exp. Path. Pharmak.* **175**, 198. *64, 72, 74, 79*

Kahlson, G. and Römer, R. (1934). Enthält das normale Blut chemisch nachweisbares Azetylcholin? *Arch. exp. Path. Pharmak.* **175**, 223. *78*

Kahn, R. H. (1912). Zur Frage der Adrenalinämie nach dem Zuckerstiche. *Pflügers Arch. ges. Physiol.* **144**, 251. *96*

Kahn, R. H. (1926). Über humorale Übertragbarkeit der Hernervenwirkung. *Pflügers Arch. ges. Physiol.* **214**, 482. *126, 147*

Kalk, H. (1929). Zur Frage der Existenz einer histaminähnlichen Substanz beim Zustandekommen des Dermographismus. *Klin. Wschr.* **8**, 64. *174*

Kapfhammer, J. and Bischoff, C. (1930). Azetylcholin und Cholin aus tierischen Organen. *Hoppe-Seyler's Z. physiol. Chem.* **191**, 179. *77*

Katz, G. (1929). Der Einfluß von Narkotika auf die Histaminwirkung am Gefäßstreifen. *Arch. exp. Path. Phramak.* **141**, 366. *29*

Katzenelbogen, S. (1929). The action of histamine on the alkali reserve. *J. Am. med. Ass.* **92**, 1240. *38*

Katzenelbogen, S. and Abramson, A. (1927). Les variations du taux de la glycémie consecutives aux injections d'histamine. *C. r. Séanc. Soc. Biol.* **97**, 240. *38*

Kauffmann, M. (1911). Über den Befund von Cholin im Ochsengehirn. *Hoppe-Seyler's Z. physiol. Chem.* **74**, 175. *110*

Keeton, R. W., Koch, F. C. and Luckhardt, A. B. (1920). The response of the stomach mucosa of various animals to gastrin bodies. *Am. J. Physiol.* **51**, 454. *33*

Kehrer, E. (1907). Wirkung von Ergotinpräparaten auf die überlebenden inneren Genitalien. *Arch. Gynaek.* **81**, 192. *39*

Kehrer, E. (1908). Der überlendende Uterus als Testobjekt für die Wichtigkeit der Mutterkornpräparate. *Arch. exp. Path. Pharmak.* **58**, 366

Kehrer, E. (1912). Die motorische Function, Function des Uterus und ihre Beeinflussung durch Wehenmittel. *Münch. med. Wschr.* **59**, 1831. *37*

Keith, N. M. (1919). Blood volume changes in wound shock and primary haemorrhage. *Spec. Rep. Ser. med. Res. Comm.,* no. **27**, *176*

Kellaway, C. H. (1930). The anaphylactic reaction of the isolated uterus of the rat. *Br J. exp. Path.* **11**, 72. *180, 184*

Kellaway, C. H. and Cowell, S. J. (1923). On the concentration of the blood and the effects of histamine in adrenal insufficiency. *J. Physiol.* **57**, 82. *35*

Kendall, A. I. (1927). The relaxation of histamine contractions in smooth muscle by certain aldehydes. *J. infect. Dis.* **40**, 689. *19*

Kendall, A. I. and Shumate, F. O. (1930). The quantitative response of intestine from sensitised guinea-pigs to homológous protein and to histamine. *J. infect. Dis.* **47**, 267. *31*

Kendall, A. I. and Varney, P. L. (1927). The physiologic action of histamine applied directly to the mucosa of the isolated surviving intestine of the guinea pig. *J. infect. Dis.* **41**, 143. *49*

Kibjakow, A. W. (1933). Über humorale Übertragung der Erregung von einem Neuron auf das andere. *Pflügers Arch. ges. Physiol.* **232**, 432. *136*

Kinoshita, T. (1910). Über den Cholingehalt tierischer Gewebe. *Pflügers Arch. ges. Physiol.* **132**, 607. *104, 106*

Kisch, B. (1927). Besprechung von L. Haberlandt: Das Hormon der Herzbewegung. Z. *Kreislaufforsch.* **19**, 355. *103*

Kitakoji, Y. (1931). Über den Einfluß von Nervengiften und die Funktionen der Gallenblase und des Oddischen Muskels. Cited in *Ber. ges. Physiol. exp. Pharm.* **63**, 213. *65*

Klaus, K. (1926). Biochemische Vorgänge bei der Menstruation. *Sborník lékarsky Jg.* **27**, 55. (Czech and German summary.) Cited in *Ber. ges. Physiol. exp. Pharm.* **37**, 189. *120*

Knoop, F. (1908). Eine Farbenreaktion des Histidins. *Ber. chem. Physiol. Path.* **11**, 356. *42*

Koch, W. F. (1913). Toxic bases in the urine of parathyroidectomized dogs. *J. biol. Chem.* **15**, 43. *46, 50, 114*

Kodera, Y. (1928). Über das Schicksal des Azetylcholins im Blute. *Pflügers Arch. ges. Physiol.* **219**, 181.

Kögl, F., Duisberg, H. and Erxleben, H. (1931). Uber das Muskarin. *Justus Liebigs Annalen Chem.* **489**, 156. *55*

Koessler, K. K. and Hanke, M. T. (1918). The synthesis of histamine. *J. Am. chem. Soc.* **40**, 1716. (Cf. also Hanke and Koessler.) *19*

Koessler, K. K. and Hanke, M. T. (1919a). Studies on proteinogenous amines. II. A micro-chemical colorimetric method for estimating imidazole derivatives. *J. biol. Chem.* **39**, 497. *41, 52*

Koessler, K. K. and Hanke, M. T. (1919b). Studies on proteinogenous amines. IV. The production of histamine from histidine by *Bacillus coli communis*. *J. biol. Chem.* **39**, 539. *52*

Koessler, K. K. and Handke, M. T. (1924). Studies on proteinogenous amines. XXI. The intestinal absorption and detoxication of histamine in the mammalian organism. *J. biol. Chem.* **59**, 889. *41, 47, 48, 52*

Koessler, K. K., Hanke, M. T. and Sheppard, M. S. (1928). Production of histamine, tyramine, bronchospastic and arteriospastic substances in blood broth by pure cultures of micro-organisms. *J. infect. Dis.* **43**, 363. *52*

Koessler, K. K. and Lewis, J. H. (1927). Determination of bronchospasm in the guinea-pig. *Archs. intern. Med.* **39**, 163. *32*

Koessler, K. K., Lewis, J. H. and Walker, J. A. (1927). Pharmacodynamic actions of bacterial poisons. *Archs. intern. Med.* **39**, 188. *44*

Komarow, S. A. (1933). Über die Anwesenheit von sekretagogen Substanzen in dem reinen Magensaft des Hundes. *Biochem. Z.* **261**, 92. *120*

Koskowski, W. and Kubikowski, P. (1929). Sécrétion du suc gastrique après injection d'histamine et presence de cette substance dans le sang. *C. r. Séanc. Soc. Biol.* **100**, 292. *48*

Kossel, A. (1885). Über eine neue Base aus dem Tierkörper. *Ber dtsch. chem. Ges.* **18**, 79. *91*

Kraut, H., Frey, E. K., Bauer, E. and Schultz, F. (1932). Zur Kenntnis des Kallikreïns. VII. *Hoppe-Seyler's Z. physiol. Chem.* **205**, 99. *118*

Kraut, H., Frey, E. K. and Werle, E. (1930a). Der Nachweis eines Kreislaufhormons in der Pankreasdrüse. IV. *Hoppe-Seyler's Z. physiol. Chem.* **189**, 97. *107, 114, 117*

Kraut, H., Frey, E. K. and Werle, E. (1930b). Über die Inaktivierung des Kallikreïns. VI. *Hoppe-Seyler's Z. physiol. Chem.* **192**, 1. *118*

Krayer, O. (1933). Zur Kreislaufwirkung der Leberpräparate des Handels. *Dt. med. Wschr.* **59**, 576. *105*

Krayer, O. and Rühl, A. (1931). Über die Wirkung einer reinen Gefäßerweiterung auf den Gesamtkreislauf. *Arch. exp. Path. Pharmak.* **162**, 70. *116*

Krayer, O. and Verney, E. B. (1934). Veränderung des Azetylcholingehaltes im Blute der Koronarvenen unter dem Einfluß einer Blutdrucksteigerung durch Adrenalin. *Klin. Wschr.* **13**, 1250. *129*

Kroetz, C. (1931). Allgemeine Physiologie der autonomen nervösen Korrelationen. In Bethe, A. (ed.): *Handb. norm. path. Physiol.* **16**, (Vol. 2), 1769. *54, 124*

Krogh, A. (1929). *Anatomy and Physiology of Capillaries.* 2nd edn, New Haven: Yale University Press. *175*

Külz, F. (1928). Zur Humoralphysiologie des Froschherzens. *Arch. exp. Path. Pharmak.* **134**, 252. *147*

Kuroda, S. (1923). Pharmakodynamische Studien zur Frage der Magenmotilität. *Z. ges. exp. Med.* **39**, 341. *64, 65, 68*

Kuschinsky, H. (1929). Über das Verhalten von Kalium und Kalzium im Blut des Hundes beim Histaminschock. *Z. ges. exp. Med.* **64**, 563. *38*

Kutscher, F. (1910). Die physiologische Wirkung einer Sekalebase und des Imidazolyläthylamins. *Zentbl. Physiol.* **24**, 163. *46*

La Barre, J. (1926a). Sur les modifications de l'alcalinité sanguine au cors du choc histaminique. *C. r. Séanc. Soc. Biol.* **95**, 238. *38*

La Barre, J. (1926b). Sur les modifications du pH du plasma lors du choc histaminique. *C. r. Seánc. Soc. Biol.* **95**, 237. *38*

La Barre, J. (1926c). A propos des variations de la glycémie lors des chocs anaphylactiques et histaminiques chez les cobayes décérébrés. *C. r. Séanc. Soc. Biol.* **95**, 855. *38*

La Barre, J. (1926d). Sur les causes de l'hyperglycémie consécutive a l'injection intraveneuse d'histamine. *C. r. Séanc. Soc. Biol.* **94**, 1021. *38*

La Barre, J. (1926e). A propos des variations de la glycémie consécutives a l'injection intraveneuse d'histamine. *C. r. Séanc. Soc. Biol.* **94**, 779. *38*

Läwen, A. (1904). Quantitative Untersuchungen über die Gefäßwirkung von Suprarenin. *Arch. exp. Path. Pharmak.* **51**, 415. *96*

Lampe, W. and Mehes, J. (1926). Aus Gefäßstudien an der überlebenden Warmblutleber. *Arch. exp. Path. Pharmak.* II. **117**, 115; III. **119**, 66; IV. **119**, 73. *28*

Lange, F. (1932). Über die blutdrucksenkende Wirkung gewisser Organextrakte. *Arch. exp. Path. Pharmak.* **164**, 417. *62, 109*

Langley, J. N. (1897). On the regeneration of pre-ganglionic and of post-ganglionic visceral nerve fibres. *J. Physiol.* **22**, 215. *167*

Langley, J. N. (1898). On the union of cranial autonomic (visceral) fibres with the nerve cells of the superior cervical ganglion. *J. Physiol.* **23**, 240. *166*

Langley, J. N. (1905). The action of nicotine and curare upon somatic nerve endings and upon skeletal muscle. *J. Physiol.* **33**, 380. *66*

Langley, J. N. (1922). The secretion of sweat. Part I. Supposed inhibitory nerve fibres in the posterior nerve roots. Secretion after denervation. *J. Physiol.* **56**, 110. *138*

Langley, J. N. (1923). Antidromic action. Part II. Stimulation of the peripheral nerves of the cat's hind foot. *J. Physiol.* **58**, 49. *139*

Langley, J. N. and Anderson, H. K. (1904a). On the union of the fifth cervical nerve with the superior cervical ganglion. *J. Physiol.* **30**, 439. *166*

Langley, J. N. and Anderson, H. K. (1904b). The union of different kinds of nerve fibres. *J. Physiol.* **31**, 365. *166, 167*

Langley, J. N. and Orbeli, L. A. (1910). Observations on the sympathetic and sacral autonomic systems of the frog. *J. Physiol.* **41**, 450. *146*

Langworthy, O. R. (1924). Problems of tongue innervation. *Johns Hopkins Hosp. Bull.* **35**, 239. *144*

Lanz, A. B. (1928). Sur la formation dans le cœur d'une substance semblable à l'adrénaline par suite de l'excitation du nerf sympathique. *Archs. néerl. Physiol.* **12**, 433; (1928) **13**, 423. *147, 152*

Laurent, L. P. E. (1935). Clinical observations on the use of prostigmin in the treatment of myasthenia gravis. *Br. med. J.* **5**, 463. *145*

Le Heux, J. W. (1919). Cholin als Hormon der Darmbewegung. *Pflügers Arch. ges. Physiol.* **173**, 8. *80, 129*

Le Heux, J. W. (1920). Cholin als Hormon der Darmbewegung. II Mitteilung; Zur Erklärung der wechselnden Wirkung des Atropins auf den Darm. *Pflügers Arch. ges. Physiol.* **179**, 177. *130*

Le Heux, J. W. (1921). Cholin als Hormon der Darmbewegung. III. Mitteilung; Die Beteiligung des Cholins an der Wirkung verschiedener organischer Säuren auf den Darm. *Pflügers Arch. ges. Physiol.* **190**, 280. *130*

Lee, F. C. (1925). The effect of histamine on cerebro-spinal fluid pressure. *Am. J. Physiol.* **74**, 317. *25*

Leethem, C. (1913). Action of certain drugs on isolated strips of ventricle. *J. Physiol.* **46**, 151. *60*

Leschke, E. (1913). Über die Beziehungen zwischen Anaphylaxie und Fieber sowie über die Wirkungen von Anaphylatoxin, Histamin, Organextrakten und Pepton auf die Temperatur. *Z. exp. Path. Ther.* **14**, 151. *37*

Levene, P. A. (1918). The structure of yeast nucleic acid. *J. biol. Chem.* **33**, 425. *93*

Levene, P. A. and Bass, L. W. (1931). *Nucleic Acids.* New York: Chemical Catalog Co. 86. *93*

Levene, P. A. and Jacobs, W. A. (1909). Über Hefenucleinsäure. *Ber. dtsch. chem. Ges.* **42**, 2703. *91*

Lewis, T. (1927). *The Blood Vessels of the Human Skin and their Responses.* London: Shaw. *23, 52, 168, 171, 173, 175, 181*

Lewis, T. and Grant, R. T. (1924). Vascular reactions of the skin to injury. *Heart* **11**, 209. *23*

Lewis, T. and Harmer, I. M. (1926). Release of vaso-dilator substances in injuries of the skin. *J. Physiol.* **62**, 11P. *174*

Lewis, T. and Harmer, I. M. (1927). Vascular reactions of the skin to injury. XI. Further evidence of the release of a histamine-like substance from the injured skin. *Heart* **14**, 19. *174*

Lewis, T. and Marvin, H. M. (1927). Observations relating to vasodilation arising from antidromic impulses, to herpes zoster and trophic effects. *Heart* **14**, 27. *23, 25, 139, 142*

Lim, R. K. S. (1924). On the relationship between the gastric acid response and the basal secretion of the stomach. *Am. J. Physiol.* **69**, 318. *34*

Lim, R. K. S., Ivy, A. C. and McCarthy, J. E. (1925). Contributions to the theory of gastric secretion. *Q. Jl exp. Physiol.* **15**, 13. *48*

Lim, R. K. S. and Liu, A. C. (1926). Ermüdung der Magensekretion. *Pflügers Arch. ges. Physiol.* **211**, 647. *34*

Lim, R. K. S. and Schlapp, W. (1923). The effect of histamine, gastrin and secretin on the gastro-duodenal secretions in animals. *Q. Jl exp. Physiol.* **13**, 393. *33*

Lindner, F. and Rigler, R. (1930). Über die Beeinflussung der Weite der Herzkranzgefäße durch Produkte des Zellkernstoffwechsels. *Pflügers Arch. ges. Physiol.* **226**, 697. *88*

Lipschütz, H. and Schilf, E. (1931). Über die Wirkung von Azetylcholin auf die Iris des Warmblüters. *Arch. exp. Path. Pharmak.* **162**, 617. *63*

Loach, J. W. V. (1934). The alleged occurrence of acetylcholine in normal ox blood. *J. Physiol.* **82**, 118. *78*

Loening, P. (1913). Beobachtungen über die vasotonierenden Eigenschaften des Blutserums. *Z. Biol.* **62**, 54. *98, 101*

Loewi, O. (1912). III. Vaguserregbarkeit und Vagusgifte. *Arch. exp. Path. Pharmak.* **70**, 351. *83*

Loewi, O. (1921). Über humorale Übertragbarkeit der Herznervenwirkung. *Pflügers Arch. ges. Physiol.* I. **189**, 239; II. **193**, 201; III. **203**, 408. *7, 124, 128, 146, 147*

Loewi, O. (1924). Über humorale Übertragbarkeit der negativ chrono- und dromotropen Vaguswirkung. *Pflügers Arch. ges. Physiol.* **204**, 629. *125*

Loewi, O. (1926). Kritische Bemerkungen zu L. Ashers Mitteilungen. *Pflügers Arch. ges. Physiol.* **212**, 695. *126, 147*

Loewi, O. and Mansfeld, G. (1910). Über den Wirkungsmodus des Physostigmins. III. *Arch. exp. Path. Pharmak.* **62**, 180. *156*

Loewi, O. and Navratil, E. (1924). Über humorale Übertragbarkeit der Herznervenwirkung. VI. Der Angriffspunkt des Atropins. *Pflügers Arch. ges. Physiol.* **206**, 123. *154, 160*

Loewi, O. and Navratil, E. (1926a). X. Über das Schicksal des Vagusstoffes. *Pflügers Arch. ges. Physiol.* **214**, 678. *81, 82, 84, 103, 122, 128, 131, 147*

Loewi, O. and Navratil, E. (1926b). XI. Über den Mechanismus der Vaguswirkung von Physostigmin und Ergotamin. *Pflügers Arch. ges. Physiol.* **214**, 689. *128, 131*

Lohmann, K. (1928a). Die Menge der leichthydrolysierbaren P-Verbindung in tierischen und pflanzlichen Zellen. *Biochem. Z.* **203**, 164. *90*

Lohmann, K. (1928b). Nachweis und Isolierung des Pyrophosphats. *Biochem. Z.* **202**, 466. *90*

Lohmann, K. (1929). Über die Pyrophosphatfraktion im Muskel. *Naturwissenschaften* **17**, 624. *91*

Lohmann, K. (1931a). Darstellung der Adenylpyrophosphorsäure aus Muskulatur. *Biochem. Z.* **233**, 460. *87, 91, 93*

Lohmann, K. (1931b). Untersuchungen über die chemische Natur des Koferments der Milchsäurebildung. *Biochem. Z.* **237**, 445. *87, 91, 93*

Lohmann, K. (1931c). Untersuchungen zur Konstitution der Adenylpyrophosphorsäure. *Biochem Z.* **254**, 381. *90, 91*

Longcope, W. T. (1922). Insusceptibility to sensitization and anaphylactic shock. *J. exp. Med.* **36**, 627. *183*

Lucas, G. H. W. (1926). Blood and urine findings in desuprarenalised dogs. *Am. J. Physiol.* **77**, 114. *42*

Luisada, A. (1928). Neue Untersuchungen über des Wirkung des Morphiums auf Blutgefäße, besonders Lungengefäße. *Arch. exp. Path. Pharmak.* **132**, 296. *28*

McCarrison, R. (1924). Effects of long-continued injection of tyramine and histamine. *Indian J. med. Res.* **11**, 1137. *36*

McDowall, R. J. S. and Thornton, J. W. (1930). A method of recording the movements of isolated bronchi. *J. Physiol.* **70**, XLIV. *44*

MacGregor, R. G. and Peat, S. (1933). The histamine–histaminase system in the isolated perfused kidney–lung preparation. *J. Physiol.* **77**, 310. *24, 50, 51, 57*

MacGregor, R. G. and Thorpe, W. V. (1933a). The extraction of histamine from tissues by electrodialysis. *J. Physiol.* **77**, 33P. *40, 95*

MacGregor, R. G. and Thorpe, W. V. (1933b). The quantitative extraction of histamine from tissues by electrodialysis. *Biochem. J.* **27**, 1394. *40, 42, 95*

McHenry, E. W. and Gavin, G. (1932). Ammonia production during histamine–histaminase reaction. *Trans. R. Soc. Can.* **26**, 321. *51*

Macht, D. I. (1926). Concerning the point of attack of pituitary extract and histamine on smooth muscle. *J. Pharmac. exp. Ther.* **27**, 389. *29*

Macht, D. I. and Ting, G. (1921). A study of antispasmodic drugs on the bronchus. *J. Pharmac. exp. Ther.* **18**, 373. *32*

McIllroy, P. T. (1928). Experimental production of gastric ulcer. *Proc. Soc. exp. Biol. Med.* **25**, 268. *34*

MacKay, M. E. (1927a). Histamine and salivary secretion. *Am. J. Phsyiol.* **82**, 546. *33*

MacKay, M. E. (1927b). The vascular reaction of the pilocarpinized submaxillary gland to histamine. *J. Pharmac. exp. Ther.* **32**, 147. *33*

MacKay, M. E. (1929). Histamine and adrenaline in relation to the salivary secretion. *J. Pharmac. exp. Ther.* **37**, 349. *33, 36*

MacKay, M. E. (1930). The action of histamine on the mobility of different parts of the intestinal tract. *Am. J. Physiol.* **95**, 527. *31, 49*

MacKay, M. E. and Baxter, S. G. (1931). Restoration of the pancreatic secretion by peptone and histamine. *Am. J. Physiol.* **98**, 42. *49, 50*

McKenney, F. D., Essex, H. E. and Mann, F. C. (1932). The action of certain drugs on the oviduct of the domestic fowl. *J. Pharmac. exp. Ther.* **45**, 113. *65*

McMichael, J. (1933). The portal circulation. II. The action of acetylcholine. *J. Physiol.* **76**, 399. *57, 63*

McSwiney, B. A. (1933). Unpublished. *156*

Magnus, R. (1930). *Lane Lectures on Experimental Pharmacology and Medicine.* Stanford University Publications. Vol. II, no. **3**. *129*

Major, R. H., Nanninga, J. B. and Weber, C. J. (1932). Comparison of properties of certain tissue extracts having depressor effects. *J. Physiol.* **76**, 487. *110, 111*

Major, R. H. and Weber, C. J. (1929). Observations on the depressor substances in certain tissue extracts. *J. Pharmac. exp. Ther.* **37**, 367; (1930). **40**, 246. *110, 111*

Manwaring, W. H. (1911). Anaphylaktischer Schock. *Z. ImmunForsch. exp. Ther.* **8**, 1 and 589. *181*

Manwaring, W. H., Hosepian, V. M., O'Neill, F. L. and Moy, H. B. (1925). Hepatic reactions in anaphylaxis. *J. Immun.* **10**, 575. *182*

Manwaring, W. H., Meinhard, A. R. and Denhart, H. L. (1916). Toxicity of foreign sera for the isolated mammalian heart. *Proc. Soc. exp. Biol. Med.* **13**, 173. *100*

Marmorston-Gottesman, J. and Peral, D. (1931). Effect of injections of cortin on resistance of suprarenalectomized rats to large amounts of histamine. *Proc. Soc. exp. Biol. Med.* **28**, 1022. *35*

Martin, L. and Morgenstern, M. (1932). Carbon dioxide changes in alveolar air and blood plasma or serum after subcutaneous histamine injection in human beings. *J. Lab. clin. Med.* **17**, 1228. *38*

Matsuda, A. (1929). Über die Wirkung des Histamins auf die Iris. *Arch. exp. Path. Pharmak.* **142**, 70. *29*

Matthes, K. (1930). The action of blood on acetylcholine. *J. Physiol.* **70**, 338. *82, 84, 85*

Mautner, H. (1924). Die Bedeutung der Venen und deren Sperrvorrichtungen für den Wasserhaushalt. *Wien. Arch. f. inn. Med.* **7**, 251. *28*

Mautner, H. and Pick, E. P. (1915). Über die durch 'Schockgifte' erzeugten Zirkulationsströrungen. *Münch. med. Wschr.* **62**, 1141. *28*

Mautner, H. and Pick, E. P. (1921). Über die durch Schockgifte erzeugten Zirkulationsstörungen. II. *Biochem. Z.* **127**, 72; (1929) III. *Arch. exp. Path. Pharmak.* **142**, 271. *28*

Meakins, J. and Harington, C. R. (1923). The relation of histamine to intestinal intoxication. II. The absorption of histamine from the intestine. *J. Pharmac. exp. Ther.* **20**, 45. *48, 49*

Méhes, J. and Wolsky, A. (1932). Untersuchungen an der quergestreiften Muskulatur des Darmes der Schleie (Tinca vulgaris). *Arb. Ung. Biol. Forsch.* **5**, 139. *64*

Mellanby, E. (1915). An experimental investigation on diarrhoea and vomiting of children. *Q. Jl Med.* **9**, 165. *48*

Mellanby, E. and Twort, F. W. (1912). On the presence of β-imidazolethylamine in the intestine wall; with a method of isolating a bacillus from the alimentary canal which converts histidine into this substance. *J. Physiol.* **45**, 53. *52*

Menten, M. and Krugh, H. K. (1928). Changes in blood dextrose in rabbits, after intravenous injections of histamine. *J. infect. Dis.* **43**, 117. *38*

Meyer, O. B. (1906). Über einige Eigenschaften der Gefäßmuskulatur mit besonderer Berücksichtigung der Adrenalinwirkung. *Z. Biol.* **48**, 352. *98*

Meyerhof, O. and Lohmann, K. (1932). Über das Kofermentsystem der Milchsäurebildung. *Naturwissenschaften* **20**, 387. *91*

Minz, B. (1932). Eine Methode zum biologischen Nachweis von Azetylcholin. *Arch. exp. Path. Pharmak.* **167**, 85; **168**, 292. *72, 79*

Mitsuda, T. (1923). Über den Mechanismus der Innervation der Muskeln und Drüsen des Dünndarms. *Z. ges. exp. Med.* **39**, 330. *65, 68*

Moir, C. (1932). The action of ergot preparations in the puerperal uterus. *Br. med. J.* **1**, 1119. *30*

Molinari-Tosatti, P. (1928). L'influensa dell' istamina sulla secrezione pancreatica. *Boll. Soc. ital. Biol. sper.* **3**, 928. *33*

Moraeus, B. (1928). Action de quelques narcotiques sur l'innervation parasympathique du cœur. *C. r. Séanc. Soc. Biol.* **98**, 807. *161*

Murakami, K. (1931). Cited in *Ber. ges. Physiol. exp. Pharm.* **61**, 555. *65*

Nakayama, K. (1925). Studie über antagonistische Nerven. Fortgesetzte Prüfung der Frage der hormonalen Ubertragung der Herznervenwirkung. *Z. Biol.* **82**, 581. *126*

Narayana, b. (1933). Vaso-constricteurs et vaso-dilatateurs coronaires. *C. r. Séanc. Soc. Biol.* **114**, 550. *62*

Navratil, E. (1927). Über humorale Übertragbarkeit der Herznervenwirkung. XII. Ergotamin und Accelerans. *Pflügers Arch. ges. Physiol.* **217**, 610. *147, 154, 161*

Neukirch, P. (1912). Physiologische Wertbestimmung am Dünndarm (nebst Beiträgen zur Wirkungsweise des Pilokarpins). *Pflügers Arch. ges. Physiol.* **147**, 153. *129*

Neymann, C. A. (1917). Changes in the blood picture after nucleic acid injections. *Johns Hopkins Hosp. Bull.* **28**, 146. *90*

O'Connor, J. M. (1912). Über den Adrenalingehalt des Blutes. *Arch. exp. Path. Pharmak.* **67**, 95. *96, 101, 102*

Oehme, C. (1913). Über die Wirkungsweise des Histamins. *Arch. exp. Path. Pharmak.* **72**, 76. *37, 50, 51*

Olivecrona, H. (1921). The action of histamine and peptone on the isolated small intestine. *J. Pharmac. exp. Ther.* **17**, 141. *31*

Oliver, G. and Schäfer, E. A. (1895). The physiological effects of extracts of the suprarenal capsules. *J. Physiol.* **18**, 230. *105*

Oppenheimer, E. T. (1929). Studies on the so-called heart hormone. *Am. J. Physiol.* **90**, 656. *26, 103*

Orias, O. (1932). Response of the nictitating membrane to prolonged stimulation of the cervical sympathetic. *Am. J. Physiol.* **102**, 87. *154*

Osborne, W. A. and Vincent, S. (1900). The physiological effects of extracts of nervous tissues. *J. Physiol.* **25**, 283. *109*

O'Shaugnessy, L. (1931). Etiology of peptic ulcer. *Lancet* **109**, 177. *35*

O'Shaugnessy, L. and Slome, D. (1935). Etiology of traumatic shock. *Br. J. Surg.* **22**, 589. *177, 178*

Ostern, P. and Parnas, J. K. (1932). Über die Auswertung von Adenosinderivaten am überlebenden Froschherz. *Biochem. Z.* **248**, 389. *87, 88, 91*

Oswald, A. (1916). Über die Wirkung der Schilddrüse auf den Blutkreislauf. *Pflügers Arch. ges. Physiol.* **164**, 506. *107*

Ott, L. and Field, G. B. W. (1878–1879). Sweat centres; the effect of muscarin and atropin on them. *J. Physiol.* **1**, 193. *138*

Otto, R. (1906). Das Theobald Smithsche Phänomen der Serumüberempfindlichkeit. 5. *Leuthold Festschrift.* (Vol. 1) *179*

Page, I. (1935). Pressor substances from the body fluids of man. *J. exp. Med.* **61**, 67. *100*

Pak, C. and Tang, T. K. (1933). Ephedrine mydriasis. *Chin. J. Physiol.* **7**, 229. *158*

Parker, G. H. (1932). *Humoral Agents in Nervous Activity.* Cambridge University Press. *124*

Parnas, J. K. (1929). Der Zusammenhang der Ammoniakbildung mit der Umwandlung des Adeninnukleotide zu Inosinsäure. *Biochem. Z.* **206**, 16. *93*

Parnas, J. K. and Mazolowski, W. (1927). Über den Ammoniakgehalt und die Ammoniakbildung im Muskel. *Biochem. Z.* **184**, 399. *93*

Pauly, H. (1904). Über die Konstitution des Histidins. *Hoppe-Seyler's Z. physiol. Chem.* **42**, 508. *41*

Phemister, D. B. and Handy, J. (1927). Vascular properties of traumatised and laked bloods. *J. Physiol.* **64**, 155. *99*

Pickering, G. W. (1933). Observations on the mechanism of headache produced by histamine. *Clin. Sci.* **1**, 77. *25*

Plattner, F. (1925). Eine Bestätigung der humoralen Übertragbarkeit der 'Herznervenwirkung'. *Z. Biol.* **83**, 544. *126*

Plattner, F. (1926). Der Nachweis des Vagusstoffes beim Säugetier. *Pflügers Arch. ges. Physiol.* **214**, 112. *74, 79, 103, 127, 128, 162*

Plattner, F. (1932). Über das Vorkommen eines azetylcholinartigen Körpers in den Skelettmuskeln. III. Mitteilung. *Pflügers Arch. ges. Physiol.* **230**, 705. *79, 164*

Plattner, F. (1933). Über das Vorkommen eines azetylcholinartigen Körpers in den Skelettmuskeln. IV. Mitteilung. *Pflügers Arch. ges. Physiol.* **232**, 342. *164*

Plattner, F. (1934). Azetylcholinartige Substanz in den Nebennieren und in einigen anderen Organen. *Pflügers Arch. ges. Physiol.* **234**, 258. *71, 79, 164*

Plattner, F. and Bauer, R. (1928). Die Wirkung von Froschblut auf Vagusstoff und Azetylcholin. *Pflügers Arch. ges. Physiol.* **220**, 180. *82*

Plattner, F. and Galehr, O. (1928). Der Einfluß verschiedener Narkotika auf die Spaltung des Azetylcholins. *Pflügers Arch. ges. Physiol.* **220**, 606. *85*

Plattner, F., Galehr, O. and Kodera, Y. (1928). Die Abhängigkeit der Azetylcholinzerstörung von der Wasserstoffionenkonzentration. *Pflügers Arch. ges. Physiol.* **219**, 678. *81*

Plattner, F. and Hintner, H. (1930a). Die Spaltung von Azetylcholin durch Organextrakte und Körperflüssigkeiten. *Pflügers Arch. ges. Physiol.* **225**, 19. *83, 84, 131*

Plattner, F. and Hintner, H. (1930b). Der Nachweis einer parasympathicomimetischen Substanz in der Iris. *Wien. klin. Wschr.* **43**, 1113. *113, 132, 163*

Plattner, F. and Hou, C. L. (1930). Der Einfluß von Paraldehyd, Chloralose und Amylenhydrat auf vagale Wirkung am Herzen. *Pflügers Arch. ges. Physiol.* **225**, 686. *85, 161*

Plattner, F. and Hou, C. L. (1931). Versuche am Embryonalherzen und am Flimmerepithel. *Pflügers Arch. ges. Physiol.* **228**, 281. *61, 69*

Plattner, F. and Krannich, E. (1932). Über das Vorkommen eines azetylcholinartigen Körpers in den Skelettmuskeln. *Pflügers Arch. ges. Physiol.* 229, 730; 230, 356. *74, 79*

Pohle, K. (1929a). Über das Vorkommen von Muskeladenylsäure und Hexosemonophosphorsäure (Lactacidogen) im Herzen. *Hoppe-Seyler's Z. physiol. Chem.* 184, 261. *91, 103*

Pohle, K. (1929b). Über das Vorkommen von Adenylsäure im Gehirn. *Hoppe-Seyler's Z. physiol. Chem.* 185, 281. *91, 110*

Poos, F. (1927). Pharmakologische und physiologische Untersuchungen an den isolierten Irismuskeln. *Arch. exp. Path. Pharmak.* 126, 307. *63*

Popielski, L. (1920a). β-Imidazolylaethylamin als mächtiger Erreger der Magendrüsen. *Pflügers Arch. ges. Physiol.* 178, 214. *32, 45, 49*

Popielski, L. (1920b). Einfluß der Säuren auf die Magensaftsekretion erregende Wirkung der Organextrakte. *Pflügers Arch. ges. Physiol.* 178, 237. *121*

Popper, M. and Russo, G. (1925). Recherches expérimentales sur la transmission humorale de l'excitation des nerfs cardiaques. *J. Physiol. Path. gén.* 23, 562. *127*

Portier, P. and Richet, C. (1902). Noveaux faits d'anaphylaxie. *C. r. Séanc. Soc. Biol.* 54, 548. *179*

Pritchard, E. A. B. (1935). 'Prostigmin' in the treatment of myasthenia gravis. *Lancet* 1, 432. *145*

Provost, J. L. and Saloz, J. (1909). Contribution a l'étude des muscles bronchiques. *Archs. int. Physiol.* 8, 327. *156*

Pyman, F. L. (1911). A new synthesis of 4- (or 5-) aminoethylglyoxaline. *J. chem. Soc.* 99, 668. *19*

Quagliariello, G. (1914). Über die Wirkung des β-Imidoazolyläthylamins und des β-Oxyphenyläthylamins auf die glatten Muskeln. *Z. Biol.* 64, 263. *30, 32*

Rehsteiner, R. (1927). Die Bedeutung der Nervenversorgung für die Azetylcholin-Kontraktur und den Zuckungsablauf des Skelettmuskels. *Pflügers Arch. ges. Physiol.* 217, 419. *66*

Reznikoff, P. (1929). Experimental leucocytosis and leucopenia. *J. clin. Invest.* 6, 16. *90*

Reznikoff, P. (1930). Nucleotide therapy in agranulocytosis. *J. clin. Invest.* 9, 381. *90*

Rich, A. R. (1921). Condition of the capillaries in histamine shock. *J. exp. Med.* 33, 287. *24*

Richards, A. N. and Plant, O. H. (1915). Urine formation by the perfused kidney. *J. Pharmac. exp. Ther.* 7, 485. *22*

Riesser, O. (1921). Untersuchungen an den Muskeln von Meerestieren. *Arch. exp. Path. Pharmak.* 91, 342; (1922) 92, 254; (1927) 120, 282; (1928) 131, 1; (1933) 172, 194. *66, 67*

Riesser, O. (1931). Beiträge zur Kenntnis des Azetylcholins. *Arch. exp. Path. Pharmak.* 161, 34. *54*

Rigler, R. (1932). Über die Ursache der vermehrten Durchblutung des Muskels während der Arbeit. *Arch. exp. Path. Pharmak.* 167, 54. *89, 171*

Rigler, R. and Schaumann, O. (1930). Beeinflussung der Weite der Herzkranzgefäße durch Produkte des Zellkernstoffwechsels. *Klin. Wschr.* 9, 1728. *89*

Rigler, R. and Singer, R. (1927). Über das Herzhormon. *Klin. Wschr.* 6, 2357. *103*

Rigler, R. and Tiermann, F. (1928). Über den Herzautomatiestoff. *Klin. Wschr.* 7, 553. *103*

Rogowicz, N. (1885). Über pseudomotorische Einwirkung der Ansa vieussenii auf die Gesichtsmuskeln. *Pflügers Arch. ges. Physiol.* 36, 1. *140*

Rona, P. and Neukirch, P. (1912). Experimentelle Beiträge zur Physiologie des Darmes. II. *Pflügers Arch. ges. Physiol.* 146, 371. *130*

Rosenau, N. J. and Anderson, J. F. (1906). Study of the cause of sudden death following the injection of horse serum. *Hygienic Lab. Bulletin Washington*, no. 29; (1907). Studies upon hypersusceptibility and immunity, no. **36**; (1907). The influence of antitoxin upon postdiphtheritic paralysis, no. **38**. *179*

Rosenblueth, A. (1932). The chemical mediation of autonomic nervous impulses as evidenced by summation of responses. *Am. J. Physiol.* **102**, 12. *151, 154*

Rosenblueth, A. and Cannon, W. B. (1932). Studies on conditions of activity in endocrine organs. XXVIII. Some effects of sympathin on the nictitating membrane. *Am. J. Physiol.* **99**, 398. *146, 148, 151*

Rosenblueth, A. and Rioch, D. M. (1933). Nature of responses of smooth muscle to adrenalin and augmentor action of cocaine for sympathetic stimuli. *Am. J. Physiol.* **103**, 681. *158*

Rosenblueth, A. and Schloßberg, T. (1931). The sensitization of vascular response to 'sympathin' by cocaine and the quantification of 'sympathin' in terms of adrenalin. *Am. J. Physiol.* **97**, 365. *146, 150, 158*

Rosenheim, O. (1909). The pressor principles of placental extracts. *J. Physiol.* **38**, 337. *15, 114*

Rosenheim, O. and Schuster, E. (1927). A new colorimeter based on the Lovibond colour system, and its application to the testing of cod-liver oil, and other purposes. *Biochem. J.* **21**, 1329. *42*

Roske, G. (1928). Über Bedingungen der Aminbildung durch Bact. coli. *Jahrb. f. Kinder* **120**, 186. *52*

Rothlin, E. (1920a). Experimentelle Untersuchungen über die Wirkungsweise einiger Substanzen auf überlebende Gefäße. *Biochem. Z.* **111**, 299. *27*

Rothlin, E. (1920b). Über die Einwirkung des Milzextraktes (Lienins) auf die Tätigkeit des Froschherzens und Säugetierherzens. *Pflügers Arch. ges. Physiol.* **185**, 111. *26, 105*

Rothmann, H. (1927). Pharmakologische Untersuchungen am isolierten Ureter. *Z. ges. exp. Med.* **55**, 776. *29*

Rothmann, H. (1930). Der Einfluß der Adenosinphosphorsäure auf die Herztätigkeit. *Arch. exp. Path. Pharmak.* **155**, 129. *88*

Rous, P. and Drury, D. R. (1929). Outlying acidosis due to functional ischemia. *J. exp. Med.* **49**, 435. *170*

Roy, C. S. and Brown, J. G. (1879–80). The blood pressure and its variations in the arterioles, capillaries and smaller veins. *J. Physiol.* **2**, 323. *169*

Rückert, W. (1930a). Über die tonischen Eigenschaften fötaler Muskeln. *Arch. exp. Path. Pharmak.* **150**, 221. *66, 67*

Rückert, W. (1930b). Die phylogenetische Bedingtheit tonischer Eigenschaften der quergestreiften Wirbeltiermuskulatur. *Pflügers Arch. ges. Physiol.* **226**, 323. *67*

Rühl, A. (1929). Über Herzinsuffizienz durch Histamin. *Arch. exp. Path. Pharmak.* **145**, 255. *26*

Rühl, A. (1930). Über Störungen des Sauerstoffdurchtritts in der Lunge. *Arch. exp. Path. Pharmak.* **158**, 282. *38*

Rydin, H. (1924). Influence de certains narcotiques sur l'action exercée par l'acétylcholine sur le cœur. *C. r. Séanc. Soc. Biol.* **91**, 1098. *61*

Rydin, H. and Backman, E. (1925). Influence de la nicotine sur l'action exercée par l'adrenaline et l'acétylcholine sur l'uterus de Lapine. *C. r. Séanc. Soc. Biol.* **93**, 1193. *65*

Rylant, P. (1927). La 'transmission de l'action des nerfs cardiaques' de Loewi chez le mammifère. *C. r. Séanc. Soc. Biol.* **96**, 1054. *127, 147*

Rylant, P. and Demoor, J. (1927). La 'transmission humorale de l'action des nerfs cardiaques' de Loewi chez le mammifère. *C. r. Séanc. Soc. Biol.* **96**, 204. *127, 147, 160*

Saad, K. (1935). The action of drugs on the splenic capsule of man and other animals. *Quart. J. Pharm.* **8**, 31. *65*

Saalfeld, E. von (1934). Humorale übertragene Vaguswirkung. *Pflügers Arch. ges. Physiol.* **235**, 15, 22. *132*

Sacks, J., Ivy, A. C., Burgess, J. P. and Vandolah, J. E. (1932). Histamine as the hormone for gastric secretion. *Am. J. Physiol.* **101**, 331. *34*

Samojloff, A. (1925). Zur Frage des Überganges der Erregung vom motorischen Nerven auf den quergestreiften Muskel. *Pflügers Arch. ges. Physiol.* **208**, 508. *144*

Santenoise, D. (1930). Sécrétion par le pancréas d'une hormone vagotonisante (vagotonine) différente de l'insuline. *C. r. Séanc. Soc. Biol.* **104**, 765. *107*

Santenoise, D. (1932). Sur l'individualité hormonale de la vagotonie. *C. r. hebd. Séanc. Acad. Sci., Paris.* **194**, 572. *107*

Santenoise, D., Brieu, T., Fuchs, G. and Vidacovitch, M. (1932). Sur l'action hypoglycémiante propre de la vagotonine. *Bull. Acad. Méd.* **107**, 302. *106*

Santenoise, D. and Penau, H. (1932). Sur la préparation de la vagotonine. *Bull. Acad. Méd.* **107**, 861. *106*

Saradjichvili, P. and Rafflin, R. (1930). Effets de l'histamine sur mouvements du chlore et de l'eau. *J. Physiol. Path. gén.* **27**, 795. *38*

Saunders, F., Lackner, J. E. and Schochet, S. S. (1931). Studies in absorption. *J. Pharmac. exp. Ther.* **44**, 169. *19, 55*

Sawasaki, H. (1925). Cholin als Hormon der Darmbewegung: Über den Cholingehalt der Muskularis und Mukosa des Dünndarms. *Pflügers Arch. ges. Physiol.* **210**, 322. *130*

Schäfer, E. A. and Moore, B. (1896). On the contractility and innervation of the spleen. *J. Physiol.* **20**, 1. *109*

Schenk, P. (1921). Über die Wirkungsweise des β-imidazolyläthylamins (Histamin) auf den menschlichen Organismus. *Arch. exp. Path. Pharmak.* **89**, 332. *37*

Schilf, E. (1932). Einfluß von Azetylcholin, Adrenalin, Histamin und Thymianextrakt auf die Bronchialschleimhautsekretion. *Arch. exp. Path. Pharmak.* **166**, 22. *68*

Schloßmann, H. (1927). Untersuchungen über den Adrenalingehalt des Blutes. *Arch. exp. Path. Pharmak.* **121**, 160. *95, 99*

Schmidt, G. (1928). Über fermentative Desaminierung im Muskel. *Hoppe-Seyler's Z. physiol. Chem.* **179**, 243. *93*

Schultz, F. (1928). *Über die Darstellung eines neuen Kreislaufstoffes.* Thesis, Munich (1928). *118*

Schultz, W. H. (1910a). The reaction of smooth muscle of the guinea-pig sensitised with horse serum. *J. Pharmac. exp. Ther.* **1**, 549. *179*

Schultz, W. H. (1910b). The reaction of smooth muscle from guinea-pigs rendered tolerant to large doses of serum. *J. Pharmac. exp. Ther.* **2**, 221. *179*

Schultz, W. H. (1912). Reaction of smooth muscle from various organs of different animals, including reaction of muscle from non-sensitized, sensitized, tolerant and immunized guinea-pigs. *Hygienic Lab. Bulletin, Washington*, no. **80**. *179*

Schwartz, C. and Lederer, R. (1908). Über das Vorkommen von Cholin in der Thymus, in der Milz und in den Lymphdrüsen. *Pflügers Arch. ges. Physiol.* **124**, 353. *105*

Schwartz, C. and Lemberger, F. (1911). Über die Wirkung kleinster Säuremengen auf die Blutgefäße. *Pflügers Arch. ges. Physiol.* **141**, 149. *169*

Secker, J. (1934a). The humoral control of the secretion by the submaxillary gland of the cat following chorda stimulation. *J. Physiol.* **81**, 81. *119*

Secker, J. (1934b). Humoral agency in salivary secretion. *J. Physiol.* **82**, 293. *119*

Shanks, W. F. (1923). The excretion of choline in the urine. *J. Physiol.* **58**, 230. *114*

Sherif, M. A. F. (1935). A cholinergic mechanism of the nerve supply to the uterus. *J. Physiol.* **85**, 298. *143*

Sherrington, C. S. (1925). Remarks on some aspects of reflex inhibition. *Proc. R. Soc. B.* **97**, 519. *153*

Shimidzu, K. (1926). Die Bildung von vegetativen Reizstoffen im tätigen Muskel. *Pflügers Arch. ges. Physiol.* **211**, 403. *144*

Siehe, H. J. (1934). Die Reaktion des denervierten Nebennierenmarkes auf humorale
Sekretionsreize. *Pflügers Arch. ges. Physiol.* **234**, 204. *58*

Simonart, A. (1930). Étude expérimentale sur la toxémie traumatique et la toxémie des
grands brulés. *Archs int. Pharmacodyn. Ther.* **37**, 269. *176, 177*

Simonson, E. (1922). Zur Kenntniss der Wirkung des Azetylcholins auf den Froschmuskel.
Arch. exp. Path. Pharmak. **96**, 284. *66, 67*

Smith, F. M., Miller, G. H. and Graber, V. C. (1926). The action of adrenalin and
acetylcholin on the coronary arteries of the rabbit. *Am. J. Physiol.* **77**, 1. *62*

Smith, M. I. (1920). Studies in anaphylaxis. The relation of certain drugs to the
anaphylactic reaction, and the bearing thereof on the mechanism of anaphylactic
shock. *J. Imun.* **5**, 239. *38*

Smith, M. I. (1928). Studies on experimental shock with especial reference to its
treatment. *J. Pharmac. exp. Ther.* **32**, 465. *178*

Smith, Theobald. (1906). Cited by Otto, R. 1906. *179*

Sollmann, T. (1905). The action of blood on the kidney. *Am. J. Physiol.* **13**, 291. *100*

Sollmann, T. and Pilcher, J. D. (1917). Endermic reactions. I. *J. Pharmac. exp. Ther.* **9**, 309.
23

Sommerkamp, H. (1927). Das Substrat des Dauerverkürzung am Froschmuskel. *Arch. exp.
Path. Pharmak.* **128**, 99. *66*

Stavraky, G. (1931). The response of the submaxillary and parotid glands of the dog to
histamine. *J. Pharmac. exp. Ther.* **43**, 265. *33*

Stedman, E. and Stedman, E. (1931). III. The inhibitory action of certain synthetic
urethanes on the activity of liver esterase. *Biochem. J.* **25**, 1147. *84*

Stedman, E. and Stedman, E. (1932). IV. The inhibitory action of certain synthetic
urethanes on the activity of esterases. *Biochem. J.* **26**, 1214. *84*

Stedman, E. and Stedman, E. (1935). The relative choline-esterase activities of serum and
corpuscles from the blood of certain species. *Biochem. J.* **29**, 2107. *82*

Stedman, E., Stedman, E. and Easson, L. H. (1932). Choline-esterase. *Biochem. J.* **26**, 2056. *82*

Stedman, E., Stedman, E. and White, A. C. (1933). Choline esterase of blood serum.
Biochem. J. **27**, 1055. *82*

Steffanutti, P. (1930). Über das Verhalten der Serumlipase im Histaminschock. *Biochem. Z.*
223, 421. *38*

Stern, L. and Rothlin, E. (1919). Action des extraits de tissues animaux sur les organes à
fibres musculaires lisses. *J. Physiol. Path. gén.* **18**, 441, 753. *105*

Stewart, G. N. and Rogoff, J. M. (1920). The relation of the epinephrin output of the
adrenals to changes in the rate of the denervated heart. *Am. J. Physiol.* **52**, 304. *148*

Stewart, G. N. and Zucker, T. F. (1913). A comparison of the action of plasma and serum
on certain objects used in biological tests for epinephrin. *J. exp. Med.* **17**, 152. *101, 102*

Stewart, H. A. and Harvey, S. C. (1912). The vasodilator and vasoconstrictor properties of
blood serum and plasma. *J. exp. Med.* **16**, 103. *100*

Strack, E., Neubaur, E. and Geißendörfer, H. (1933). Über den Cholingehalt bzw.
Azetylcholingehalt tierischer Gewebe. *Hoppe-Seyler's Z. physiol. Chem.* **220**, 217. *15, 104*

Strömbeck, J. P. (1932). Der Einfluß des Histamins auf die Magensekretion beim
Kaninchen nach längerer Anwendung. *Scand. Arch. Physiol.* **65**, 92. *34*

Suda, G. (1924). Experimentelle Untersuchungen über den Innervationsmechanismus der
Magendrüsen. *Arch. path. Anat.* **251**, 56. *65*

Suma, K. (1931). Cited in *Ber. ges. Physiol. exp. Pharm.* **60**, 672. *64*

Sumbal, J. J. (1924). The action of pituitary extracts, acetylcholine and histamine upon
the coronary arteries of the tortoise. *Heart* **11**, 285. *26, 27, 62*

Szakall, A. (1932). Über die Wirkung intravenöser Kallikreinzufuhr auf den tierischen
Organismus. *Arch. exp. Path. Pharmak.* **166**, 301. *116*

Szczygielski, J. (1932). Die adrenalinabsondernde Wirkung des Histamins und ihre Beeinflussung durch Nikotin. *Arch. exp. Path. Pharmak.* **166**, 319. *36, 45*

Tang, C. (1932). Einfluß des Histamins auf die arterielle Sauerstoffbindungskurve. *Arch. exp. Path. Pharmak.* **168**, 274. *38*

Tangl, H. and Recht, S. (1928). Die Wirkung des Histamins auf den Cholesteringehalt des Blutes normaler und entmilzter Hunde. *Biochem. Z.* **200**, 190. *38*

Teschendorf, W. (1921). Über die Gefäßwirkung organischer Kationen und ihre Beeinflussung durch anorganische Ionen. *Biochem. Z.* **118**, 267. *59*

Thannhauser, S. J. (1919). Isolierung der Kristallisierten Adenosinphosphorsäure. *Hoppe-Seyler's Z. physiol. Chem.* **107**, 157. *91*

Thornton, J. W. (1931). Reactions of isolated bronchi. *Q. Jl exp. Physiol.* **21**, 305. *44*

Thornton, J. W. (1934). The liberation of acetylcholine of vagus nerve endings in isolated perfused lungs. *J. Physiol.* **82**, 14P. *132*

Thorpe. W. V. (1928). Vasodilator constituents of tissue extracts. Isolation of histamine from muscle. *Biochem. J.* **22**, 94. *42, 46*

Thorpe, W. V. (1930). The isolation of histamine from the heart. *Biochem. J.* **24**, 626. *46, 47, 103*

Tidmarsh, C. J. (1932). The action of histamine on the motility of large intestine. *Q. Jl exp. Physiol.* **22**, 33. *31*

Titone, F. P. (1914). Über die Funktion der Bronchialmuskeln. *Pflügers Arch. ges. Physiol.* **155**, 77. *32*

Tomaszewski, Z. (1918). Über die chemischen Erreger der Magendrüsen. *Pflügers Arch. ges. Physiol.* **171**, 1. *121*

Tournade, A., Chabrol, M. and Malmejac, J. (1926). Au sujet de l'hormone vagale. Echec des tentatives faites pour la découvrir dans le sang de la circulation coronaire chez le chien. *C. r. Séanc. Soc. Biol.* **95**, 1358. *127*

Trendelenburg, P. (1910). Bestimmung des Adrenalingehaltes im normalen Blut. *Arch. exp. Path. Pharmak.* **63**, 161. *96*

Trendelenburg, P. (1916). Über die Adranalinkozentration im Säugetierblut. *Arch. exp. Path. Pharmak.* **79**, 154. *96*

Trevan, J. W. (1925). The micrometer syringe. *Biochem. J.* **19**, 1111. *45*

Trevan, J. W., Boock, E., Burn, J. H. and Gaddum, J. H. (1928). The pharmacological assay of digitalis by different methods. *Q. Jl Pharm. and Pharmacol.* **1**, 6. *75*

Tschannen, F. (1933). Fortgesetzte Untersuchungen über humorale Übertragungen nach Reizung autonomer Nerven. *Z. Biol.* **93**, 459. *147*

Tutkewitsch, L. M. (1929). Vegetatives Nervensystem und Blutlipoide. *Arch. exp. Path. Pharmack.* **144**, 55. *69*

Underhill, F. P. and Fisk, M. E. (1930). Studies on the mechanism of water exchange in the animal organism. The composition of edema fluid resulting from a superficial burn. *Am. J. Physiol.* **95**, 330. *176*

Underhill, F. P. and Ringer, M. (1921). Studies on the physiological action of some protein derivatives. IX. Alkali reserve and experimental shock. *J. biol. Chem.* **48**, 533. *38*

Underhill, F. P. and Roth, S. C. (1922). The influence of water deprivation, pilocarpin and histamine upon changes in blood concentration in the rabbit. *J. biol. Chem.* **54**, 607. *38*

Vairel, J. (1933). Action de l'adrénaline et de l'acétylcholine sur la rate. *J. Physiol. Path. gen.* **31**, 42. *65*

Vartiainen, A. (1933). The sensitization of leech muscle to barium by eserine. *J. Physiol.* **80**, 21P. *155*

Vartiainen, A. (1934). Acetylcholine and the cardiac vagus. *J. Physiol.* **82**, 282. *163*

Velhagen, K. (1930). Einleitende Untersuchungen über das Vorkommen aktiver und neurotroper Substanzen im Auge. *Arch. Augenheilk.* **103**, 424. *132*

Velhagen, K. (1931a). Alkaliempfindlichkeit als Unterschiedungsmerkmal von Cholin und Azetylcholin. *Arch. exp. Path. Pharmak.* 161, 697. *55*

Velhagen, K. (1931b). Über das Vorkommen depressorischer Substanzen im Auge. *Arch. Augenheilk.* 104, 546. *113*

Velhagen, K. (1932). Zur Frage der vagotropen Substanzen im Auge. *Arch. Augenheilk.* 105, 573. *113, 163*

Verney, E. B. and Starling, E. H. (1922). On secretion by the isolated kidney. *J. Physiol.* 56, 353. *100*

Viale, G. (1929). Préexistence d'une substance vagotrope dans le sang, la lymphe et le liquide céphalorachidien et son identité probable avec la substance vagale de Loewi. *C. r. Séanc. Soc. Biol.* 100, 118. *129*

Villaret, M., Justin-Besançon, L. and Cachera, R. (1934). *Recherches Expérimentales sur quelques Esters de la Choline.* Paris: Masson. *54*

Villaret, M., Schiff-Wertheimer, A. and Justin-Besanọn, L. (1928). Effets de l'injection sous-cutanée d'acétylcholine sur l'artère rétinienne de l'homme. *C. r. Séanc. Soc. Biol.* 98, 909. *56*

Vincent, S. and Cramer, W. (1904). The nature of the physiologically active substances in extracts of nervous tissues and blood, with some remarks on the methods of testing for choline. *J. Physiol.* 30, 143. *110*

Vincent, S. and Sheen, W. (1903). The effects of intravascular injections of extracts of animal tissues. *J. Physiol.* 29, 242. *104, 109*

Viotti, C. (1924). Action de l'histamine sur le cœur et importance de l'atropine à cet égard. *C. r. Séanc. Soc. Biol.* 91, 1085. *27*

Voegtlin, C. and Dyer, H. A. (1924). Natural resistance of albino rats and mice to histamine, pituitary and certain other poisons. *J. Pharmac. exp. Ther.* 24, 101. *25, 177*

Vogelfanger, I. (1932). Azetylcholin im Rinderblut. *Hoppe-Seyler's Z. physiol. Chem.* 214, 109. *78*

Volhard, F. (1931). Die Veränderungen am Herzen und am Gefäßapparat: Mechanismus der Blutdrucksteigerung. In von Bergmann and Staehelin: *Handbuch der inneren Medizin.* Bd. vi. (Vol. 1), *409.*

Voss, O. (1926). Über Wirkungsänderungen des Azetylcholins. *Arch. exp. Path. Pharmak.* 116, 367. *59*

Wachholder, K. and von Ledebur, J. F. (1930). Die Umklammerungshaltung des Frosches und die Schutzhaltung der Schildkröte. *Pflügers Arch. ges. Physiol.* 225, 627. *66*

Wachholder, K. and von Ledebur, J. F. (1932). Rote und weiße Muskeln, Verhalten im Winterschlaf. *Pflügers Arch. ges. Physiol.* 229, 657. *66*

Wachholder, K. and Matthias, F. (1933). Einfluß verschieden zusammengesetzter Ringerlösung auf das Kontrakturvermögen von Froschmuskeln. *Pflügers Arch. ges. Physiol.* 232, 159. *67*

Walker, M. B. (1934). Treatment of myasthenia gravis with physostigmin. *Lancet* 1, 1200. *145*

Waud, R. A. (1928). Viscosity of the blood in histamine shock. *Am. J. Physiol.* 84, 563. *38*

Weber, C. H., Nanninga, J. B. and Major, R. H. (1933). Isolation of crystalline depressor substance from the brain. *Proc. Soc. exp. Biol. Med.* 30, 513. *112*

Weber, E. (1914). Über experimentelles Asthma und die Innervation der Bronchialmuskeln. *Arch. Anat. Physiol.* 6, 63. *32*

Wedd, A. M. (1931). The action of adenosine and certain related compounds on the coronary flow of the perfused heart of the rabbit. *J. Pharmac. exp. Ther.* 41, 355. *89*

Webb, A. M. and Fenn, W. O. (1933). The action on cardiac musculature and the vagomimetic behaviour of adenosine. *J. Pharmac. exp. Ther.* 47, 365. *59, 62, 88*

Wedum, A. G. and Gabauer-Fuelnegg, E. (1932). Histamine and acetylcholine contraction-ratio in the surviving intestinal strip. *Proc. Soc. exp. Biol. Med.* **29**, 888. *64, 75*

Weese, H. (1933). Zur biologischen Auswertung des Kallikreïns. *Arch. exp. Path. Pharmak.* **173**, 36. *117*

Weichardt, W. (1927). Über Spezifität. *Klin. Wschr.* **6**, 1555. *103*

Weiland, W. (1912). Zur Kenntnis der Entstehung der Darmbewegung. *Pflügers Arch. ges. Physiol.* **147**, 171. *129*

Weiss, O. (1897). Ein Nachtrag zu den Untersuchungen über die Wirkung von Blutseruminjektionen ins Blut. *Pflügers Arch. ges. Physiol.* **68**, 348. *99*

White, A. C. and Stedman, E. (1931). On the physostigmine-like action of certain synthetic urethanes. *J. Pharmac. exp. Ther.* **41**, 259. *70, 84*

Wilson, W. C. (1929). The tannic acid treatment of burns. *Spec. Rep. Ser. med. Res. Counc.* no. 141. London. *176*

Windaus, A. and Vogt, W. (1907). Synthese des Imidazolyläthylamins. *Ber. dtsch. chem. Ges.* **40**, 3691. *5, 18, 19*

Winterberg, H. (1907). Über die Wirkung des Physostigmins auf das Warmblüterherz. *Z. exp. Path. Ther.* **4**, 636. *140*

Witanowski, W. R. (1925). Über humorale Übertragbarkeit der Herznervenwirkung. VIII. Mitteilung. *Pflügers Arch. ges. Physiol.* **208**, 694. *74, 79, 103, 128, 136, 162*

Wrede, F. and Keil, W. (1931). Azetylcholin im Rinderblut. *Hoppe-Seyler's Z. physiol. Chem.* **194**, 229. *78*

Wyman, L. C. (1928). The relative importance of cortex and medulla in the susceptibility to histamine of suprarenalectomised rats. *Am. J. Physiol.* **87**, 29, 42. *35*

Yanagawa, H. (1916). On the vasoconstrictive action of serum on the coronary vessels of the mammalian heart. *J. Pharmac. exp. Ther.* **8**, 89. *100*

Yoshida, Y. (1931). Pharmakologische Studien über den Erreger im Magensaft, besonders über Histamin. *Nagasaki Igak. Zass.* **9**, 1005. Cited in *Ber. ges. Physiol. exp. Pharm.* (1932) **65**, 240. *50*

Yoshimura, K. (1910). Über Fäulnisbasen (Ptomaine) aus gefaulten Sojabohnen (Glycine hispida). *Biochem. Z.* **28**, 16. *46*

Young, J. Z. (1930). The pupillary mechanism of the teleostean fish *Uranoscopus scaber*. *Proc. R. Soc. B.* **107**, 464. *63*

Young, J. Z. (1932). Comparative studies on the physiology of the iris. II. *Uranoscopus* and *Lophius. Proc. R. Soc. B.* **112**, 242. *63*

Zimmerman, W. (1929). Über eine spezifische Farbenreaktion für Histamin. *Hoppe-Seyler's Z. physiol. Chem.* **186**, 260. *42*

Zipf, K. (1927). Die Austauschbindung als Grundlage der Aufnahme basischer und saurer Fremdsubstanzen in die Zelle. *Arch. exp. Path. Pharmak.* **124**, 259. 286. *171*

Zipf, K. (1931a). Über die physiologische und pharmakologische Bedeutung kreislaufwirksamer, intermediärer Stoffwechselprodukte. *Klin. Wschr.* **10**, 1521. *94*

Zipf, K. (1931b). Die chemische Natur der 'depressorischen Substanz' des Blutes. *Arch. exp. Path. Pharmak.* **160**, 579. *91, 99, 101*

Zipf, K. and Hülsmeyer, P. (1933). Histaminähnliche Stoffe und Spätgift im Blut. *Arch. exp. Path. Pharmak.* **173**, 1. *95, 102*

Zipf, K. and Wagenfeld, E. (1930). Über die pharmakologische Wirkung des frisch defibrinierten Blutes. *Arch. exp. Path. Pharmak.* **150**, 70; (1930) **156**, 91. *97, 99*

Zucker, K. (1923). Die Wirkung des Physostigmins auf den quergestreiften Muskel. (Ein Beitrag zur Tonusfrage.) *Arch. exp. Path. Pharmak.* **96**, 29. *143, 156*

Zuelzer, G. (1908). Spezifische Anregung der Darmperistaltik durch intravenöse Injektion des 'Peristaltikhormons'. *Berl. klin. Wschr.* **45**, 2065. *105*

Zuelzer, G. (1911). Das Peristaltikhormon 'Hormonal'. *Ther. Gegenw.* 197. *105*

Zuelzer, G. (1928). Zum gegenwärtigen Stand der Herzhormonfrage. *Med. Klin.* **24**, 571. *104*

Zunz, E. and Govaerts, P. (1924). Action hypotensive du sang carotidien recueilli pendant l'excitation du pneumogastrique. *C. r. Séanc. Soc. Biol.* **91**, 389. *127*

Zunz, E. and György, P. (1914). A propos de l'action de la morphine sur l'intestin. *Archs. int. Physiol.* **14**, 221. *130*

Zunz, E. and La Barre, J. (1926). Tension superficielle et coagulation du plasma lors du choc histaminique du chien. *C. r. Séanc. Soc. Biol.* **95**, 722. *38*

Index

271